Artificial Intelligence Technologies for Engineering Applications

This book enables the readers to design, optimize, and control complex systems with greater precision and efficiency. It further provides practical insights and presents case studies for readers interested in exploring the intersections between artificial intelligence and industry. This book discusses important topics such as algorithmic design, mathematical modeling, natural language processing, machine learning, and computer vision.

This book:

- Explores practical applications of artificial intelligence in engineering, including optimization, predictive modeling, decision-making, and control systems.
- Provides real-world examples of the applications of artificial intelligence in engineering, drawing from a range of industries, including aerospace, automotive, and manufacturing.
- Discusses technologies such as machine learning and computer vision for aircraft design optimization, fault diagnosis, and autonomous navigation.
- Explains natural language processing for analyzing and optimizing building systems, while robotics can be used for construction automation.
- Presents artificial intelligence technologies for optimization of manufacturing processes, predictive maintenance, and quality control.

This book is primarily written for senior undergraduates, graduate students, and academic researchers in the fields of electrical engineering, electronics and communications engineering, computer science and engineering, and information technology.

Artificial Intelligence for Sustainable Engineering and Management

Sachi Nandan Mohanty
School of Computer Science & Engineering (SCOPE), VIT-AP University, Amaravati, Andhra Pradesh, India
Deepak Gupta

Artificial intelligence is shaping the future of humanity across nearly every industry. It is already the main driver of emerging technologies like big data, robotics and IoT, and it will continue to act as a technological innovator for the foreseeable future. Artificial intelligence is the simulation of human intelligence processes by machines, especially computer systems. Specific applications of AI include expert systems, natural language processing, speech recognition and machine vision. The future of business intelligence combined with AI will see the analysis of huge quantities of contextual data in real-time. So, the tool will quickly capture customer needs and priorities and do what is needed.

Green Metaverse for Greener Economies
Edited by Sukanta Kumar Baral, Richa Goel, Tilottama Singh, and Rakesh Kumar

Healthcare Analytics and Advanced Computational Intelligence
Edited by Sushruta Mishra, Meshal Alharbi, Hrudaya Kumar Tripathy, Biswajit Sahoo, and Ahmed Alkhayyat

AI in Agriculture for Sustainable and Economic Management
Edited by Sirisha Potluri, Suneeta Satpathy, Santi Swarup Basa, and Antonio Zuorro

Deep Learning in Biomedical Signal and Medical Imaging
Edited by Ngangbam Herojit Singh, Utku Kose, and Sarada Prasad Gochhayat

Sustainable Development Using Private AI: Security Models and Applications
Edited by Uma Maheswari V and Rajanikanth Aluvalu

Sustainable Farming through Machine Learning Enhancing Productivity and Efficiency
Suneeta Satpathy, Bijay Paikaray, Ming Yang, and Arun Balakrishnan

Artificial Intelligence Technologies for Engineering Applications
G. Sucharitha, Anjanna Matta, M. Srinivas, and Sachi Nandan Mohanty

https://www.routledge.com/AI-for-Sustainable-Engineering-and-Management-series/book-series/AISEM

Artificial Intelligence Technologies for Engineering Applications

Edited by
G. Sucharitha, Anjanna Matta, M. Srinivas,
and Sachi Nandan Mohanty

CRC Press
Taylor & Francis Group
Boca Raton London New York

CRC Press is an imprint of the
Taylor & Francis Group, an **informa** business

First edition published 2025
by CRC Press
2385 NW Executive Center Drive, Suite 320, Boca Raton FL 33431

and by CRC Press
4 Park Square, Milton Park, Abingdon, Oxon, OX14 4RN

CRC Press is an imprint of Taylor & Francis Group, LLC

ISBN: 978-1-032-76581-5 (hbk)
ISBN: 978-1-032-93357-3 (pbk)
ISBN: 978-1-003-56552-9 (ebk)

DOI: 10.1201/9781003565529

Typeset in Sabon
by codeMantra

Contents

Preface

Welcome to the intersection of technology and sustainability. In an era marked by rapid industrialization and environmental concerns, the role of artificial intelligence (AI) in shaping sustainable engineering and management practices is more critical than ever. This book delves into the dynamic landscape where the cutting-edge AI techniques meet the pressing challenges of sustainable development. In recent years, AI has emerged as a powerful tool with the potential to revolutionize various sectors, including engineering and management. Its ability to process vast amounts of data, identify patterns, and make predictions has opened up new avenues for optimizing resource utilization, minimizing environmental impact, and enhancing overall efficiency. From renewable energy systems and smart infrastructure to eco-friendly manufacturing processes and waste management, AI holds the promise of driving transformative change toward a more sustainable future. However, harnessing the full potential of AI for sustainability requires a multidisciplinary approach. It demands collaboration between engineers, environmental scientists, policymakers, and business leaders to integrate AI solutions seamlessly into existing frameworks while addressing ethical considerations and ensuring equitable outcomes. This book aims to facilitate such collaboration by providing insights into the latest AI technologies, case studies highlighting their real-world applications, and discussions on the opportunities and challenges they present. Whether you are an AI researcher, an environmental enthusiast, a business strategist, or a policymaker, this book offers valuable perspectives on leveraging AI for sustainable engineering and management. By exploring innovative solutions and best practices, we hope to inspire readers to embark on a journey toward building a more resilient, inclusive, and environmentally conscious society. We extend our gratitude to the contributors, whose expertise and dedication have made this book possible. We also thank the readers for their interest in this crucial intersection of AI and sustainability. Together, let us explore the transformative potential of AI in shaping a better world for generations to come.

About the editors

G. Sucharitha is working as a Professor in the Department of Computer Science and Engineering at the School of Engineering, Anurag University, Hyderabad. Electronics and Communication Engineering at the Institute of Aeronautical Engineering, Hyderabad. Her research areas are Biomedical Image Processing, Image Retrieval, and Machine and Deep Learning. She has more than 13 years of teaching experience. She is an expert of teaching Image Processing, Artificial Intelligence, Machine and Deep Learning, Computer Organization & Architecture, Microprocessors, and Internet of Things. She has published more than 20 research articles in SCI & Scopus journals. She has also presented many research papers in national and international conferences, holds six patents, and has published nine book chapters, one edited book, and one authored book. She has received CIEMA "Outstanding Thesis & Dissertation Award-2022" in the month of March, 2022.

Anjanna Matta is working as an Associate Professor in the Department of Mathematics, Faculty of Science and Technology (ICFAITECH), IFHE, Hyderabad. He has published several articles in various international journals. He is authoring few books and book chapters. He received PhD degree from IIT Hyderabad, India, completed MTech (IMSC) from IIT Madras, India, and MSc (Mathematics) from NIT Warangal, India.

M. Srinivas is working as an Assistant Professor in the Department of Computer Science and Engineering at the National Institute of Technology (NIT), Warangal, India. He received his postdoc from ETS, Canada, and Academia Sinica, Taiwan, in 2018 and 2016. He received his PhD degree from IIT Hyderabad in the year 2014. He has published many papers in reputed Scopus/SCI journals, conference, book chapters, and books. He has received many travel grants to attend international conferences, conclaves, and summits. He received Global Distinguished Young Scientist award from IEEE IAS Global Conference in 2023. He is reviewer for several journals of international repute. His research areas include Biomedical Image Analysis, Machine and Deep Learning, and Artificial Intelligence.

Sachi Nandan Mohanty received his postdoc from IIT Kanpur in the year 2019 and PhD from IIT Kharagpur, India, in the year 2015, with MHRD scholarship from the government of India. He has authored/edited 32 books, published by international publishers. His research areas include Data Mining, Big Data Analysis, Cognitive Science, Fuzzy Decision Making, Brain-Computer Interface, Cognition, and Computational Intelligence. Prof. S. N. Mohanty has received four Best Paper Awards during his PhD at IIT Kharagpur from the International Conference at Beijing, China, and the other at International Conference on Soft Computing Applications organized by IIT Roorkee in the year 2013. He has been awarded the first prize for the best thesis by Computer Society of India in the year 2015. He has guided nine PhD scholars. He has published 120 international journals of international repute and has been elected as fellow of the Institute of Engineers, European Alliance Innovation (EAI) Springer, and senior member of IEEE Computer Society Hyderabad Section. He is also the reviewer of the *Journal of Robotics and Autonomous Systems*, *Computational and Structural Biotechnology Journal*, *Artificial Intelligence Review*, and *Spatial Information Research*.

Contributors

Abdulsattar Abdullah Hamad
Dijlah University College
Baghdad, Iraq

Ganji Adithya
Department of Computer Science
    and Engineering (Cyber Security)
Institute of Aeronautical
    Engineering
Hyderabad, Telangana, India

Golla Angel
Information Technology Institute
    of Aeronautical Engineering
Hyderabad, Telangana, India

Thota Anjushree
Computer Science and Engineering
    (AI & ML)
Institute of Aeronautical
    Engineering
Hyderabad, Telangana, India

E. Bharat Babu
Department of Electronics and
    Communication Engineering
B V Raju Institute of Technology
Narsapur, Telangana, India

D. Celesty Bliss Rufus
Department of CSE
Institute of Aeronautical
    Engineering
Hyderabad, Telangana, India

G. Chandra Sekhar
Department of CSE
Institute of Aeronautical
    Engineering
Hyderabad, Telangana, India

V. Deekshitha
Department of Computer Science
    and Engineering (Cyber Security)
Institute of Aeronautical
    Engineering
Hyderabad, Telangana, India

Mamata Garanayak
Kalinga Institute of Social Sciences
Deemed to be University
Bhubaneswar, Odisha, India

E. Goutham
Information Technology Institute
    of Aeronautical Engineering
Hyderabad, Telangana, India

Anita Hembram
Kalinga Institute of Social Sciences
Deemed to be University
Bhubaneswar, Odisha, India

Indu P.V.
Department of Data Science
CHRIST University
Lavasa, Maharashtra, India

N. Jaya Krishna
Department of Computer Science
and Engineering (Cyber Security)
Institute of Aeronautical
Engineering
Hyderabad, Telangana, India

Pooja Jena
KIIT School of Management
KIIT University
Bhubaneswar, Odisha, India

Rahul Joshi
Department of Journalism & Mass
Communication, School of
Media Studies & Humanities
Manav Rachna International
Institute of Research & Studies
Faridabad, Haryana, India

Vijay Kumar Reddy Julakanti
Indian Institute of Technology
Hyderabad, Telangana, India

G. Kavya Yadav
Department of Electronics and
Communication Engineering
B V Raju Institute of Technology
Narsapur, Telangana, India

Parikshita Khatua
KISS University
School of Management
Bhubaneswar, Odisha, India

A. Krishna Chaitanya
Department of Computer Science
and Engineering (Cyber Security)
Institute of Aeronautical
Engineering
Hyderabad, Telangana, India

E. Krishna Rao Patro
Computer Science and Engineering
Institute of Aeronautical
Engineering
Hyderabad, Telangana, India

Suman Kumari
Department of Journalism & Mass
Communication, School of
Media Studies & Humanities
Manav Rachna International
Institute of Research & Studies
Faridabad, Haryana, India

P. Lohitha
Department of Computer Science
and Engineering (Cyber Security)
Institute of Aeronautical
Engineering
Hyderabad, Telangana, India

C. Mahananda Reddy
Department of CSE
Institute of Aeronautical
Engineering
Hyderabad, Telangana, India

Abhiraj Malia
KISS University
School of Management
Bhubaneswar, Odisha, India

M. Manideep
Department of Computer Science
   and Engineering (Cyber Security)
Institute of Aeronautical
   Engineering
Hyderabad, Telangana, India

Y. Manohar Reddy
Department of Computer Science
   and Engineering (Cyber Security)
Institute of Aeronautical
   Engineering
Hyderabad, Telangana, India

S. Mrudula
Department of Electronics and
   Communication Engineering
B V Raju Institute of Technology
Narsapur, Telangana, India

Preethi Nanjundan
Department of Data Science
CHRIST University
Lavasa, Maharashtra, India

K. Naveena
Department of Computer Science
   and Engineering (Cyber Security)
Institute of Aeronautical
   Engineering
Hyderabad, Telangana, India

P. Neha
BVRIT Hyderabad College of
   Engineering for Women
Department of Computer Science
   and Engineering
Hyderabad, Telangana, India

Rachamalla Nikhitha
Department of Computer Science
   and Engineering
Vignana Bharathi Institute of
   Technology
Ghatkesar, Hyderabad, Telangana,
   India

T. Nithin
Information Technology Institute
   of Aeronautical Engineering
Hyderabad, Telangana, India

B. Padmaja
Computer Science and Engineering
   (AI & ML)
Institute of Aeronautical
   Engineering
Hyderabad, Telangana, India

Bijay Kumar Paikaray
Centre for Data Science
Siksha 'O' Anusandhan, Deemed
   to be University
Bhubaneswar, Odisha, India

Liji Panda
KISS University
School of Management
Bhubaneswar, Odisha, India

Krishna Pandey
Department of Journalism & Mass
   Communication, School of
   Media Studies & Humanities
Manav Rachna International
   Institute of Research & Studies
Faridabad, Haryana, India

Suresh K. Peddoju
University of Houston
Houston, Texas

Sri Sai Madhuvani Petla
Department of CSE (AI & ML)
Institute of Aeronautical
   Engineering
Hyderabad, Telangana, India

S. Pranitha Pradhan
BVRIT Hyderabad College of
  Engineering for Women
Department of Computer Science
  and Engineering
Hyderabad, Telangana, India

Mudimanchi Pranay Kumar
Department of Computer Science
  and Engineering (Cyber Security)
Institute of Aeronautical
  Engineering
Hyderabad, Telangana, India

Sugyanta Priyadarshini
Kalinga Institute of Industrial
  Technology Deemed to be
  University
Bhubaneswar, Odisha, India

Sukanya Priyadarshini
Berhampur University
Korapalli, Odisha, India

M. Purushotham Reddy
Information Technology Institute
  of Aeronautical Engineering
Hyderabad, Telangana, India

Siripothula Rahul
Department of Computer Science
  and Engineering
Vignana Bharathi Institute of
  Technology
Ghatkesar, Hyderabad,
  Telangana, India

K. Rajendra Prasad
Department of Computer Science
  and Engineering (Cyber Security)
Institute of Aeronautical
  Engineering
Hyderabad, Telangana, India

Mahesh Ratnaparkhe
Department of Information
  Technology
Institute of Aeronautical
  Engineering
Hyderabad, Telangana, India

Rajesh Saturi
K. Vikas Ghatkesar
Hyderabad, Telangana, India

Amjan Shaik
Department of Computer Science
  and Engineering
St. Peters Engineering College
Maisammaguda, Telangana, India

Shanmuga Sundari M.
BVRIT Hyderabad College of
  Engineering for Women
Department of Computer Science
  and Engineering
Hyderabad, Telangana, India

Ravipalli Shreya
Department of CSE
Institute of Aeronautical
  Engineering
Hyderabad, Telangana, India

Zuha Siddiqui
Department of Computer Science
  and Engineering
Vignana Bharathi Institute of
  Technology
Ghatkesar, Hyderabad, Telangana,
  India

Vedururu Sireesha
Department of CSE
Gandhi Institute of Technology and
  Management
Hyderabad, Telangana, India

U. Sivaji
Department of Information
  Technology
Institute of Aeronautical
  Engineering
Hyderabad, Telangana, India

G. Sucharitha
Department of Computer Science
  and Engineering
School of Engineering, Anurag
  University
Hyderabad, Telangana, India

G. Sucharitha Reddy
Anurag University
Hyderabad, India

Thurimella Sujitha
BVRIT Hyderabad College of
  Engineering for Women
Department of Computer Science
  and Engineering
Hyderabad, Telangana

Lijo Thomas
CHRIST University
Lavasa, Maharashtra, India

Sachin Turreraa
Department of Information
  Technology
Institute of Aeronautical
  Engineering
Hyderabad, Telangana, India

Tejaswini Vemulapalli
Department of CSE (AI & ML)
Institute of Aeronautical
  Engineering
Hyderabad, Telangana, India

K. Vikas
Department of Electronics and
  Communication Engineering
B V Raju Institute of Technology
Narsapur, Telangana, India

Kummari Vikas
University of Houston
Houston, Texas, USA

Sirra Yashwanth
Department of Information
  Technology
Institute of Aeronautical
  Engineering
Hyderabad, Telangana, India

Chapter 1

Navigating the ethical landscape of artificial intelligence

Challenges, frameworks, and responsible deployment

Preethi Nanjundan, Indu P.V., and Lijo Thomas

1.1 INTRODUCTION

This introduction section provides an overview of the complex ethical issues related to the use of artificial intelligence (AI), as discussed in the book *"Navigating the Ethical Landscape of Artificial Intelligence: Challenges, Frameworks, and Responsible Deployment"* The thorough investigation of the writers delves into the various issues that follow AI's pervasive incorporation into various spheres of society. The opening establishes the scene by outlining the moral obligations that come with this technological revolution while acknowledging the revolutionary possibilities of AI in their overview of the complexities surrounding ethical issues. The authors stress the importance of addressing biases present in training data, maintaining decision-making processes' transparency, and protecting user privacy. The introduction highlights the significance of developing reliable and responsible models as AI systems become essential to important decision-making. The introduction highlights the importance of responsible deployment, emphasizing transparency, interpretability, and ongoing monitoring to uphold accountability and user trust. As the ethical landscape of AI continues to unfold, this introduction serves as a compass, guiding readers through the challenges and ethical frameworks that shape the responsible deployment of AI in modern society. The central theme of the introduction is the ethical dimensions of AI, which call for a holistic approach that goes beyond technical advancements and incorporates frameworks for responsible development and deployment.

1.2 INTRODUCTION TO MACHINE LEARNING

The fundamental idea of machine learning (ML), a paradigm that allows computers to learn from experience without explicit programming, is crucial to AI. This broad area includes several techniques, the two main ones

DOI: 10.1201/9781003565529-1

being supervised and unsupervised learning. When using supervised learning, the model is trained on input–output pairs, and the algorithm generates predictions based on labelled data. For many applications, this labelled data is essential since it enables the algorithm to generalize and make well-informed predictions on data that hasn't been seen yet. Conversely, unsupervised learning explores unlabelled data with the goal of finding naturally occurring patterns and structures in the data.

In the field of supervised learning, the analysis of labelled example is critical. During the procedure, the model is trained on a dataset with predetermined results, allowing it to discover the connections between input feature labels and related features. In applications such as image identification, natural language processing (NLP), and predictive analytics, this supervised technique is frequently used. The calibre and representativeness of the labelled examples used in training have a significant impact on the model's accuracy and effectiveness. In contrast, unsupervised learning makes use of clustering algorithms in order to find patterns within unlabelled data. This experimental method is especially useful when dealing with large datasets, where labelling could be expensive or difficult. Unsupervised learning contributes to a deeper knowledge of complicated datasets by revealing linkages and insights that may not be immediately evident by detecting underlying structures.

ML gains a dynamic component with reinforcement learning (RL), which takes cues from human learning via trial and error. In this paradigm, an agent interacts with its surroundings, takes feedback, and modifies its behaviour to optimize the sum of rewards. In applications such as gaming, robotics, and autonomous systems, RL is widely used to teach agents optimum tactics through ongoing interaction and feedback loops. The growing significance of AI has made ethical questions and responsible AI deployment crucial. The necessity of appropriate deployment procedures and the importance of addressing ethical challenges in AI are emphasized in the beginning. As part of this, AI systems must be made transparent, equitable, and accountable in order to reduce the possibility of prejudice and guarantee the moral use of technology. The introduction also emphasizes how important high-quality data is to ML applications. The reason why ML models work well is highly reliant on the training set of data, highlighting the need of representative, diversified, and carefully selected datasets. In addition to improving model performance, high-quality data helps ensure that AI technologies are developed and used in an ethical and responsible manner.

1.2.1 Supervised learning and unsupervised learning

Supervised learning and unsupervised learning are two [1] of the most important methods in ML, a revolutionary idea in AI. By training algorithms on labelled datasets, supervised learning enables them to form conclusions or predictions based on patterns found throughout the training phase. For tasks where the model learns to correlate input properties with preset outcomes, such as regression and classification, this organized approach is essential. Unsupervised learning, in contrast, is concerned [2] with identifying patterns in unlabelled data. This exploratory approach frequently reveals underlying patterns and organic groupings, providing insights into the underlying complexity of the data.

1.2.2 Reinforcement learning: learning through interaction and feedback

The dynamic and potent field of RL in ML is revolutionizing the way algorithms learn. Through trial and error, it emulates how humans learn, enabling agents to gradually modify and improve their tactics [3]. This flexibility is especially useful in situations that are dynamic and complex. Applications of RL are widespread in domains where adaptive decision-making is essential, such as robots and gaming. Robots in robotics are able to optimize their movements and responses in real time by learning from their interactions with the environment. Reinforcement [4] learning has proven crucial in helping players reach superhuman levels of performance in challenging games. RL principles provide a flexible framework for tackling difficult problems in several domains, such as resource allocation optimization in business processes and improving the effectiveness of self-governing automobiles [5]. The uses of RL are anticipated to grow as technology progresses, helping to create increasingly intelligent and adaptable systems.

1.2.3 Importance of quality data and data pre treatment

Effective ML relies heavily on high-quality data, which is essential to the dependability and capacity for generalization of models. Within ML, the proverb "garbage in, garbage out" is applicable, highlighting the direct correlation between the calibre of input data and the calibre of the model's predictions. Accuracy, relevance, and representativeness are traits of high-quality data. The concept of accuracy guarantees that the data accurately depicts the actual state of the phenomenon being modelled, whereas relevance guarantees that the data has direct application to the issue at hand. In order [6] for the model to effectively generalize to new, unknown cases, representativeness guarantees that the dataset captures the diversity and variability found in the larger population [7].

1.3 NEURAL NETWORKS AND DEEP LEARNING

The inventive applications of deep learning and neural networks in AI are examined in this chapter, with an emphasis on the latter's complex architectures and revolutionary effects on a range of practical uses. Drawing inspiration from the intricate network of neurons found in the human brain, neural networks are effective instruments for tasks such as pattern identification and judgement. Understanding how these structures allow machines to learn, adapt, and digest information in a way that is similar to human cognition is made possible by exploring the intricacies of these networks, including nodes, layers, and connections. An important factor behind the unparalleled progress in a number [8] of fields, including image and speech recognition, is deep learning, which is defined by the integration of numerous hidden layers. The practical applications of these technologies are across a range of industries such as financial forecasting, NLP, healthcare diagnostics, and autonomous systems. Neural networks and deep learning are clearly more than just theoretical ideas; they are driving forces behind innovation that are expanding the bounds of what is possible and taking AI to new heights.

1.3.1 Simulating human brain with neural networks

In order to provide machines cognitive capacities, artificial neural network research attempts to replicate the structure and functions of the human brain. The complex connection-building and communication processes present in biological neurons are intended to be replicated by these networks. Attempting to decipher the secrets of cognition and convert them into computing models, this endeavour [8] is both technological and philosophical. Neural networks function as information processing and transmission units, just like neurons do. Features can be abstracted hierarchically thanks to layers, which reflect the brain's layered organization. Similar to synapses, connections distribute weighted information among nodes to support decision-making and learning. Due to this interaction, machines are able to learn from data patterns, generalize, and make decisions [9].

Neural networks have proven to be highly adaptable in a variety of fields, such as banking and healthcare. Neural networks are used in healthcare to identify patterns in medical imaging data, which increases the accuracy of diagnosis [10]. By identifying intricate market patterns, they improve predictive analytics in the financial industry. The insights gleaned from comparing artificial neural networks to the human brain underscore their adaptability and potential to transform a multitude of technical domains. The applications on these networks will keep pushing the limits of AI as they develop.

1.3.2 Deep learning: multi-layered neural networks

AI systems can now do much more thanks to deep learning, which is a sophisticated method to ML [11] that is encapsulated by multi-layered neural networks. Several layers of interconnected nodes make up the hierarchical architecture of multi-layered neural networks, sometimes referred to as deep neural networks or deep learning models. Deep neural networks, in contrast to conventional ML models, have the ability to learn hierarchical representations of data on their own, which enables them to recognize complex patterns and characteristics. These networks are especially good at tasks involving pattern recognition, meaningful representation extraction, and decision-making based on subtle information because of their depth, which allows them to capture and model intricate relationships within the data [12].

The ability of multi-layered neural networks to automatically learn and adapt to the underlying structures of the data they are trained on is largely responsible for their success. Deeper layers of the network capture more intricate and abstract properties as it learns to extract various levels of abstraction from the input data. The model's efficacy in tasks including picture recognition, speech processing, and natural language understanding is attributed to its ability to comprehend and represent complex relationships through hierarchical feature learning. In order to push the limits of what is possible [12] with these potent neural networks, researchers are investigating new [13] topologies, optimization strategies, and applications as the discipline of deep learning continues to develop.

1.3.3 Applications of deep learning in speech and image recognition

Deep learning has improved computer comprehension of complex material, dramatically changing the fields of speech and image recognition. Advances in automatic voice recognition systems have resulted 11] from the demonstration of unparalleled accuracy and efficiency in speech recognition by technologies such as Recurrent Neural Networks (RNNs) and Long Short-Term Memory networks (LSTMs). Applications such as interactive voice response systems, transcription [8] services, and voice-controlled virtual assistants have been made possible by this. The comprehension of accents, context, and varied speech patterns by deep learning models has greatly improved user experience across a range of applications.

Convolutional Neural Networks (CNNs) are a key technique in image recognition, with exceptional performance in semantic segmentation, object detection, and image categorization. Applications in the automotive, entertainment, security, and healthcare sectors have increased the opportunities for creative applications in domains where visual data is essential. Beyond only improving accuracy [10], deep learning has a significant impact on speech and image recognition, creating new opportunities for more intuitive and natural human–computer interactions [6]. The future of intelligent systems will be shaped by deep learning, as evidenced by the more complex applications that these technologies' constant progress and continuous research and development promise.

1.4 PRACTICAL CONSIDERATIONS IN MACHINE LEARNING

To guarantee their [14] dependability, effectiveness, and moral application, ML systems need to be carefully considered. Preprocessing and high-quality data are necessary for precise forecasts, while model evaluation and selection are necessary to choose the best model for a given task. In production settings, scalability is critical for models because they must manage massive datasets and growing workloads. Techniques such as feature importance analysis and model-agnostic interpretability methodologies add to the need for interpretability and explainability [2] in delicate fields such as healthcare and finance.

Due to the potential for ML models to reinforce biases in training data, ethical considerations and bias mitigation are crucial. The key to reducing these problems is to implement algorithms and methods that are fairness-aware and to continuously monitor and evaluate results for bias. Techniques such as federated learning and differential privacy secure user data, making security and privacy essential [1]. Integration with current systems, real-time performance monitoring, updating, and retraining are all part of deployment and maintenance. Adherence to regulatory frameworks and data protection legislation is imperative in order to avert legal issues [4] and guarantee proper utilization of data. The appropriate and successful implementation of ML systems in a variety of applications requires a comprehensive strategy that incorporates these elements [15].

1.4.1 Feature engineering

Feature engineering is essential for improving ML models' performance since it converts unprocessed, raw data into a more readable format. Improving the model's understanding and forecast accuracy both depend on this process. It takes domain knowledge to find key variables and correlations in the dataset in order to extract pertinent data and produce useful features. The groundwork for efficient model training is laid by this first phase [9]. Feature engineering makes use of a number of approaches to maximize the performance of the model. To capture the combined influence of two or more factors, one popular method is to use interaction terms. Furthermore, in order to guarantee that the model is reliable and capable of handling situations in which the data may be lacking, it is imperative to manage missing data converting and encoding in categories Moreover, categorical variables are essential because they let the model handle and understand various kinds of input in an efficient manner [7].

Reducing training time and resource needs can be achieved through effective feature engineering. The model gains efficiency and improves its ability to generalize to new data by concentrating on the most pertinent features.

It can be difficult to strike a balance between simplicity and complexity since features that are too complex can result in both underfitting and overfitting. In order to optimize the features' impact on model performance and fine-tune them, data scientists and domain experts must collaborate and engage in iterative testing [2]. A crucial component of ML is feature engineering, which calls for a blend of domain and technical understanding. It entails gathering relevant data, developing useful features, and maximizing their influence on model performance. Data scientists and domain experts can iteratively modify the characteristics by carefully experimenting and working together to find the ideal balance between simplicity and complexity in order to obtain the highest level of predicted accuracy [14].

1.4.2 Model assessment

A critical stage in the ML process is model assessment, which entails assessing how well-trained models perform in order to make sure they generalize well to new data. This procedure entails a thorough examination of numerous measures, including precision, recall, F1 score, and area under the receiver operating characteristic curve (AUC-ROC), in addition to just measuring accuracy. The type of problem will determine which assessment metrics are used, and it is crucial to weigh the trade-offs between them in light of the application's particular objectives. By training and assessing a model on several subsets of the data, cross-validation—a key technique in model assessment—provides a reliable estimate of the model's performance. By doing so, the model's dependability across various data partitions is guaranteed and possible problems such as overfitting or underfitting are helped to identify. In order to understand the behaviour of the model and identify possible areas for development, model assessment also entails closely examining learning curves, confusion matrices, and other visualization tools [7].

Additionally, analysing many models and choosing the one that most closely matches the project's goals is part of the assessment step. This could entail experimenting with various algorithms, adjusting hyperparameters, or investigating ensemble techniques. Data scientists can iteratively refine [6] models, resolving flaws and strengthening strengths to create a more reliable and efficient machine through thorough model assessment.

1.4.3 Balancing variance and bias

In the field of ML, creating models that successfully generalize to new data requires finding a careful balance between two essential components: variance and bias. Variance quantifies the model's responsiveness to modifications in the training set, whereas bias signifies the errors caused by oversimplifying actual scenarios. Reducing bias could require investigating more complex techniques, adding more pertinent data, or making the model more complex. Methods to reduce bias and improve the accuracy of the model include feature engineering, increasing model complexity, and utilizing complex architectures. On the other hand, overfitting—which is associated with high variance—presents a problem when a model grows unnecessarily complicated and starts to perceive noise in the training set as real patterns. The use of regularization techniques combats overfitting. These techniques punish overly complicated models; dropout in neural networks and pruning in decision trees are two examples. Furthermore, increasing the training dataset's size or using strategies like cross-validation can help with variance management.

It takes trial and error with various feature sets, regularization strategies, and model complexity to find the best trade-off between bias and variance. Models must be continuously monitored and corrected during the training process to keep them from deviating towards simplicity or complexity and to guarantee that they function as best they can on fresh, untested data. Keeping bias and building robust models that generalize well beyond the training data requires finding the proper equilibrium, which is a fundamental difficulty in ML. To sum up, managing the complexities of bias and variance in ML calls for a sophisticated strategy that entails ongoing testing and optimization. Practitioners can achieve the perfect balance between variance and bias by carefully monitoring the model's performance and selectively adding varied strategies. This will ultimately result in models that are robust and have good generalization capabilities on new datasets.

1.5 RESPONSIBLE DEPLOYMENT OF MACHINE LEARNING MODELS

ML model deployment responsibly is a multifaceted process that takes into account social, legal, and ethical considerations in addition to performance measures. To ensure that decision-making processes are comprehensible and interpretable, transparency is essential. Moreover, good documentation of the model architecture, data sources, and feature engineering techniques builds user and stakeholder trust. Methods such as LIME (Local Interpretable Model-agnostic Explanations) assist people in comprehending the rationale behind a given choice. When using ML models ethically, ethical factors must be taken into account. It's critical to recognize and reduce biases in training data and algorithms since they can reinforce social injustices. Models that comply with ethical norms are a result of employing fairness-aware methodologies, continuously [12] checking for differential impacts, and integrating multiple perspectives into the creation process.

Another important ethical consideration is protecting user privacy. Users' trust is increased when data usage is disclosed openly, privacy-preserving measures are put in place, and data protection rules are followed. When it comes to data utilization and consent for data collection and model deployment, responsible deployment entails open and honest communication with consumers. Reliability in deployment implies a sustained effort that goes beyond the first model release; it includes frequent upgrades, situational awareness monitoring, and user input integration. Organizations can utilize these rules to make sure their use of ML technology is morally just and benefits society [11].

1.5.1 Necessity of responsible model deployment

Given the profound effects ML models have on people, societies, and organizations, responsible model deployment is essential. With the growing integration of ML models into everyday life, there is a greater chance of unforeseen outcomes. It is crucial to address any biases in training data and algorithms in order to optimize ML's benefits while avoiding hazards and ethical problems. Prejudicial results and societal injustices can be sustained by biased models. Users and stakeholders need to be able to comprehend how ML models make decisions, which requires transparency and interpretability. As part of responsible deployment, model predictions must be explained, confidence must be built, and users must be able to comprehend and challenge the judgements made by automated systems. Since ML models frequently rely on big datasets comprising delicate data. Informed permission, strong privacy-preserving methods, and adherence to data protection laws are all necessary for the responsible deployment of models. All things considered, appropriate model deployment is necessary to guarantee favour able societal outcomes and safeguard user privacy [10].

1.5.2 Ethical considerations in machine learning applications

In order to make sure that ML applications adhere to social norms and values, ethical concerns are essential. The possibility of skewed results, which frequently result from biased training data, is one major worry. In order to avoid making biased decisions and maintain current inequities, it is imperative to address these prejudices. In order to practice ethical ML, one must recognize and reduce biases, aim for justice, and divide rewards fairly [11]. Ethics demands transparency and interpretability since users and other stakeholders should be able to comprehend the decision-making process used by ML models. Accountability and trust are increased when model predictions are explained in detail and procedures are made comprehensible. Transparency in data gathering and user consent is also crucial [10].

Another ethical component is the responsible handling of sensitive information, which includes adhering to privacy-preserving methods, secure data storage, and rules pertaining to data privacy that are essential to moral ML applications. Employing ML models may have social repercussions; therefore, organizations need to plan ahead and deal with any unexpected effects [9].

1.6 FUTURE TRENDS IN MACHINE LEARNING

The future of ML is full with interesting possibilities since it develops quickly. One major breakthrough is the combination of ML with the cutting-edge technologies such as edge computing and quantum computing. Using the ideas of quantum mechanics, quantum ML in particular offers a promising approach to solving complicated problems more quickly. This could perform faster and more computationally efficient than conventional ML techniques. In addition, edge computing aims to analyse data nearer to the source in order to minimize latency and facilitate instantaneous decision-making. It is anticipated that the fusion of these technologies would open up new avenues in a number of fields, such as autonomous systems, finance, and healthcare, where quick and informed decision-making is critical [7]. An further noteworthy development influencing ML in the future is the increased focus on interpretability and explainability of model outputs. Increasing integration of AI systems into crucial decision-making procedures necessitates that these models be as clear and understandable as possible. Researchers and academics are working hard to find ways to make complex models easier to understand so that people can understand why AI makes the decisions that it does. This movement emphasizes the significance of responsibility in the application of ML solutions, which is in line with ethical concerns and a larger social push for responsible AI [5].

Additionally, the increasing use of AI in decision-making processes emphasizes the necessity of clear and intelligible models, particularly in delicate fields like banking and healthcare. Achieving interpretability makes it possible for end users and domain experts to have confidence in the choices made by ML models. For AI technology to be widely adopted and accepted across a variety of industries, this trust is essential. The combination of cutting-edge technologies and a greater emphasis on interpretability and explainability will define ML in the future [6]. Quantum and edge computing together have the potential to completely transform a number of industries, and the focus on transparent and understandable models is in line with social demands for responsible AI as well as ethical considerations. Together, these tendencies are forming a future where ML not only offers sophisticated skills but also does so in a way that inspires trust and confidence in its users [3].

1.6.1 Issues and challenges in machine learning

Even if ML has a bright future, there are still certain difficulties. The ethical and societal implications of AI applications are ongoing problems. ML models have the ability to introduce bias into both algorithmic decision-making and training data, which could exacerbate already-existing societal inequities. Research and development must continue in order to address these biases and guarantee justice in AI systems, which continue to be significant problems. Other challenges are scalability and resource constraints. The need for processing power and storage capacity rises with the complexity of ML models and the size of datasets. It will take advances in distributed computing architectures, hardware, and algorithmic efficiency to overcome these scalability issues.

Another difficulty that lies ahead is interdisciplinary teamwork. ML in the future will increasingly need when knowledge in a variety of domains, including domain-specific expertise, morality, and legal issues. Reducing the distance between non-technical stakeholders and technical experts is essential for prudent and successful ML deployments. In spite of these obstacles, overcoming them offers chances for creativity and advancement. Resolving ethical issues, improving scalability, and encouraging interdisciplinary cooperation are critical areas of concentration to guarantee ML's sustained success and beneficial effects in the years to come [3].

1.6.2 Emerging trends in ML research and applications

NLP, edge AI, RL, and transfer learning are some of the revolutionary trends that are transforming ML. RL has demonstrated efficacy in robotics, gaming, and self-governing systems by means of trial and error model training. By investigating methods to enhance the effectiveness and practicality of RL algorithms, research is breaking new ground in the field of decision-making. ML models may become more adaptive and capable of generalizing to a wider range of applications through the process of transfer learning, which entails training models on one task and applying the knowledge gained to perform better on a related task [5]. NLP and ML are being combined by NLP to enhance human–computer interactions. More sophisticated language models, such as GPT (Generative Pre-trained Transformer) models, demonstrate the possibility of robots comprehending and produce text that seems human. As devices at the edge of the network take on ML activities locally to cut down on latency and reliance on centralized cloud services, edge AI is becoming more and more popular [3].

These tendencies do, however, offer present opportunities as well as difficulties. Concerns about bias reduction, ethical issues, and the social effects of ML applications must always be on the minds of researchers and practitioners. The future of ML will be shaped by the further investigation of these tendencies, which will also have an impact on how AI systems interact with and enhance human capabilities in a variety of disciplines.

1.7 CONCLUSION

The field of AI ethics is intricate and multidimensional, involving the appropriate application of AI technologies, moral conundrums, and developing frameworks. The obstacles include possible social consequences and the existence of intrinsic biases in training data, which call for an anticipatory approach to ethical issues. In order to reduce risks and promote ethical AI activities, this project has justice, openness, and privacy as major emphasis points. To effectively navigate the complexities of AI ethics, new ethical frameworks that place a premium on user permission, accountability, and openness must be established. These frameworks provide an organized method for navigating moral dilemmas in the rapidly changing field of AI. They serve as guiding principles for a variety of stakeholders, including legislators, developers, and decision-makers. The responsible application of AI is essential to converting moral precepts into workable processes. It entails making sure AI applications are visible, comprehensible, and closely supervised in addition to being morally right. AI systems must be transparent and comprehensible in order to uphold responsibility and foster trust among users and the general public. A collective dedication to responsible deployment is essential if we are to fully realize AI's revolutionary potential in a way that aligns with human values and the well-being of society. A balanced combination of moral concerns, the creation of ethical frameworks, and an unrelenting commitment to guide AI towards a future that values justice, accountability, and beneficial societal effect are necessary to achieve responsible deployment of AI. Stakeholders may uphold appropriate deployment methods and confront ethical issues head-on make sure AI technologies respect human values and help create a more morally upright and prosperous future for society.

REFERENCES

1. Graepel, T., Lauter, K., & Naehrig, M. (2012). ML Confidential: Machine Learning on Encrypted Data. *International Conference on Information Security and Cryptology (ICISC 2012): Information Security and Cryptology*, Seoul, Korea, November 28–30, 2012.
2. King, D. E. (2009). Dlib-ml: A Machine Learning Toolkit. *Journal of Machine Learning Research*, 10, 1755–1758.
3. Endo, A., Kuroda, M., & Tanzawa, K. (1976). Competitive Inhibition of 3-Hydroxy-3-Methylglutaryl Coenzyme A Reductase by ML-236A and ML-236B Fungal Metabolites, Having Hypocholesterolemic Activity. *FEBS Letters*, 72, 323–326.
4. Milner, R. (n.d.). *The Definition of Standard ML: Revised*. MIT Press. Rediff Books.
5. Heintze, N. (1994). Set-Based Analysis of ML Programs. *ACM SIGPLAN Lisp Pointers*, 7, 306–317.
6. Endo, A., Kuroda, M., & Tsujita, Y. (1976). ML-236A, ML-236B, and ML-236C, New Inhibitors of Cholesterogenesis produced by *Penicillium citrinium*. *The Journal of Antibiotics*, 29 (12), 1346–1348.
7. Pottier, F., & Simonet, V. (2002). Information flow inference for ML. *POPL '02: Proceedings of the 29th ACM SIGPLAN-SIGACT Symposium on Principles of Programming Languages.* 319–330.

8. MacQueen, D. (1984). Modules for standard ML. *LFP '84: Proceedings of the 1984 ACM Symposium on LISP and Functional Programming,* Austin, TX, USA, August 6–8, 1984.

9. Pottier, F., & Simonet, V. (2003). Information Flow Inference for ML. *ACM Transactions on Programming Languages and Systems (TOPLAS),* 25(1), 117–158.

10. Gee, M. W., Siefert, C. M., Hu, J. J., Tuminaro, R. S., & Sala, M. G.

11. Leroy, X. (1990). The ZINC experiment: An economical implementation of the ML language (RT-0117). INRIA.

12. Kalinowski, S. T., Wagner, A. P., & Taper, M. L. (2006). A Computer Program for Maximum Likelihood Estimation of Relatedness and Relationship. *Molecular Ecology Notes,* 6 (2), 576–579.

13. Rubin, D. B., & Thayer, D. T. (1982). EM algorithms for ML factor analysis. *Psychometrika,* 47 (1), 69–76.

14. Milner, R. (1983). *A Proposal for Standard ML.* University of Edinburgh.

15. Roddam, A. W., Duffy, M. J., Hamdy, F. C., Ward, A. M., Patnick, J., Price, C. P., ... Allen, N. E. (2005). Use of Prostate-Specific Antigen (PSA) Isoforms for the Detection of Prostate Cancer in Men with a PSA Level of 2–10 ng/ml: Systematic Review and Meta-Analysis. *European Urology,* 48, 386–399.

Chapter 2

Opportunities of intelligent machine learning techniques for sustainable development

*K. Rajendra Prasad, K. Naveena,
V. Deekshitha, P. Lohitha, and G. Sucharitha*

2.1 INTRODUCTION

Artificial intelligence (AI)-driven solutions [1,2] offer instruments for risk assessment, scenario analysis, and predictive modeling in sustainable management. This chapter looks at how AI can help optimize waste management, energy use, and supply chains, all of which can result in increased productivity and smaller environmental footprints. Adaptive planning, resilient strategy creation, and real-time monitoring are made possible by the integration of intelligent algorithms into management systems. Furthermore, the abstract explores the moral issues and difficulties related to the advancement of AI in sustainable engineering and management. It is crucial to address problems such as bias, which are vital for maintaining good governance in the public sector, in order to guarantee the proper use of modern technologies.

Due to its unmatched capacity for data extraction, AI can be used as a means of connection in order to sustain advancement. It will enable green investors to evaluate weather risks and serve as the foundation for innovative solutions in the industry. AI can also be used to track the advancement of sustainable development and mobilize a sizable portion of society to support the cause. The summary concludes by highlighting the revolutionary potential of AI and ML foundations in promoting sustainable engineering and management practices. Adopting intelligent systems helps to create a more resilient and environmentally responsible future in addition to increasing efficiency. This investigation lays the groundwork for more study and advancement, directing the incorporation of AI technologies in engineering and management fields in the direction of a sustainable paradigm. The integration of AI with sustainable engineering and management has become a crucial factor in tackling the intricate problems presented by our swiftly changing global environment. The foundational field of machine learning (ML), which forms the basis for intelligence decision-making and optimization procedures, is at the center of this fascinating field. Understanding the foundations of ML is essential for maximizing the potential of AI

DOI: 10.1201/9781003565529-2

applications in the pursuit of sustainable development [3]. ML enables systems to learn from data, spot patterns, and make predictions or decisions without the need for explicit programming. In the context of sustainable engineering and management, this introduction seeks to shed light on the importance of computer learning foundations. ML is an innovation accelerator in the field of sustainable engineering [4], where resource optimization and minimizing environmental effect are critical. Practitioners may create intelligent systems that are capable of adaptive learning, which will optimize workflows, forecast results, and allocate resources more effectively by grasping the fundamentals of ML.

2.1.1 Application of AI in sustainable administrator and engineering

Optimize Resource Allocation: AI algorithms are able to analyze vast volumes of data in order to find inefficiencies and optimize the distribution of resources, which reduces waste and boosts output.

Boost Predictive Capabilities: By using ML models to forecast changes in the environment, energy usage trends, and infrastructure maintenance requirements, proactive risk management and decision-making are made possible.

Enhance Decision Support Systems (DSSs): AI-driven DSSs enable stakeholders to receive real-time insights and recommendations, facilitating well-informed decision-making that strikes a balance between social, environmental, and economic goals.

Encourage Innovation: By utilizing AI methods such as deep learning and reinforcement learning, scholars and professionals can create novel answers to challenging sustainability issues, propelling breakthroughs in technology and revolutionary shifts in society.

2.2 LITERATURE STUDY

There is an equally strong connection between ML and sustainable management. Predictive modeling, data-driven decision-making, and real-time monitoring are made possible by ML algorithms, which give enterprises the ability to adopt flexible and adaptable tactics. Gaining an understanding of ML principles is crucial for implementing intelligent technologies that optimize energy usage, boost supply chain management, and increase waste management procedures. We will examine supervised and unsupervised learning, reinforcement learning, and the rapidly developing topic of deep learning as we delve into the complexities of ML basics. These ideas are the fundamental building blocks upon which intelligent systems are built, enabling the proactive, data-driven development of strategies and the extraction of insightful information from massive datasets.

Nonetheless, it is imperative to recognize the ethical aspects linked to the incorporation of ML within sustainable environments. In order to ensure the ethical and fair development of these technologies, we must address the challenges of bias, transparency, and accountability as we explore the possibilities of AI for sustainable engineering and management [5]. This investigation of the foundations of ML in the context of AI for sustainable engineering and management opens the door to a future in which intelligent systems [6,7] will play a major role in the worldwide pursuit of sustainability. Through acquiring a fundamental comprehension of ML, professionals can set out on a path to harness its potential, ultimately influencing a more robust, effective, and ecologically aware globe.

System optimization for materials obtained from biomass is aided by ML. Data difficulties for sustainability assessments are addressed by ML [8,9]. Modern neutral network models and ensemble models both perform and generalize well.

Energy Management: Examine how ML approaches might help integrate renewable energy sources more easily, optimize energy usage, and increase energy efficiency. Observation of the Environment and Discussion: Talk about how AI is used to track environmental changes and aid in conservation initiatives. [10,11]. Management of a sustainable supply chain: Energy management: Examine how ML approaches might help integrate renewable energy sources more easily, optimize energy usage, and increase energy efficiency. Observation of the Environment and Discussion: Talk about how AI is used to track environmental changes and aid in conservation initiatives [10,11].

Within the larger field of AI, ML has become a game-changer, allowing computers to understand patterns and make data-driven judgments without the need for explicit programming. This literature background study explores the basic ideas of ML, following its historical evolution, comprehending its essential algorithms, and investigating its modern applications. Our goal in offering this thorough review is to set the foundation for a more in-depth comprehension of the theoretical foundations and real-world applications of ML. The foundation for ML was established in the middle of the 20th century by trailblazers such as Alan Turing and Arthur Samuel. But ML didn't see a comeback until the past 10 years, thanks to advances in algorithmics, computing power, and data abundance. Later advancements in the subject were built upon the foundation of classical ML, which is commonly divided into supervised and unsupervised learning.

2.2.1 Core machine learning algorithms

A solid understanding of ML's underlying algorithms is necessary. Algorithms such as support vector machines for the classification and linear regression for regression tasks are examples of supervised learning, which is typified by labeled training data. When working with unlabeled data, unsupervised learning employs dimensionality reduction strategies such as principal component analysis (PCA) and grouping methods such as k-means. In domains such as robotics and gaming, reinforcement learning—in which agents learn by interacting with an environment—has become increasingly popular. Theory of Statistical Learning: Statistical learning theory, which offers a strict framework for comprehending the procedures involved in learning from data, is the foundation of ML. When evaluating ML models, concepts such as generalization error, overfitting, and the bias-variance tradeoff are essential. To determine the circumstances in which learning algorithms converge and produce precise predictions, researchers delve into theoretical details, adding to the dependability and robustness of ML applications.

ML is revolutionizing industries and solving difficult problems across a wide range of fields. Using algorithms for image analysis and predictive modeling, ML helps healthcare providers diagnose illnesses and create individualized treatment regimens. Financial organizations increase security and efficiency by using ML for fraud detection and risk assessment. Advances in computer vision and natural language processing (NLP) have been impressive, with large-scale models such as OpenAI's GPT-3 demonstrating the promise.

2.3 CHALLENGES AND OPPORTUNITIES

2.3.1 Data quality and accessibility

Talk about the difficulties in obtaining, analyzing, and integrating data while discussing AI applications for sustainable engineering and management. Implications for Society and Ethics: Examine moral issues including algorithmic prejudice, privacy problems, and how AI will affect jobs in sustainable businesses. restrictions, such as those pertaining to data privacy and environmental standards, that control the application of AI in sustainable management and engineering.

2.3.2 Case studies and best practices

Showcase effective applications of AI and ML in engineering and sustainable management initiatives. Highlight the cutting-edge techniques and technologies that have had positive effects on society and the environment.

2.3.3 Future directions and research opportunities

Determine new directions in research that need to be explored, such as the use of computer vision, NLP, and reinforcement learning in sustainable engineering and management.

Talk about the value of knowledge exchange between government, business, and academia as well as multidisciplinary collaborations.

2.3.4 Problem identification and definition

Clearly state the issue or challenge in engineering or sustainable management that AI and ML can help with. This could involve cutting waste, increasing resource efficiency, and optimizing energy use, among other things.

2.3.5 Data collection a preprocessing

Determine which data sources—such as sensor data, historical documents, satellite imaging, etc.—are pertinent to the issue at hand. To assure consistency and quality, handle missing numbers, and eliminate noise from the data, clean and preprocess it.

2.3.6 Feature selection and engineering

Determine the variables or pertinent elements in the data that are crucial for modeling. If additional features are needed to improve the model's prediction ability, develop them.

2.3.7 Model selection and training

Choose the right ML models and algorithms for the task, taking into account variables such as complexity, size, and kind of data.

Train the chosen models with the preprocess data, and assess their performance using methods such as cross-validation.

2.3.8 Model evaluation and validation

Use relevant measures, such as accuracy, precision, recall, or F1-score, to assess the trained models. To make sure the models can be applied to new data, validate them using different test datasets or methods such as k-fold cross-validation.

2.3.9 Application of AI in sustainable
management and engineering

Optimize Resource Allocation: AI systems are able to examine enormous volumes of data in order to find inefficiencies and optimize the use of resources, which lowers waste and raises output.

Boost Predictive Capabilities: By using ML models to forecast changes in the environment, energy usage trends, and infrastructure maintenance

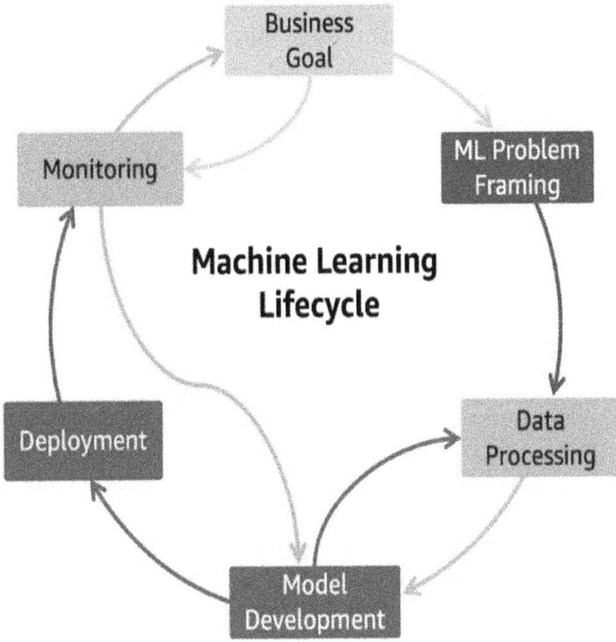

Figure 2.1 Sustainable development model reference in AI.

requirements, proactive risk management and decision-making are made possible.

Enhance Decision Support Systems: AI-driven DSSs enable stakeholders to receive real-time insights and recommendations, facilitating well-informed decision-making that strikes a balance between social, environmental, and economic goals.

Encourage Innovation: By utilizing AI methods such as deep learning and reinforcement learning, scholars and professionals can create novel answers to challenging sustainability issues, propelling breakthroughs in technology and revolutionary shifts in society.

2.4 PROPOSED STEPS TOWARD VARIOUS SUSTAINABILITY APPLICATIONS

A robust framework that is adapted to various settings is necessary for the efficient integration of ML basics into the sustainable management of AI/ML. This technique outlines effective strategies for implementation and highlights potential benefits in order to pinpoint specific areas and scenarios where AI/ML can improve sustainable practices. The following actions are suggested for the different applications of sustainable development.

Use ML techniques to examine past energy usage trends and optimize energy use instantly. By preventing equipment breakdowns, predictive maintenance models can save downtime and energy usage. Use reinforcement learning to promote effective resource allocation and adaptive energy utilization in dynamic contexts. Maintain a close eye on and evaluate the efficacy and efficiency of used AI/ML solutions. Make use of key performance indicators (KPIs) that are in line with sustainability objectives, such as waste reduction and energy savings.

To improve the sustainability of the supply chain, use ML algorithms for inventory management, route optimization, and demand forecasting. Utilize real-time data to dynamically modify supply chain procedures, minimizing waste of resources and the impact on the environment. In order to ensure the resilience of the supply chain, incorporate predictive analytics to foresee possible interruptions and take proactive measures to address them.

To maximize fuel efficiency, cut emissions, and optimize waste collection routes, use smart waste management systems with ML algorithms. To support proactive waste management initiatives, use predictive analytics to forecast trash generation patterns. Optimize water use in urban settings, industrial processes, and agricultural by implementing ML techniques. Use predictive modeling to prepare for sustainable consumption and to foresee resource availability. Utilize sensor data and ML to optimize yields through precision agriculture, reducing water and resource consumption.

Use ML techniques to improve occupant comfort, optimize energy use, and intelligent building management systems. Reduce energy consumption related to equipment breakdowns by using predictive analytics to forecast maintenance needs. Make dynamic adjustments to lighting, (Heating, Ventilation, and Air Conditioning) HVAC, and other systems through real-time monitoring to ensure effective resource use.

2.5 CONCLUSION AND SCOPE OF THE RESEARCH

In summary, there is great potential for improving sustainable engineering and management processes through the incorporation of AI into ML. We can solve difficult problems pertaining to resource optimization, environmental preservation, and societal well-being by methodically applying AI technology. In order to achieve sustainable development goals, more resilient, egalitarian, and efficient systems must be created. This can be done through the synthesis of AI and ML approaches. The convergence of AI, ML, and sustainable management calls for interdisciplinary research and knowledge sharing across disciplines, which speeds up the transition to sustainability. In conclusion, a comprehensive strategy that strikes a balance between technical innovation and ethical principles, societal values, and environmental stewardship is needed for the successful integration of AI in ML for sustainable management and engineering. We can build a more resilient and sustainable future for future generations if we responsibly and cooperatively use AI.

REFERENCES

1. J. Liu et al., "Artificial Intelligence in the 21st Century," In *IEEE Access*, vol. 6, pp. 34403–34421, 2018, Doi: 10.1109/ACCESS.2018.2819688.
2. Y. Xin et al., "Machine Learning and Deep Learning Methods for Cybersecurity," In *IEEE Access*, vol. 6, pp. 35365–35381, 2018, Doi: 10.1109/ACCESS.2018.2836950.
3. "Sustainable Engineering, Energy and Environment [Table of Contents]," 2013 *African*, Pointe aux Piments, Mauritius, 2013, pp. 1–21, Doi: 10.1109/AFRCON.2013.6757875.
4. "Management for Sustainable Development," In *Management for Sustainable Development*, River Publishers, 2016, pp.-xviii.
5. S. Soni and U. Yadav, "Sustainable Supply Chain Management: Research Review and its Future," *2022 International Conference on Computing, Communication, and Intelligent Systems (ICCCIS)*, Greater Noida, India, 2022, pp. 930–936, Doi: 10.1109/ICCCIS56430.2022.10037713.
6. H. Qu, P. Ling and L. Wu, "Electricity Consumption Analysis and Applications Based on Smart Grid Big Data," *2015 IEEE 12th Intl Conf on Ubiquitous Intelligence and Computing and 2015 IEEE 12th Intl Conf on Autonomic and Trusted Computing and 2015 IEEE 15th Intl Conf on Scalable Computing and Communications and Its Associated Workshops (UIC-ATC-Com)*, Beijing, China, 2015, pp. 923–928, Doi: 10.1109/UIC-ATC-ScalCom-CBDCom-IoP.2015.176.
7. H. N. Saha et al., "Waste Management Using Internet of Things (IoT)," *2017 8th Annual Industrial Automation and Electromechanical Engineering Conference (IEMECON)*, Bangkok, Thailand, 2017, pp. 359–363, Doi: 10.1109/IEMECON.2017.8079623.
8. D. F. Mujtaba and N. R. Mahapatra, "Ethical Considerations in AI-Based Recruitment," *2019 IEEE International Symposium on Technology and Society (ISTAS)*, Medford, MA, USA, 2019, pp. 1–7, doi: 10.1109/ISTAS48451.2019.8937920.
9. C. Cichy and S. Ras, "An Overview of Data Quality Frameworks," In *IEEE Access*, vol. 7, pp. 24634–24648, 2019, Doi: 10.1109/ACCESS.2019.2899751.
10. V. Sitharamulu, K. Rajendra Prasad, K. S. Reddy, A. V. K. Prasad, and M. V. Dass, "Hybrid Classifier Model for Big Data by Leveraging Map Reduce Framework," In *International Journal of Data Mining, Modelling and Management, InderScience*, vol. 16, Issue 1, pp. 23–48, 2024.
11. K. Rajendra Prasad, K. Narasimhulu, C. N. Santhosh Kumar, and N. Ramanjaneya Reddy, "Sampling-Based Fuzzy Speech Clustering Systems for Faster Communication with Virtual Robotics toward Social Applications," *Journal of Soft Computing, Springer*, pp. 1–9, 2023, Doi: 10.1007/s00500-023-08273-y.

Chapter 3

Disease prediction based on drug reviews using TF-IDF in natural language processing

Anita Hembram, Mamata Garanayak, and Bijay Kumar Paikaray

3.1 INTRODUCTION

The amount of routinely gathered information in the healthcare industry has increased exponentially since the last 10 years. The medical literature now often references procedures for managing and analyzing massive information set, such as machine learning (ML), which have grown in popularity as an outcome. These procedures have occasionally shown remarkable outcomes in challenging tasks such as categorization and natural language interpretation [1]. However, ML procedures frequently do not outperform conventional statistical procedures in terms of prediction [2], are not well documented [3], and give rise to issues with interpretability and generalizability [4]. Healthcare prototypes will advance by starting with clinical academics, addressing both the potential and limitations of modern data science. Clinicians are particularly positioned to spot chances for ML to assist patients [5]. Despite this, big data analytics training is rarely given to physicians. The chapter aims to offer readers knowledge of how popular ML procedures and natural language processing (NLP) can be utilized to analyse text passages provided by patients as reviews of medications they have taken for various purposes.

Review text may be a huge place of origin of information to support decision-making and improve quality. Examples of written text include medical records, patient reviews, evaluations of physicians' work, and remarks on social media. It is theoretically feasible to gather text-relied data from the internet such as social media evaluations of healthcare services via a technique termed web scraping, which is simple to carry out with open source software [6]. Specific text from websites such as comments on patient forums may be identified and downloaded by utilizing web scraping software, which then stores the data in databases so it can be examined later. The analysis of drug review dataset by the patients is the main emphasis of this chapter.

As these open text databases are so large, it would be unfeasible to choose qualitative research approaches for manually synthesizing all of the

DOI: 10.1201/9781003565529-3

relevant materials. Text passages can be transformed into comprehensible datasets for analysis by statistical and ML prototypes through a series of steps termed natural language processing (NLP) [7]. In order to automatically convert clinical text into structured clinical data that can be analyzed directly by utilizing ML procedures, it is imperative that NLP techniques be developed. NLP is being more widely utilized in the healthcare setting for a variety of purposes, including as identifying biological concepts from radiology reports, nursing paperwork, and discharge summaries [8]. However, there is a lack of widespread utilization of NLP-dependent frameworks for clinical narratives in clinical settings to assist workflows or decision support prototypes.

The chapter is organized in the following way: The fundamental ideas discussed in Section 3.2 are necessary to comprehend the chapter's work and the illness prediction using NLP. The Literature Review is presented by the authors in Section 3.3. The earlier research on illness prediction by utilizing various NLP approaches is discussed in this section. Motivation and goals are emphasized in Section 3.4 and are met in this article. Problem statements are discussed in Section 3.5. In order to build a new strategy for the virtual machine, this part describes the problem statements and discusses issues that were present in the prior system. Section 3.6 reviews proposed work. This chapter discusses the proposed work, which utilizes the suggested strategy to increase system efficiency. The authors provide a description of the implementation and outcomes in Section 3.7. Section 3.8 provides the conclusion and recommendations for the future.

3.2 BASIC CONCEPTS

3.2.1 Machine learning

In relation to computers, AI, ML as well as additionally data analysis, have expanded quickly in recent years, usually enabling the applications to perform in an intelligent manner [9]. ML is often considered as the most popular current technology of the Fourth Industrial Revolution. It typically gives prototypes the aptitude for acquiring knowledge to study and upgrade from experience automatically without being manually designed. Four main categories of learning procedures may be distinguished in this domain: supervised, un-supervised, semi-supervised, and reinforcement learning [10].

Type and qualities of the dataset as well as the functionality of the learning techniques determine how successful and efficient a ML solution is. To efficiently create data-driven prototypes, ML procedures can be employed in the areas of classification, regression, clustering of data, feature engineering and dimensional reduction, and reinforcement ML [11].

3.2.2 Natural language processing

NLP has garnered a lot of attention lately as a procedure for computationally expressing and understanding human language. Its utilizations have spread to the extensive range of sectors, including machine translation, question answering, information extraction, summarization, and the medical sector [12]. The main focus of NLP is making words and phrases comprehensible to computers. It falls into two categories: NLU, or linguistics, that was developed to reduce person's workloads and satisfy the aspiration to speak with the computer in natural languages, and Natural Language Generation, which advances the task of comprehending and producing text [13]. Furthermore, NLG, also termed text generation, involves generating coherent and contextually relevant linguistic outcome from data that is structured in nature or internal representations.

3.2.2.1 Components of natural language processing

NLP is subcategorized into two parts: Natural Language Generation and Natural Language Understanding. NLP advances the process of text generation and comprehension. General categorization of NLP is shown in Figure 3.1.

3.3 NAÏVE BAYES

A batch of classification procedures based on Bayes' theorem is termed Naïve Bayes classifiers. Naïve Bayes makes simplifying assumptions, as indicated by the term "Naive." Given the class label, the classifier makes the assumption that the characteristics utilized to characterize an observation are conditionally independent. In real-world scenarios, Naïve Bayes'

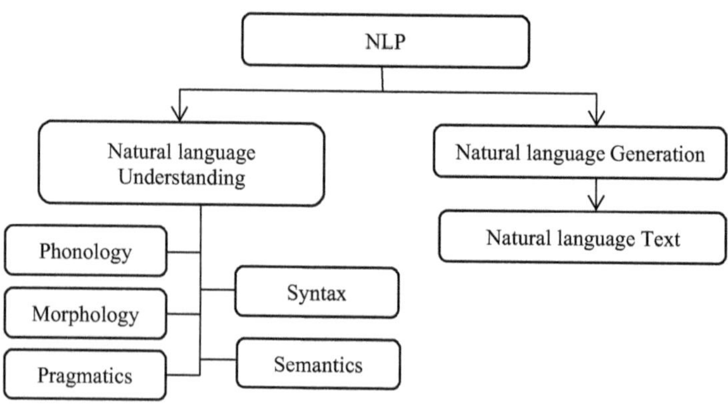

Figure 3.1 Components of NLP.

assumptions are typically incorrect. The independence assumption is really never true, even though it frequently functions effectively in real life [14]. Three varieties of Naïve Bayes models exist, which are as follows:

3.4 GAUSSIAN NAÏVE BAYES CLASSIFIER

Continuous values connected to every feature in Gaussian Naïve Bayes are taken to follow a Gaussian distribution.

3.5 MULTINOMIAL NAÏVE BAYES

A feature vector illustrates how usually a multinomial distribution has produced a given set of events. This event model is commonly employed in the process of classifying documents.

3.6 BERNOULLI NAÏVE BAYES

Features in the event of multivariate Bernoulli prototype are not dependent on variables that are binary in nature, or Booleans that describe inputs. Similar to the multinomial prototype, this prototype is well-liked for document classification problems when term frequencies are not as useful as binary term occurrence [15].

3.7 TF-IDF

Term frequency denotes the frequency of occurrence of a specific term within the essay. In the experimental part, terms are represented as n grams, which can consist of character, word, part-of-speech tag, or a part of blend thereof.

$$w\!f_{t,e} = \begin{cases} \dfrac{1 + \log\left(t\, f_{t,e}\right) \quad \text{if } t\!f_{t,e} > 0}{0 \qquad\qquad \text{Other case}} \end{cases} \tag{3.1}$$

In the equation (3.1), $w\, f_{t,e}$ is the weight and $t\, f_{t,e}$ is the frequency term t in e essay.

The idea that a phrase that appears often in essay is not an effective discriminator and must be obtained minimizes weight than the one that happens in fewer essays is quantified by Inverse Document Frequency (IDF) [16].

$$i\, df\left(t_i\right) = \log \frac{N}{n_i} \tag{3.2}$$

In equation (3.2), N is number of essays in the corpus and t_i happens in n_i of them. IDF provides a brand new weight when merged with TF to obtain TF-IDF. It merges the weights of TF and IDF by multiplying both of them. TF produces maximum weight to a frequent term in an essay and IDF down scales the weight if the term happens in many essays. Equation (3.3) gives the final weight that each term of the essay gets before the normalization.

In equation (3.2), N represents the total no. of essays in corpus, and the term t_i happens in n_i of them. IDF introduces a new weight when merged along with TF to obtain TF-IDF. It merges weights of TF and IDF by multiplying both. TF assigns greater weight to a term that appears frequently in the essay, while IDF minimizes the weight if term is prevalent across many essays. Equation (3.3) illustrates the final weight assigned to each and every term within an essay before the normalization.

$$w_{i,\,e} = \left(1 + \log\!\left(tf_{t,\,e}\right) \right) x \log\!\left(N/n_i\right) \tag{3.3}$$

3.7.1 Literature review

The literature review covers earlier research on drug recommendation prototypes that make utilization of NLP. This section examines the many options that researchers have put up to address this issue.

A prototype for a drug recommendation system was created by Nayak et al. [17] using symptoms entered by the user to suggest medications with possible side effects. The illnesses are predicted using four different prototypes. The reviews are analyzed using NLP-based sentiment analysis and the Vader tool. Lastly, weighted average and probabilistic approaches are used to suggest the drugs. The study of Kumar et al. [18] makes a significant addition to the area of AI-based medical diagnostics. Their NLP-DL method presents a viable path towards creating accessible and non-invasive speech analysis-based illness prediction systems. Users record voice descriptions of their symptoms. NLP methods are utilized to translate speech to text and extract pertinent keywords and phrases. To train a Deep Learning model for illness prediction, extracted characteristics are utilized. The authors compare and contrast different ML prototypes like ensemble voting classifier, SVM, as well as logistic regression. In order to predict illnesses from unstructured voice data, a deep learning prototype including an embedding layer, a LSTM layer, as well as dense layer has been suggested. It achieves a promising accuracy of 98.94%.

An innovative methodology was presented by Yu [19] to investigate the viability of integrating NLP and ML for symptom-based data-based disease prediction. A reliable NHS website is used to scrape the data about symptoms. In the framework, powerful NLP and ML techniques are used to extract and process 298 disease-related data. To determine the frequency and co-occurrence of symptoms, statistical analysis is done. The project predicts the illness by using Naïve Bayes that is Multinomial Algorithm, Keens, and NLP techniques. Correctness of suggested framework is 93%.

FarSight is presented by Gangavarapu et al. [20] as a means of long-term illness forecasting. Using unstructured clinical nursing notes is especially interesting since it makes use of an important, but frequently underutilized, source of information in the medical field. FarSight predictive models have great potential to improve healthcare outcomes by facilitating early diagnosis and action based on insights from several of data sources. In order to determine the most representative feature space, this approach makes use of vector space and NMF (Nonnegative Matrix Factorization) topic modelling. This information is then utilized to activate reliable illness forecasting by utilizing deep NN.

Heo et al. [21] examined data from two datasets containing radiology reports of individuals who had acute ischemic stroke (AIS). NLP approaches were used by the authors to extract pertinent data, such as the location, size, and severity of the infarct, from the reports. Using this extracted data, deep learning and conventional ML systems were then taught to predict a poor functional result 90 days after the stroke. When it comes to forecasting subpar functional outcomes, NLP-based ML models outperformed traditional models by a large margin. The multi-CNN method (0.805) had the greatest classification performance out of all the ML classifiers, with CNN (0.799) coming in second.

The results of Bacchi et al. [22] show that deep learning NLP is effective at precisely identifying the cerebrovascular origin of TIA-like manifestations. The work demonstrates how these cutting-edge computational techniques may be used in clinical decision-making, providing a viable path towards enhancing patient care and diagnostic accuracy. Among the classifier prototypes tested on complaint history, CNN had highest prediction capacity. The authors recognize that the field of AI applications in healthcare is constantly changing and stress the significance of additional validation and incorporation of these models into standard clinical practice.

Yasashvini et al. [23] investigates the analysis and classification of retinal pictures according to the presence and severity of diabetic retinopathy (DR) by utilizing CNN with a hybrid deep learning technique. Three deep learning architectures were used by the authors: a hybrid CNN along with Residual Network, a freestanding CNN, as well as a hybrid CNN and DenseNet 2.1. For DR classification, the hybrid CNN along with DenseNet 2.1 had the best accuracy (96.22%), much surpassing both the standalone CNN (75.61%) and the hybrid CNN along with ResNet (93.18%). Accuracy, sensitivity, specificity, and AUC-ROC measurements were used to gauge performance.

In their analysis of text data from a text-based programme for mentally ill patients who had recently been released from the hospital, Cook et al. [24] concentrated on answers to the open ended questions, "How do you feel today?" To extract language elements like word choice, mood, and negation, they used NLP approaches. The authors made a comparison between logistic regression forecasting prototypes with structured data and NLP-relied prototypes with an unstructured query. Based only on answers to a straightforward general mood question, NLP-relied prototypes were able to produce pretty good forecasting ranks (both suicidal ideation and high GHQ-12 scores). The research by Kulshrestha et al. [25] adds to the growing body of knowledge on predictive analytics in the medical field by showing how the NLP may be utilized to derive valuable insights from unstructured clinical narratives in order to forecast catastrophic chest injuries. Between 2014 and 2018, clinical records from a trauma centre were reviewed. The inclusion criteria were satisfied for 6,891 traumas. Total documents in the data corpus were 473,694. When NLP and ML were combined; the first 8 hours of clinical record could be used to distinguish between instances of serious chest damage with great accuracy.

3.8 MOTIVATION AND OBJECTIVES

Using specialized engines that can sift through massive amounts of unstructured data, healthcare NLP finds previously unnoticed or incorrectly classified medical problems. NLP may be utilized by medical professionals to determine a drug's potency, frequency, form, and duration. NLP will eventually greatly enhance patient outcomes, doctor decision-making, and the delivery of healthcare.

Following are the various objectives of this chapter:

1. Provide a paradigm for evaluating the effectiveness of integrating NLP & ML techniques in a disease prediction prototype relied on patients' reviews of drugs.
2. A drug review dataset with NLP features is what are employed and also carefully reviewed the data by utilizing lemmatization and stop word analysis.
3. Several ML procedures, such as Naïve Bayes and Passive Aggressive Classifier, have been employed to determine the accuracy of the prototypes.
4. The popular procedure in NLP is TFIDF, assesses term significance inside a document in connection to a class of documents, or in connection to corpus.
5. Computed the accuracy of the various prototypes by utilizing the Naïve Bayes with the TF-IDF procedure. At the end, we assess how accurate each prototype is. The TF-IDF prototype has the highest accuracy (98%), out of all the procedures.

3.9 PROBLEM STATEMENT

TF-IDF is an important methodology which provides recommendation of drugs relied on patients' review. In the architecture, data gathering, data preprocessing, and applying of several prototypes are involved in the drug recommendation. This chapter grasps the challenges in research work with NLP TF-IDF methodologies on recommendation relied on patient's review. Particularly, the research problems are as follows:

- **How to gather the data and analyze the data?**
  It's a challenge to gather an appropriate data of reviews of drugs given by the patients and analyze that data appropriately. Data is collected from the kaggle site and analyzed by utilizing several ML approaches.
- **How to prepare the data for recommendation?**
  The data is prepared for recommendation by replacing the wrong column name with NaN values, deleting the missing values, selecting the patient's condition with number of drugs, importing the stop words, wordnet, and performing word processing.
- **How to design TF- IDF drug recommendation prototype based on drugs review given by the patients?**
  First, the prototype is designed by applying the logistic regression and Naïve Bayes prototype with counter vectorizer and then creates TF-IDF versions of the count vectorizer and fit both the logistic regression on the TF-IDF data. After that by comparing both the Prototype chooses the Best one. Select top 5 conditions and reviews from the merged cleaned train dataset and dominant topic dataset and clean the review column by applying lemmatization to get the cleaned review and finally recommend the top 3 drugs as well as predict the disease from the review
  The following are the fundamental issues that this chapter resolves:
- Predict the disease first relied on patients' review.
- Provides a better drug recommendation prototype relied on reviews given by the patients after using that drug with an accuracy of 98%.

3.10 DATA COLLECTION

Data collection is a crucial component of research projects because it allows investigators to make judgments relied on current information and assess the value of information that will support ongoing investigations.

In this article the data is collected from kaggle site which contain two sets, one is train set and another is the test set. The train dataset consists of 16, 1,297 data with 7 features with 3,436 types of drugs and 885 conditions of patients (Table 3.1) and the test dataset consists of 53, 766 data (Table 3.2) with seven features with 2,637 types of drugs and 709 conditions of patients.

3.10.1 Proposed work

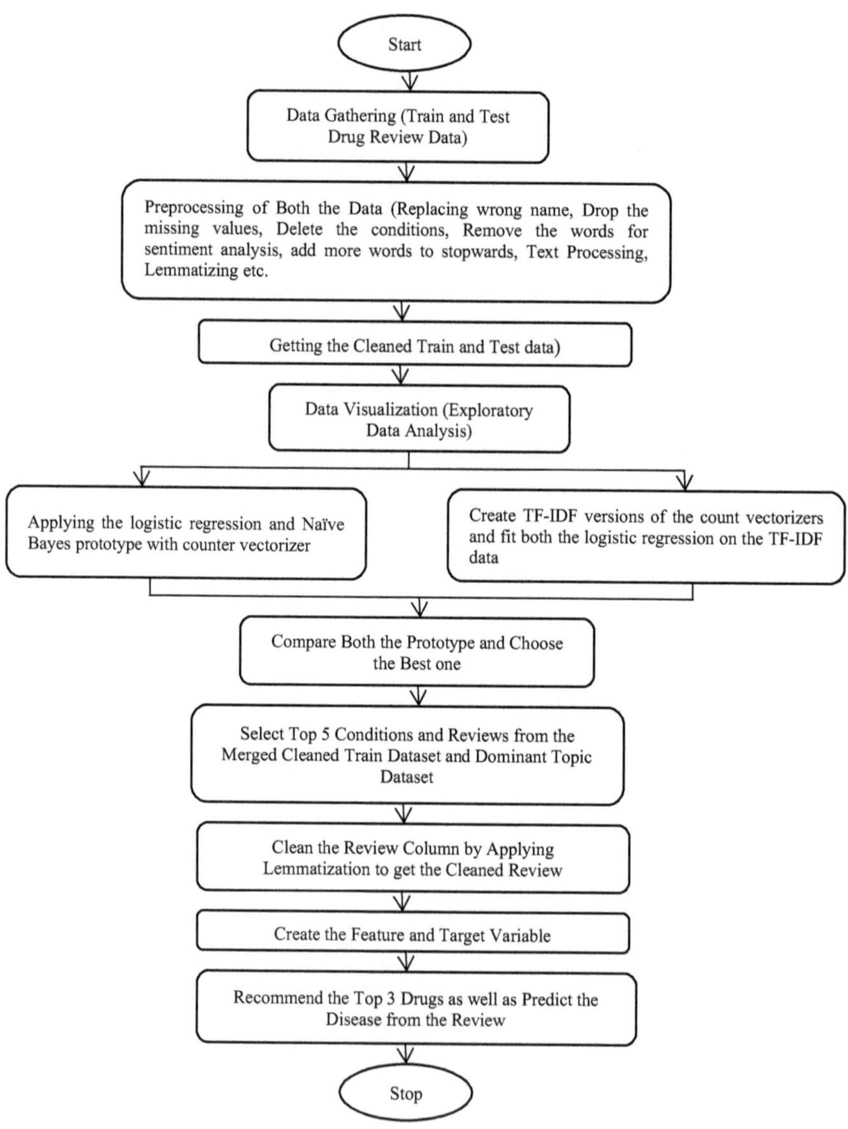

Figure 3.2 Proposed work.

Table 3.1 Train dataset

S.No	Unique-ID	Drug- name	Conditions	Reviews	Ratings	Dates	Useful-counts
0	206461	Valsartan	Left Ventricular Dysfunction	"It has no side effects, I…	9	20-May-12	27
1	95260	Guanfacine	ADHD	"My son is half way through his 4th weeks…	8	27-Apr-10	192
2	92703	Lybrel	Birth Control	"I am used to take other oral contraceptive…	5	14-Dec-09	17
3	138000	Ortho Evra	Birth Control	"It is my 1st time utilizing…	8	3-Nov-15	10
4	35696	Buprenorphine/naloxone	Opiate Dependence	"Suboxone has completely changed my life…	9	27-Nov-16	37
…	…	…	…	…	…	…	…
161292	191035	Campral	Alcohol Dependence	"I wrote my first of…	10	31-May-15	125
161293	127085	Metoclopramide	Nausea/Vomiting	"This IV was given me before surgery. I…	1	1-Nov-11	34
161294	187382	Orencia	Rheumatoid Arthritis	"Limited improvement after 4 months, developed…	2	15-Mar-14	35
161295	47128	Thyroid desiccated	Underactive Thyroid	"I & #39; have been on thyroid medication 49…	10	19-Sep-15	79
161296	215220	Lubiprostone	Constipation, Chronic	"I & #39; have had chronic constipation…	9	13-Dec-14	116

Table 3.2 Test dataset

	Unique ID	Drug name	Condition	Review	Rating	Date	Useful count
0	163740	Mirtazapine	Depression	"I' have tried some antidepressants…	10	28-Feb-12	22
1	206473	Mesalamine	Crohn Disease, Maintenance	"My son has Crohn & #39; disease…	8	17-May-09	17
2	159672	Bactrim	Urinary Tract Infection	"Quick reduction of symptoms"	9	29-Sep-17	3
3	39293	Contrave	Weight Loss	"Contrave merges drugs which are…	9	5-Mar-17	35
4	97768	Cyclafem 1/35	Birth Control	"I have been on this birth control…	9	22-Oct-15	4
…	…	…	…	…	…	…	…
53761	159999	Tamoxifen	Breast Cancer, Prevention	"I have used Tamoxifen for 5 years…	10	13-Sep-14	43
53762	140714	Escitalopram	Anxiety	"I' ve been taking Lexapro (escitaploprgra…	9	8-Oct-16	11
53763	130945	Levonorgestrel	Birth Control	"I & #39; married, 34 years…	8	15-Nov-10	7
53764	47656	Tapentadol	Pain	"I was adviced Nucynta for highest neck…	1	28-Nov-11	20
53765	113712	Arthrotec	Sciatica	"It works!!!"	9	13-Sep-09	46

3.11 DATA PREPROCESSING

1. **Replace wrong name in condition or not listed condition column with NaN**

   After observing the dataset, it is observed that the dataset contain the condition feature in which not listed conditions are also there as shown in Table 3.3, which contains total 214 not listed conditions.

   Therefore, we replaced these not listed conditions with NaN values and drop the rows with missing/NaN values in condition as in Table 3.4. We prefer to drop rows with still missing values in condition as we got 100 rows which contain missing values and 100 rows=0.0006% of total data.

Table 3.3 Not listed condition

Condition	
Not listed	214
Pain	200
Birth control	172
High BP	140
Acne	117
......	.....
Sepsis	1
72 users got this comment helpful	1
Microscopic polyangiitis	1
Short stature	1
Epicondylitis, tennis elbow	1

Table 3.4 Data after removing not listed condition

Condition	
Pain	200
Birth control	172
High BP	140
Acne	117
Depression	105
......	.....
Post-cholecystectomy diarrhea	1
Breast cancer, palliative	1
Postoperative increased intraocular pressure	1
Postpartum breast pain	1
Cachexia	1

```
Index(['Uveitis', 'Postpartum Depression', 'Burns, External',
       'Benign Essential Trem', 'Paranoid Disorde', 'Pancreatic Cance',
       'Social Anxiety Disorde', 'Cervical Dystonia', 'Autism',
       'Chronic Fatigue Syndrome',
       ...
       'Performance Anxiety', 'Peritonitis', 'Pertussis',
       'Platelet Aggregation Inhibition', 'Portal Hypertension',
       'Post-Cholecystectomy Diarrhea', 'Breast Cancer, Palliative',
       'Postoperative Increased Intraocular Pressure',
       'Postpartum Breast Pain', 'Cachexia'],
      dtype='object', name='condition', length=625)
```

Figure 3.3 Condition with less than 10 drugs.

2. **Select and delete the conditions with less than 10 drugs**

   After deletion of the not listed conditions, we have selected the conditions having less than 10 drugs as in Figure 3.3 and delete that drugs from the dataset.

3. **Remove the words for sentiment analysis from stopwords and add more words to stopwords**

   We first import the regular expression, stopwords, wordnet, and punkt to delete the words such as "aren't", "couldn't", "didn't", "don't", "doesn't", "hadn't", "hasn't", "haven't", "isn't", "mightn't", "mustn't", "needn't", "no", "nor", "not", "shalln't", "shouldn't", "wasn't", "weren't", "wouldn't" for sentiment analysis from stopwords and add more words like 'mg', 'week', 'month', 'day', 'january', 'february', 'march', 'april', 'may', 'june', 'july', 'august', 'september', 'october', 'november', 'december', 'iv', 'oral', 'pound', 'lb', 'month', 'day', 'night' to stopwords.

4. **Performed text processing**

   In the text processing step numbers, capital letters and punctuations are removed as in Table 3.5.

   Lemmatization is a text pre-processing procedure that assists NLP prototypes to find similarities by reducing a term to its most basic meaning. A lemmatization procedure, for instance, would reduce the word better to its meaning, or good as in Table 3.6.

   By following the above same five steps for the test dataset we got the cleaned test data as represented in Table 3.7.

3.12 EXPLORATORY DATA ANALYSIS (EDA)

Visualization of data is act of displaying data by utilizing well-known visual like plot, chart, as well as animation. These illustrative visualization aids simplify complex information connections and information driven insights. Additionally, it offers a simple procedure for identifying and comprehending patterns, trends, and outliers in data. To visualize the data distribution, we first merge the cleaned train data and test data as shown in Table 3.8.

Table 3.5 Dataset after preprocessing

	Drug-names	Conditions	Reviews	Ratings	Useful-counts
1	Guanfacine	ADHD	[my, son, is, halfway, …	8	192
2	Lybrel	Birth control	[i, used, to, take, Contraceptive pill…	5	17
3	Ortho Evra	Birth control	[this, is, my, first, time,…	8	10
5	Cialis	Benign prostatic hyperplasia	[nd, day, on, mg, roc…	2	43
6	Levonorgestrels	Emergency contraceptions	[he, pulled,…	1	5
…	…	…	…	…	…
161291	Junel 1.5/30	Birth control	[this, would, be, my,…	6	0
161293	Metoclopramide	Nausea/ vomiting	[i, was, given, this,…	1	34
161294	Orencia	Rheumatoid arthritis	[limited, improvement, after, months, develope…	2	35
161295	Thyroid desiccated	Underactive thyroid	[i, ve, been, on, thyroid, medication, years,…	10	79

Table 3.6 Cleaned train dataset

	Drug name	Condition	Rating	Useful count	Review
1	Guanfacine	ADHD	8	192	son half-way 4th intuniv became concerned be…
2	Lybrel	Birth Control	5	17	used take another contraceptive pill cycle hap…
3	Ortho Evra	Birth control	8	10	1st time using form…
5	Cialis	Benign prostatic hyperplasia	2	43	nd started work rock hard erection however exp…
6	Levonorgestrel	Emergency contraception	1	5	pulled cummed bit took plan b hour later took…
…	…	…	…	…	…
161291	Junel 1.5/0	Birth control	6	0	would second junel birth control year changed…
161293	Metoclopramide	Nausea/vomiting	1	34	given surgey immediately became anxious could…
161294	Orencia	Rheumatoid arthritis	2	35	limited improvement month developed bad rash m…
161295	Thyroid desiccated	Underactive thyroid	10	79	thyroid medication year spent first synthroid…
161296	Lubiprostone	Constipation, chronic	9	116	chronic constipation adult life tried linz wor…

Table 3.7 Cleaned test dataset

	Drug name	Condition	Rating	Useful count	Review
0	Mirtazapine	Depression	10	22	tried antidepressant year citalopram fluoxetin…
1	Mesalamine	Crohn's diseases	8	17	son crohn diseases done well asacol no complain…
2	Bactrim	Urinary tract infection	9	3	quick reduction symptom
3	Contrave	Weight loss	9	35	contrave combine…
4	Cyclafem 1/35	Birth control	9	4	birth control are one cycle…
…	…	…	…	…	…
53760	Apri	Birth control	9	18	started taking apri month ago breats got notic…
53762	Escitalopram	Anxiety	9	11	taking lexapro escitaploprgram since first lik…
53763	Levonorgestrel	Birth control	8	7	married year old no kid taking pill hassle dec…
53764	Tapentadol	Pain	1	20	prescribed nucynta…
53765	Arthrotec	Sciatica	9	46	work

Table 3.8 Merged train and test dataset

	Drug name	Condition	Rating	Useful count	Review
0	Guanfacine	ADHD	8	192	son halfway fourth intuniv became concerned be…
1	Lybrel	Birth control	5	17	used take another contraceptive pill cycle hap…
2	Ortho Evra	Birth control	8	10	first time utilizing form birth…
3	Cialis	Benign prostatic hyperplasia	2	43	nd started work rock hard erection however exp…
4	Levonorgestrel	Emergency contraception	1	5	pulled cummed bit took plan b hour later took…
…	…	…	…	…	…
49170	Apri	Birth control	9	18	started taking apri month ago breats got notic…
49171	Escitalopram	Anxiety	9	11	taking lexapro escitaploprgram since first lik…
49172	Levonorgestrel	Birth control	8	7	married a years old no kids…
49173	Tapentadol	Pain	1	20	prescribed nucynta severe…
49174	Arthrotec	Sciatica	9	46	work

To know the distribution of rating in the dataset we draw the count plot of rating attribute and got that in the dataset the lowest rating is 1 and the highest rating is 10 as in Figure 3.4.

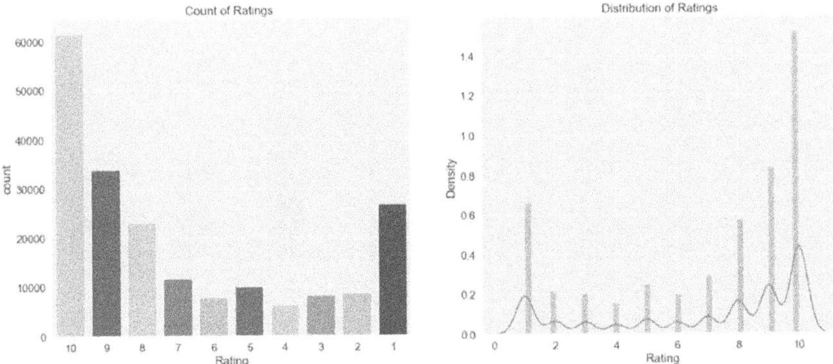

Figure 3.4 Distribution of ratings.

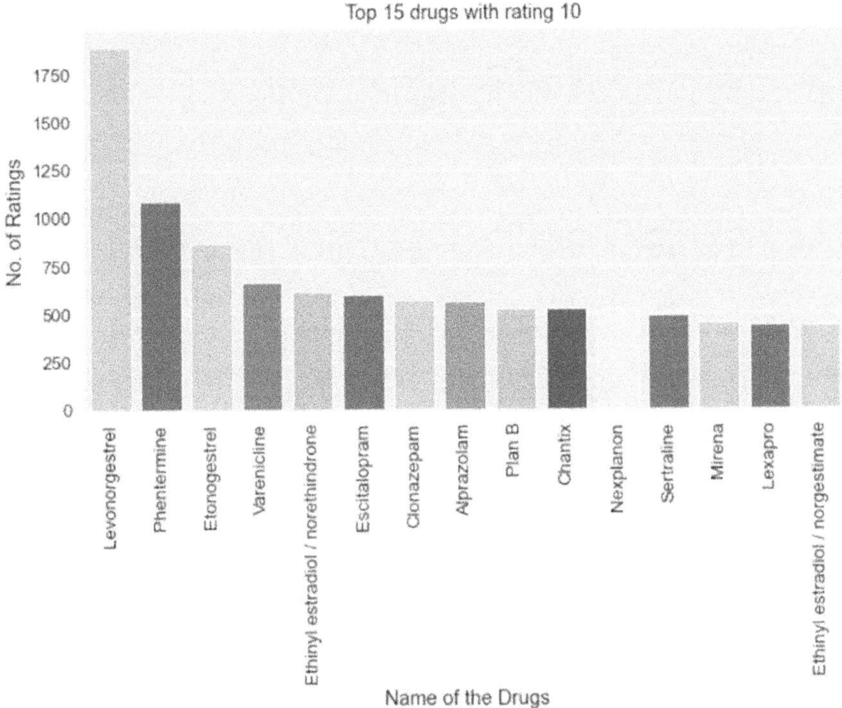

Figure 3.5 Top 15 drugs with the 10/10 ratings.

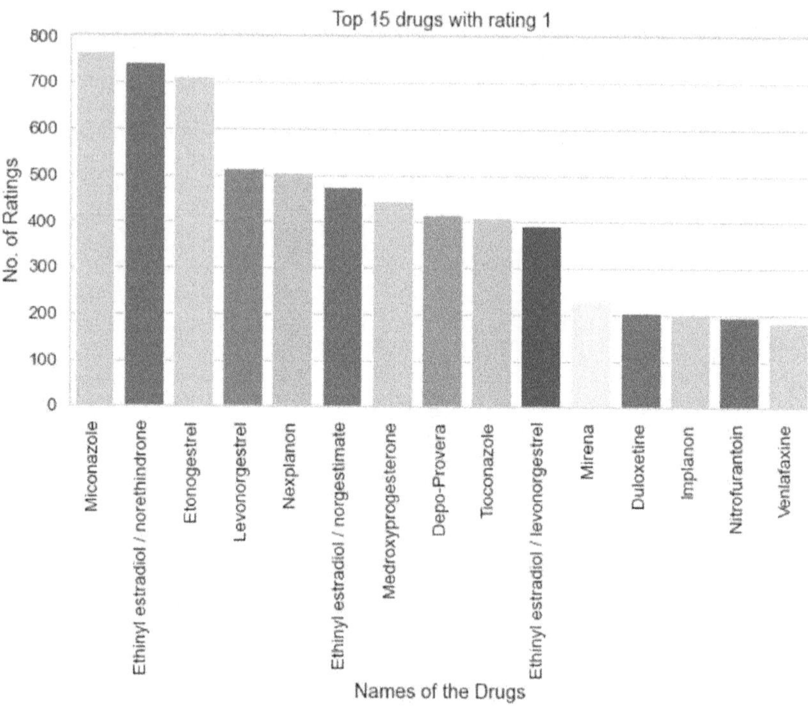

Figure 3.6 Top 15 drugs with the 1/10 ratings.

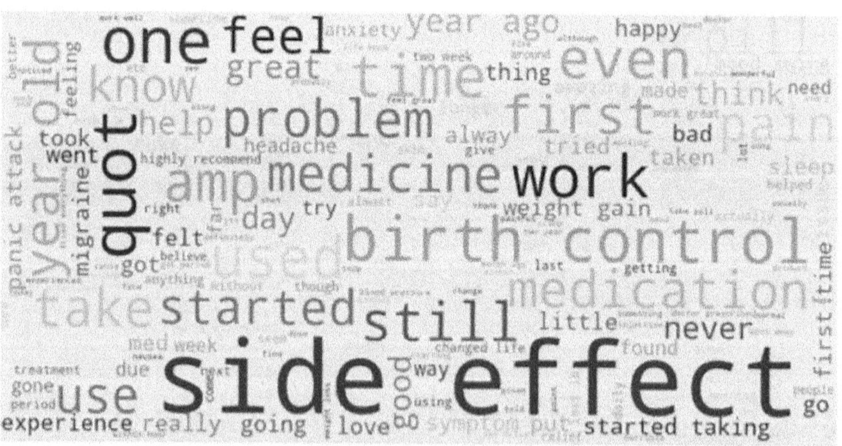

Figure 3.7 Word clouds of the reviews with rating equal to 10.

Figure 3.8 Word clouds of the reviews with rating equal to 1.

Figures 3.9–3.12 shows the highest ten conditions with which the patients are suffering, highest ten drugs that are utilized for the highest conditions that are birth control, distribution of the useful count, and number of reviews per year respectively.

Figures 3.13–3.16 display the top 20 unigrams, top 20 bigrams, top 20 trigrams, and the removal of stop words before plotting, respectively, based on the rating.

3.12.1 Drug recommendation

3.12.1.1 Drug recommendation using NLP supervised

Algorithm

Input: Train and test dataset

Output: TFIDF performs well

Step 1: Calculate the rating value of train dataset.

Step 2: Set the value 0 for the ratings <7 and value 1 for the ratings >=7.

Step 3: Import the stopwords.

Step 4: Set CountVectorizer(stop_words=stop, ngram_range=(1, 1), min_df=10, max_df=0.7).

Step 5: Apply logistic regression and Naïve Bayes with count vectorizer.

Step 6: Create TF-IDF versions of the count vectorizers, fit the both logistic regression on the TF-IDF data.

Step 7: Compare the models to get the better one.

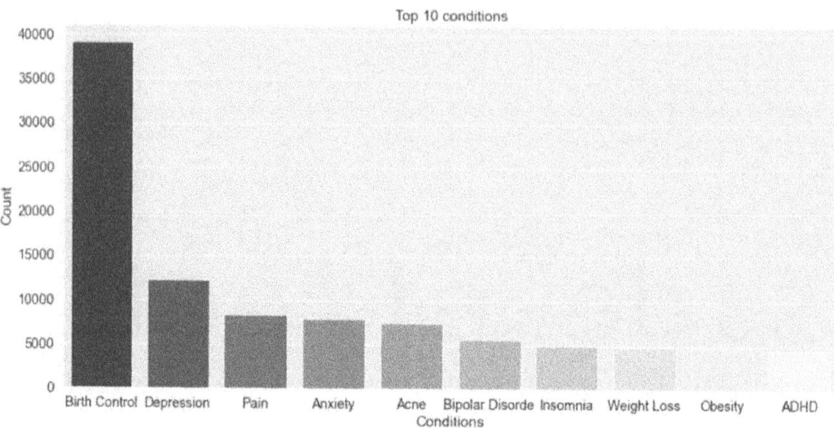

Figure 3.9 Top 10 conditions the people are suffering.

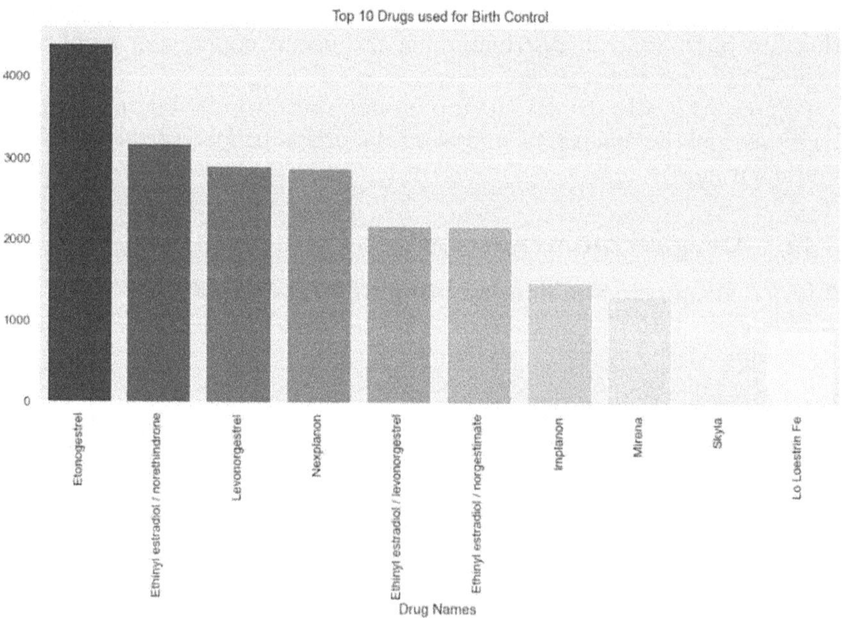

Figure 3.10 Top 10 drugs that are utilized for the top condition like birth control.

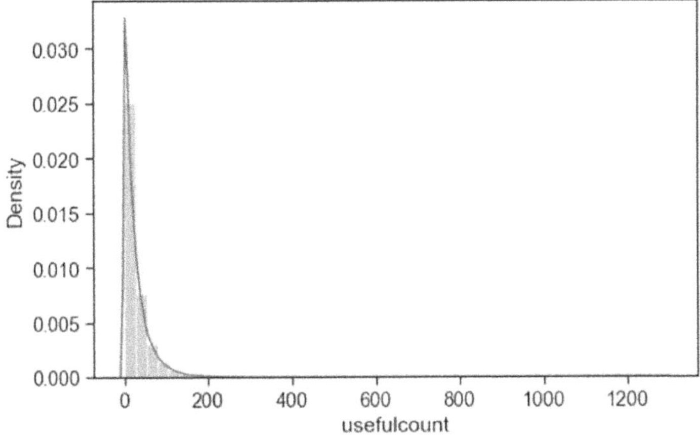

Figure 3.11 Distribution of the useful count.

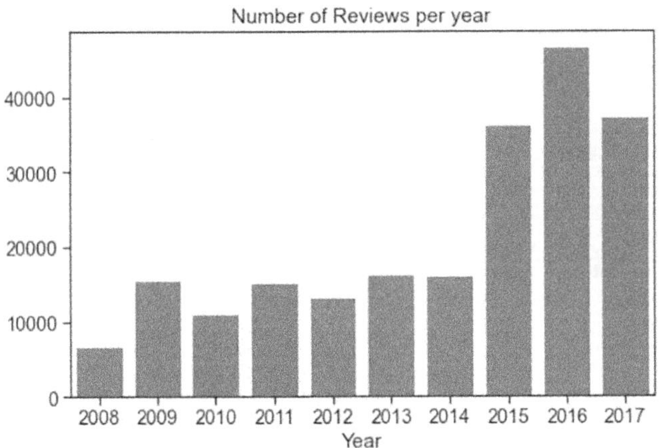

Figure 3.12 Number of reviews per year.

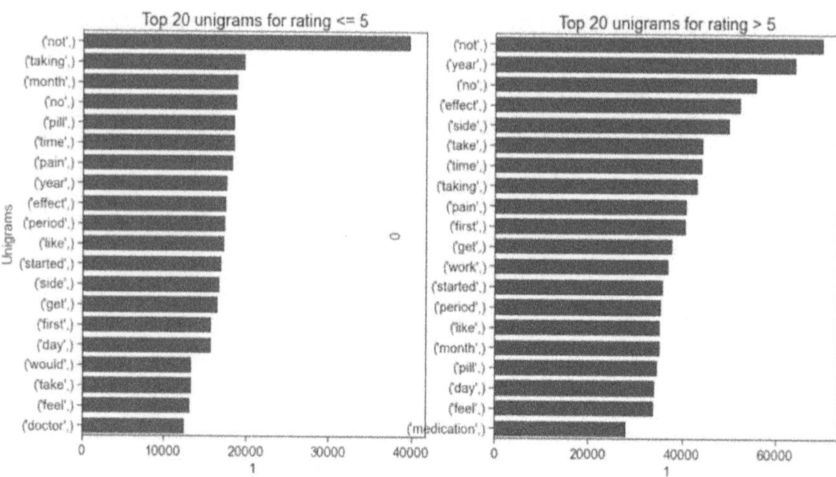

Figure 3.13 Top 20 unigrams as per rating.

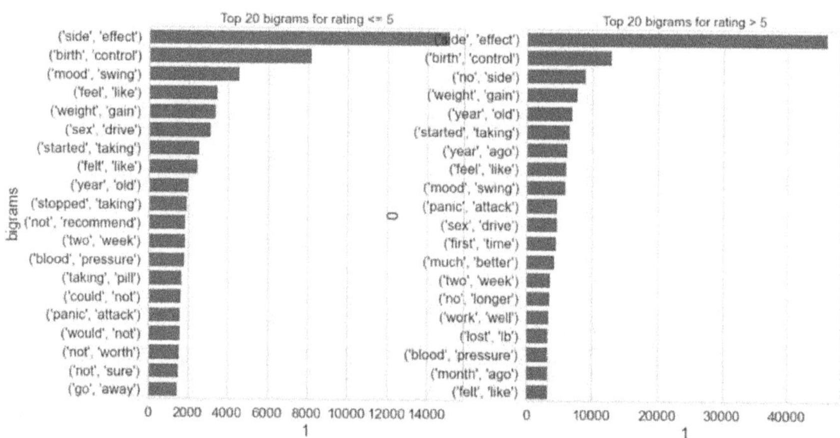

Figure 3.14 Top 20 bigrams as per rating.

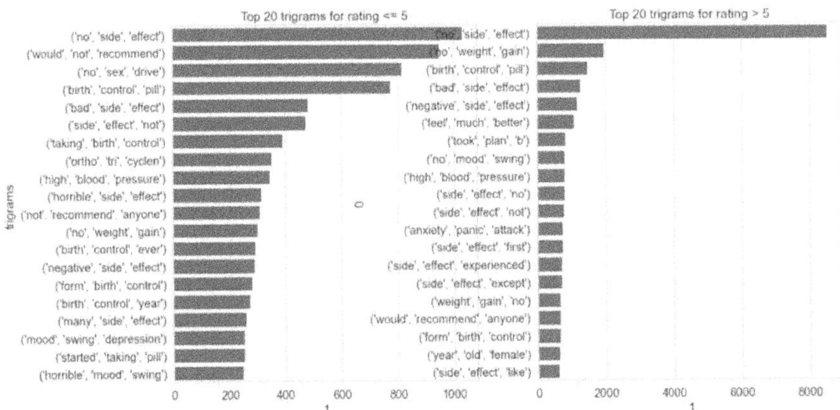

Figure 3.15 Top 20 trigram as per ratings.

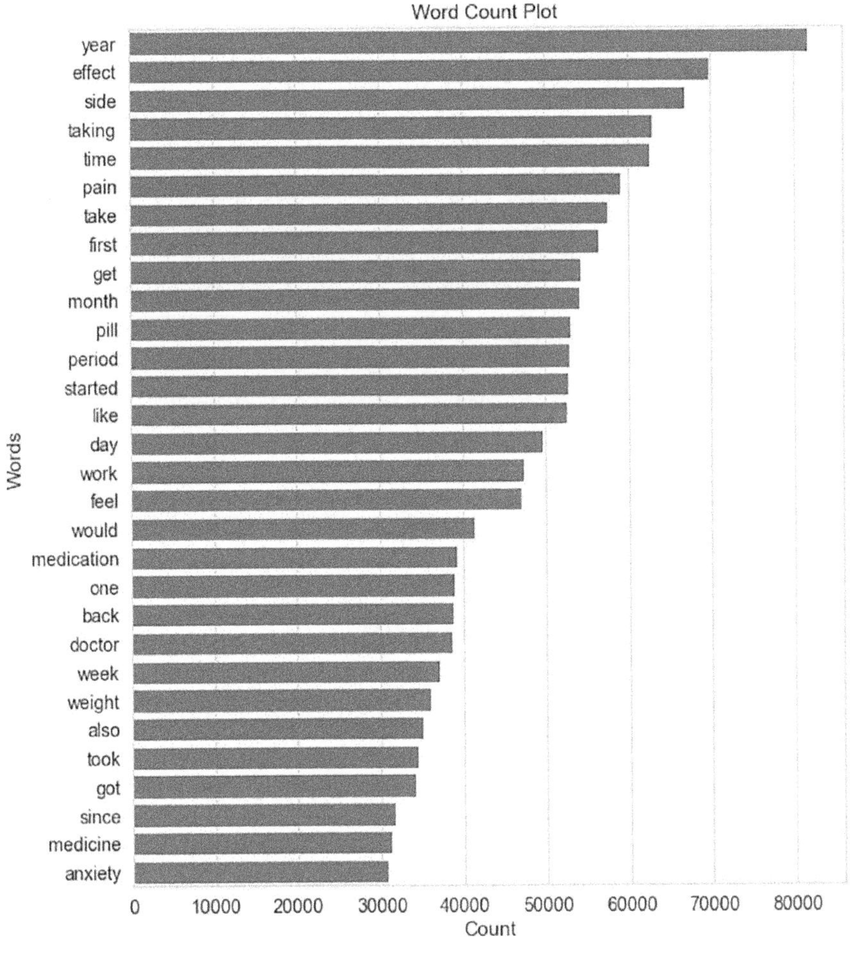

Figure 3.16 Removing stop words before plotting.

Table 3.9 Rating value calculation

Rating	Rating value
10	0.310609
9	0.170993
1	0.134743
8	0.117450
7	0.058981
5	0.050402
2	0.043906
3	0.041318
6	0.039802
4	0.031795

First we calculated the rating value as shown in Table 3.9.

We map the ratings greater than or equal to 7 to 1 and other ratings as 0 for both the train and test data. By doing this we get that the dataset has mostly positive reviews as in Tables 3.10 and 3.11.

After applying the logistic regression and Naïve Bayes prototype with counter vectorizer we get the following result as in Table 3.12.

Now we create TF-IDF versions of the count vectorizers and fit both the logistic regression on the TF-IDF data and got the following outcome as in Table 3.13.

3.13 TOP 5 CONDITION

For the recommendation of drug, here we use cleaned train dataset (Table 3.6) and a dataset called topic dataset which contain topic, contribution, keyword and review (Table 3.14).

First we merge the two datasets to get the merged dataset and deleted the review attribute as in Tables 3.15 and 3.16.

3.14 DISEASE FORECASTS & DRUG RECOMMENDATIONS

We predict diseases & recommend the top 3 drugs from the review attribute present in the dataset. To do this, first we cleaned the review attribute by applying Lemmitization to get the cleaned review (Table 3.17) and break the data into feature and target variable.

Finally, we apply the prototype to get the predicted disease (Table 3.18) and top 3 drug recommendation (Figure 3.17) as an outcome.

Table 3.10 7–10 star reviews to positive (1), the rest to negative (0) for train set

	Drug-names	Conditions	Ratings	Useful-counts	Reviews	Sentimen
0	Guanfacine	ADHD	8	192	son half-way 4th intuniv became concerned be…	1
2	Ortho Evra	Birth control	8	10	1st time using form…	1
3	Cialis	Benign prostatic hyperplasia	2	43	nd started work rock hard erection however exp…	0
4	Levonorgestrel	Emergency contraception	1	5	pulled cummed bit took plan b hour later took…	0
5	Aripiprazole	Bipolar disorde	10	32	abilify changed life hope zoloft clonidine fir…	1
…	…	…	…	…	…	…
148382	Junel 1.5/30	Birth control	6	0	would second junel birth control year changed…	0
148383	Metoclopramide	Nausea/vomiting	1	34	given surgey immediately became anxious could…	0
148384	Orencia	Rheumatoid arthritis	2	35	limited improvement month developed bad rash m…	0
148385	Thyroid desiccated	Underactive thyroid	10	79	thyroid medication year spent first synthroid…	1
148386	Lubiprostone	Constipation, chronic	9	116	chronic constipation adult life tried linz wor…	1

Table 3.11 7–10 star reviews to positive (1), the rest to negative (0) for test set

	Drug name	Condition	Rating	Useful count	Review	Sentiment
0	Mirtazapine	Depression	10	22	tried antidepressant year citalopram fluoxetin…	1
1	Mesalamine	Crohn's disease, maintenance	8	17	son crohn diseases done…	1
2	Bactrim	Urinary tract infection	9	3	quick reduction symptom	1
3	Contrave	Weight loss	9	35	contraves combined drug…	1
4	Cyclafem 1/35	Birth control	9	4	birth control one…	1
…	…	…	…	…	…	…
49170	Apri	Birth control	9	18	started taking apri month ago breats got notic…	1
49171	Escitalopram	Anxiety	9	11	taking lexapro escitaploprgram since first lik…	1
49172	Levonorgestrel	Birth control	8	7	married a years old no kids…	1
49173	Tapentadol	Pain	1	20	prescribe nucynta…	0
49174	Arthrotec	Sciatica	9	46	work	1

Table 3.12 Accuracy of logistic regression and Naïve Bayes prototype

	LogReg1	LogReg2	Naïve Bayes1	Naïve Bayes2
Accuracy	0.842	0.877	0.798	0.846
Precision	0.865	0.884	0.857	0.895
Recall	0.914	0.946	0.848	0.880
F1- Score	0.889	0.914	0.852	0.887

Table 3.13 Accuracy of logistic regression on the TF-IDF data

	LogReg1	LogReg2	Naïve Bayes1	Naïve Bayes2	LR1-TFIDF	LR2-TFIDF
Accuracy	0.842	0.877	0.798	0.846	0.845	0.918
Precision	0.865	0.884	0.857	0.895	0.862	0.932
Recall	0.914	0.946	0.848	0.880	0.924	0.951
F1 Score	0.889	0.914	0.852	0.887	0.892	0.941

Table 3.14 Dominant topic dataset

	Dominant_Topic	Perc_Contribution	Topic_Keywords	Review
0	0.0	0.25	acne no weight gain no, effect no weight gain…	son halfway fourth intuniv became concerned be…
1	0.0	0.25	acne no weight gain no, effect no weight gain…	used take another contraceptive pill cycle hap…
2	0.0	0.25	acne no weight gain no, effect no weight gain…	1st time utilizing form…
3	0.0	0.25	acne no weight gain no, effect no weight gain…	nd started work rock hard erection however exp…
4	0.0	0.25	acne no weight gain no, effect no weight gain…	pulled cummed bit took plan b hour later took…
…	…	…	…	…
148381	0.0	0.25	acne no weight gain no, effect no weight gain…	would second junel birth control year changed…
148382	0.0	0.25	acne no weight gain no, effect no weight gain…	given surgey immediately became anxious could…
148383	0.0	0.25	acne no weight gain no, effect no weight gain…	limited improvement month developed bad rash m…
148384	0.0	0.25	acne no weight gain no, effect no weight gain…	thyroid medication year spent first synthroid…
148385	0.0	0.25	acne no weight gain no, effect no weight gain…	chronic constipation adult life tried linz wor…

Table 3.15 Merged cleaned train dataset and dominant topic dataset.

	Drug-Names	Conditions	Ratings	Useful-Counts	Dominant_Topic	Perc_Contribution	Topic_Keywords
0	Guanfacine	ADHD	8	192	0.0	0.25	acne no weight gained no, effect no…
1	Lybrel	Birth Control	5	17	0.0	0.25	acne no weight gain no, …
2	Ortho Evra	Birth Control	8	10	0.0	0.25	acne no weight gain no, …
3	Cialis	Benign Prostatic Hyperplasia	2	43	0.0	0.25	acne no weight gain no, effect no weight gain…
4	Levonorgestrel	Emergency Contraception	1	5	0.0	0.25	acne no weight gain no, effect no weight gain…
…	…	…	…	…	…	…	…
148381	Tekturna	High Blood Pressure	7	18	0.0	0.25	acne no weight gain no, effect no weight gain…
148382	Junel 1.5 / 30	Birth Control	6	0	0.0	0.25	acne no weight gain no, effect no weight gain…
148383	Metoclopramide	Nausea/Vomiting	1	34	0.0	0.25	acne no weight gain no, effect no weight gain…
148384	Orencia	Rheumatoid Arthritis	2	35	0.0	0.25	acne no weight gain no, effect no weight gain…
148385	Thyroid desiccated	Underactive Thyroid	10	79	0.0	0.25	acne no weight gain no, effect no weight gain…

Table 3.16 Selected top 5 conditions

Condition	Dominant topic	Value
Acne	0.0	0.997861
	1.0	0.001604
	2.0	0.000357
	3.0	0.000178
Birth control	0.0	0.977385
	3.0	0.010239
	1.0	0.008476
	2.0	0.003899
Depression	0.0	0.996728
	3.0	0.001854
	1.0	0.000982
	2.0	0.000436
High blood pressure	0.0	0.996600
	3.0	0.001700
	1.0	0.000850
	2.0	0.000850
Pain	0.0	0.997428
	3.0	0.001125
	1.0	0.000804
	2.0	0.000643

3.14.1 Conclusion & future work

The chapter proposes a methodology to predict disease and recommend the drugs from the reviews given by the patients. First we collect the datasets and preprocessed the data by utilizing several data preprocessing procedures to make the data in our required format. We apply the logistic regression 1, logistic regression 2, Naïve Bayes 1, Naïve Bayes 2 with counter vectorizers and got the accuracies of the prototypes 84.2%, 87.7%, 79.85 and 84.6% respectively. To enhance the outcome of the prototype, we utilized the TF-IDF version of the counter vectorizers and fit both the logistic regression on the TF-IDF data and got the accuracies of the prototypes 84.5% and 91.8%, which is the better one as compare to the previous prototype's outcome. With this prototype, we utilize NLP's lemmatization text processing procedure to clean the review attribute of the dataset and predicted the disease from the review as well as recommended the top 3 drugs from the review given by the patients.

Table 3.17 Cleaned reviews

	Condition	Review	Review_clean
161273	Birth control	I have had the Nexplanon since Dec. 27, 2016 \r\r\nI got my first period at the end of January and it lasted about a month and a half. In March of 2017 I didn't bleed for close to 3 weeks and then started bleeding again March 28th and have been bleeding every since. I have gained about 13 lbs so far since getting the birth control. Although for now the weight gain isn't a deal breaker for me but the bleeding is… I am trying to be very patient to see how my body adjusts to the implant. It has been 3 months so far and I have my fingers crossed that my cycle will go away for a while.	nexplanon since dec got first period end january lasted month half march bleed close 3 week started bleeding march th bleeding every since gained lb far since getting birth control although weight gain deal breaker bleeding trying patient see body adjusts implant 3 month far finger crossed cycle go away awhile
161278	Diabetes, Type 2	I diagnosed with type two. My doctor adviced Invokana & metformin from the starting. My blood sugars went minimize to normal by the 2nd week. I am minimizing more weights. No other side effects till now. Miracle medicine…	got diagnosed type doctor prescribed invokana metformin beginning sugar went normal second week losing much weight side effect yet miracle medicine
161286	Depression	This is the 3rd med I'have tried for anxiety & mild depression. Been on it for 7 days and I hate it so much. I am so dizzy,	third med tried anxiety mild depression week hate much dizzy major diarrhea feel worse started contacting doc changing asap
161290	High blood pressure	I have only been on Tekturna for 9 days. The effect was immediate. I am also on a calcium channel blocker (Tiazac) and hydrochlorothiazide. I was put on Tekturna because of palpitations experienced with Diovan (ugly drug in my opinion, same company produces both however)…	tekturna day effect immediate also calcium channel blocker tiazac hydrochlorothiazide put tekturna palpitation experienced diovan ugly drug opinion company produce however palpitation pretty bad diovan hour monitor ekg etc. day substituting tekturna diovan palpitation
161291	Birth control	This would be my 2nd month on June. I & #39; have been on Birth Control for about 10 years now. I changed due to spotting and increased mood swings with my previous birth control…	would 2nd month junel birth control year changed due spotting increased mood swing previous birth control since switch shorter period day gained major weight increased appetite switched regular exercise routine still managed drop extra lb

Table 3.18 Disease predictions from reviews

	Test_Sent	Prediction
0	Tekturna day effect immediate also calcium channel blocker tiazac hydrochlorothiazide put tekturna palpitation experienced diovan ugly drug opinion company produce however palpitation pretty bad diovan hour monitor ekg etc. day substituting tekturna diovan palpitation	High Blood Pressure
1	Third med tried anxiety mild depression week hate much dizzy major diarrhea feel worse started contacting doc changing asap	Depression
2	Got diagnosed type doctor prescribed invokana metformin beginning sugar went normal second week losing much weight side effect yet miracle medicine	Diabetes, Type 2

```
text: This is the third med I&#039;ve tried for anxiety and mild depression. Been on it for a week and I hate it so much. I am
so dizzy, I have major diarrhea and feel worse than I started. Contacting my doc in the am and changing asap.
Condition: Depression
Top 3 Suggested Drugs:
Sertraline
Zoloft
Viibryd
text: I just got diagnosed with type 2. My doctor prescribed Invokana and metformin from the beginning. My sugars went down to
normal by the second week. I am losing so much weight. No side effects yet. Miracle medicine for me
Condition: Diabetes, Type 2
Top 3 Suggested Drugs:
Victoza
Canagliflozin
Invokana
text: I have only been on Tekturna for 9 days. The effect was immediate. I am also on a calcium channel blocker (Tiazac) and hy
drochlorothiazide. I was put on Tekturna because of palpitations experienced with Diovan (ugly drug in my opinion, same company
produces both however). The palpitations were pretty bad on Diovan, 24 hour monitor by EKG etc. After a few days of substitutin
g Tekturna for Diovan, there are no more palpitations.
Condition: High Blood Pressure
Top 3 Suggested Drugs:
Losartan
Aldactone
Spironolactone
```

Figure 3.17 Top 3 drug recommendation from reviews.

In future work, the authors will try to consider different category of prototypes, which can be more powerful and will give the better accuracies in the disease prediction and drug recommendation from the patient's review.

REFERENCES

1. Nadkarni, Prakash M., Lucila Ohno-Machado, and Wendy W. Chapman. "Natural language processing: an introduction." *Journal of the American Medical Informatics Association* 18, no. 5 (2011): 544–551.
2. Gravesteijn, Benjamin Y., et al. "Machine learning algorithms performed no better than regression models for prognostication in traumatic brain injury." *Journal of Clinical Epidemiology* 122 (2020): 95–107.
3. Christodoulou, Evangelia, et al. "A systematic review shows no performance benefit of machine learning over logistic regression for clinical prediction models." *Journal of Clinical Epidemiology* 110 (2019): 12–22.

4. Doshi-Velez, Finale, and Been Kim. "Considerations for evaluation and generalization in interpretable machine learning." In *Explainable and Interpretable Models in Computer Vision and Machine Learning*, Hugo Jair Escalante, Sergio Escalera, Isabelle Guyon, Xavier Baró, Yağmur Güçlütürk, Umut Güçlü, and Marcel van Gerven (eds.), (2018): 3–17. Springer.

5. Harrison, Conrad J., and Chris J. Sidey-Gibbons. "Machine learning in medicine: a practical introduction to natural language processing." *BMC Medical Research Methodology* 21, no. 1 (2021): 158.

6. Iacus, Stefano M. "Automated data collection with R – a practical guide to web scraping and text mining." *Journal of Statistical Software* 68 (2015): 1–3.

7. Sidey-Gibbons, Jenni A. M., and Chris J. Sidey-Gibbons. "Machine learning in medicine: a practical introduction." *BMC Medical Research Methodology* 19 (2019): 1–18.

8. Pfob, André, Sheng-Chieh Lu, and Chris Sidey-Gibbons. "Machine learning in medicine: a practical introduction to techniques for data pre-processing, hyperparameter tuning, and model comparison." *BMC Medical Research Methodology* 22, no. 1 (2022): 1–15.

9. Sarker, Iqbal H. "Machine learning: algorithms, real-world applications and research directions." *SN Computer Science* 2, no. 3 (2021): 160.

10. Mohammed, Mohssen, Muhammad Badruddin Khan, and Eihab Bashier Mohammed Bashier. *Machine Learning: Algorithms and Applications*. CRC Press, 2016.

11. Azuaje, Francisco. *Witten IH, Frank E: Data mining: Practical machine learning tools and techniques*. 2nd edition, 2006.

12. Khurana, Diksha, Aditya Koli, Kiran Khatter, and Sukhdev Singh. "Natural language processing: state of the art, current trends and challenges." *Multimedia Tools and Applications* 82, no. 3 (2023): 3713–3744.

13. Benson, Edward, Aria Haghighi, and Regina Barzilay. "Event discovery in social media feeds." In *Proceedings of the 49th Annual Meeting of the Association for Computational Linguistics: Human Language Technologies*, Portland, OR, USA, pp. 389–398, June 19–24, 2011.

14. Murphy, Kevin P. "Naive Bayes classifiers." *University of British Columbia* 18, no. 60 (2006): 1–8.

15. Bayes, Thomas. "Naive Bayes classifier." *Article Sources and Contributors* (1968): 1–9.

16. Kim, Sang-Woon, and Joon-Min Gil. "Research paper classification systems based on TF-IDF and LDA schemes." *Human-Centric Computing and Information Sciences* 9 (2019): 1–21.

17. Nayak, Suvendu Kumar, Mamata Garanayak, Sangram Keshari Swain, Sandeep Kumar Panda, and Deepthi Godavarthi. "An intelligent disease prediction and drug recommendation prototype by using multiple approaches of machine learning algorithms." *IEEE Access* 11 (2023): 99304–99318.

18. Kumar, Rahul, Sushant Pradhan, Tejaswi Rebaka, and Jay Prakash. "Disease prediction from speech using natural language processing and deep learning method." In *Congress on Intelligent Systems: Proceedings of CIS 2020 held at* New Delhi, India, September 5–6, 2020, Volume 2, pp. 407–413. Springer Singapore, 2021.

19. Yu, Hong Qing. "Experimental disease prediction research on combining natural language processing and machine learning." In *2019 IEEE 7th International Conference on Computer Science and Network Technology (ICCSNT)*, 19-20 Oct. 2019, pp. 145–150. IEEE, Dalian, China, 2019.

20. Gangavarapu, Tushaar, Gokul S. Krishnan, Sowmya Kamath, and Jayakumar Jeganathan. "FarSight: long-term disease prediction using unstructured clinical nursing notes." *IEEE Transactions on Emerging Topics in Computing* 9, no. 3 (2020): 1151–1169.

21. Heo, Tak Sung, Yu Seop Kim, Jeong Myeong Choi, Yeong Seok Jeong, Soo Young Seo, Jun Ho Lee, Jin Pyeong Jeon, and Chulho Kim. "Prediction of stroke outcome using natural language processing-based machine learning of radiology report of brain MRI." *Journal of Personalized Medicine* 10, no. 4 (2020): 286.

22. Bacchi, Stephen, Luke Oakden-Rayner, Toby Zerner, Timothy Kleinig, Sandy Patel, and Jim Jannes. "Deep learning natural language processing successfully predicts the cerebrovascular cause of transient ischemic attack-like presentations." *Stroke* 50, no. 3 (2019): 758–760.

23. Yasashvini, R, Sarobin M. Vergin Raja, Panjanathan Rukmani, Jasmine S. Graceline, and Anbarasi L. Jani. "Diabetic retinopathy classification using CNN and hybrid deep convolutional neural networks." *Symmetry* 14, no. 9 (2022): 1932.

24. Cook, Benjamin L., Ana M. Progovac, Pei Chen, Brian Mullin, Sherry Hou, and Enrique Baca-Garcia. "Novel use of natural language processing (NLP) to predict suicidal ideation and psychiatric symptoms in a text-based mental health intervention in Madrid." *Computational and Mathematical Methods in Medicine* 2016 (2016): 8708434.

25. Kulshrestha, Sujay, et al. "Prediction of severe chest injury using natural language processing from the electronic health record." *Injury* 52, no. 2 (2021): 205–212.

Chapter 4

Investigating the potential of ChatGPT in substituting or assisting teachers in the digital era

Sugyanta Priyadarshini and Sukanya Priyadarshini

4.1 BACKGROUND STUDY

The 21st century has marked the exponential growth of digital technology, making it a part and parcel of our daily lives. It has significantly affected lives by seamless penetration of technology in multiple sectors like education [1], shopping [2], navigation, medical science [3], and many more. Moreover, with the rapid advances in digital technology and rampant use of mobile phones [4,5], available information can be collected, processed, accessible and transmitted at any time to all groups of people. The genesis of integration of technology into education dates back to the age of first-generation computer [6]. Computers have been used for teaching since the early 1960s by educators and computer scientists. Due to the computer's limited level of human engagement at first, it was used to read and type text to provide instructions on how to operate the device. Later, it was employed to solve various time-consuming tasks. However, the introduction of reasonably priced microcomputers and the incorporation of text, graphics, and colour led to a quick uptake of computers in offices, schools, and households [7]. Since then, teachers were seen utilising computers to lecture, research, innovate teaching practices, and record pupils' grades. Similarly, students have also vivaciously used computers to learn, investigate, and solve difficulties, use software tools instead of pen and paper to make presentations and projects [8,9]. Along with rampant development the digital revolution took a major leap with the emergence of big data and AI [10]. AI can be defined as "the science and engineering of making intelligent machines" or "a machine that behaves in a way that could be considered intelligent if it were a human being." It can also be defined as a set of techniques aiming at using machines to approximate some feature of human or animal cognition [11]. Now it is regarded as the electricity of new age [1]. It is a rapidly emerging branch of computer science concerned with the development of intelligent systems and the solution of challenges resembling human intelligence [12]. In this contemporary world, AI has become a cutting-edge technology. Being ubiquitous in nature [13], AI has delved deep into each and every field [14] and especially in the field of education.

DOI: 10.1201/9781003565529-4

AI has rapidly expanded over the past few decades and changed many facets of society. AI has established itself as a leader in research areas including medicine, robotics, education, and autonomous vehicles. AI in education (AIEd) is one such rapidly developing fields of educational technology that has acclaimed a lot of attention recently. Education is more than just a finished good; it is also a process in which learning extends beyond the merely intellectual. Looking from this aspect, AI holds tremendous potential as it could aid to promote individualised education by adjusting to the interests and wants of each student [15]. AI has been used more and more into the educational sector, offering creative ways to improve teaching and learning. AI's educational potential is broad and is still expanding. AI is applied in every possible part of education sector starting from personalised learning experiences to effective administrative procedures. Adaptive learning systems, intelligent tutoring systems, learning analytics, and educational data mining are a few examples of how AI might be applied in education [16]. One such recently developed AI chatbot created by OpenAI is ChatGPT (Generative Pre-trained Transformer). ChatGPT can be defined as a conversational AI interface that uses natural language processing (NLP) to converse realistically and can even "answers follow-up questions, admits its mistakes, challenges incorrect premises, and rejects inappropriate requests." ChatGPT interacts with users in a human-like manner while offering them directions and apt information. This technology is capable of performing a wide array of tasks such as organising data for research, translating software and papers, inscribing mails and essays, debug and annotate code, generate poetry, and respond to questions of any genre. It is believed that it might hold tremendous potential to make the process of teaching and learning easier. ChatGPT signals towards a paradigm shift in educational sphere [17] as unlike other chatbots it has a humongous data set as training data, greater reliance on deep learning that produce realistic human like text generation, upgraded capabilities and extremely fine-tuning [18].

ChatGPT has posed itself as a double-edged sword. There has been a plethora of research papers, blogs, and articles that highlight the advantages of implementing ChatGPT in the educational landscape but some have also suggested the risks, and future challenges that come along with it [19–22]. Moreover, many have also suggested the definite ways and modes of learning by which ChatGPT can be implemented in classrooms to augment the teaching and learning capabilities [17,23–24]. However, despite the advantages ChatGPT has also instigated new challenges and threats into the educational landscape. There are rising concerns that ChatGPT can replace teachers and augment automation [25]. Further, there are concerns about AI-assisted cheating because it may be used to produce written assignments and exams on behalf of students and can respond specifically to user requests [26]. Additionally, many educators have raised concerns regarding the outsourcing of homework by students by using ChatGPT [27]. Hence, ethical and responsible approach is desired to minimise the challenges and maximise the benefits of ChatGPT. And in order to achieve this, a deeper understanding is prerequisite.

4.2 REVIEW OF LITERATURE

A new historical stage in the evolution of intelligent technology is being ushered in by the genesis of AI represented by ChatGPT. This technology is not only fundamentally transforming society and mankind as a whole, but also substantially influencing and shaping the modes of production, life, and communication of the entire society. As this technology has immense potential to revolutionise every sector, deep research on its every aspect including both strengths and challenges has become quintessential. As we focus on the impact of ChatGPT on educational landscape, the potential challenges and the requisite actions to address them are discussed in detail.

4.2.1 Leveraging ChatGPT in teaching, learning, and protecting teaching job

It is quite evident that technology is rapidly changing and diversifying the varied dimensions of education. The inception of ChatGPT in the educational practice is quite new. And hence, there isn't enough time to perform long-term studies and gather useful data on the impact of ChatGPT on education [25]. But with the available data and research it is clearly evident that ChatGPT can greatly affect the traditional methodologies of teaching and learning [28,29]. As we are getting more inclined towards the online platform, ChatGPT with its intelligent algorithms can act as a boon by providing personalised learning experience to every student. It acts as a viable educational tool with several learning advantages. The efficiency of teaching and learning is increased because this AI tool's ability to tailor activities and content to the individual needs of each learner. In other words, it gives students continual help as they acquire knowledge and allows them to adjust their learning rate. Moreover, ChatGPT makes learning more customised as well as individualised. This in turn augments student's motivation, dedication, confidence, and commitment. In other words, ChatGPT complements a basic educational support for the enhancement of the teaching and learning process. At the same time, ChatGPT encourages teamwork by enabling students to investigate topics together and improve their communication and teamwork abilities, which in turn aids learning process. As ChatGPT can now aid, students with little or no prior knowledge of a certain subject to write projects without difficulty in the long run; therefore, teachers must consider new teaching philosophies that they might use to evaluate their students.

ChatGPT may be able to offer students real-time answers to their questions and customised feedback, but it can't take the place of a teacher in helping students develop empathy, creativity, and critical thinking. To keep in pace with the evolving educational landscape, educators should upgrade their practises and competences to match these new changes and expectations. For effective applicability of ChatGPT in learning environment, teachers must be equipped with apt technological skills. This will boost technology, teachers, and teaching methodologies to go hand in hand. Further, by equipping teachers with advanced technical skills and knowledge, they will never stand as an impediment in the implementation of rising AI tools. Along with understanding and learning the developing technology, the teachers must also understand the ethical and pedagogical challenges that come with it [15]. The onus lies on the teachers to effectively implement the technological skills and make their students aware of the possibilities and limitations of AI tools [30]. This can be the only way to create a technically advanced, healthy and responsible classroom environment.

However, it is important to understand that the knowledge of ChatGPT is limited and need to be updated [31,32]. Hence, ChatGPT may generate inaccurate information and might give answers that aren't [33] reliable or correct at times. Such situations pose as a challenge before the students, creating hallucinations and adversely affecting their learning journey. Therefore, there is a chance that the chatbot will produce responses that may seem clear, coherent and cohesive but actually contain inaccurate or unsupported information. And this may negatively affect the learning and understanding quotient of the students. In such cases teachers can play a predominant role by making students understand the shortcomings of a chatbot. Only the teachers with advanced technological knowledge and skills can make students aware of the limitations and reliability of using a chatbot like ChatGPT. Hence, more research is required to delve deep into the impact AI chatbots in the field of teaching and learning. It is vital to strike a balance between the advantages and disadvantages of this technology and to make sure that it is used in favour of all students and enhance their educational process. Therefore, in order to give all students a more inclusive, accessible, and useful educational experience, educational institutions should keep exploring and creating AI applications while keeping in mind that such AI chatbots (e.g., ChatGPT) can never replace the pedagogical methodologies, guidance, and expertise of teachers.

4.2.2 State of ChatGPT in present day education system

In this digital world, new technology is continually emerging and frequently disappearing over time. But only a few technological breakthroughs are incorporated into the educational system. And such AI generative tool is ChatGPT. AI chatbot ChatGPT has posed itself as a game changer in the evolving era of education. It has bought a radical transformation in every aspect of education by revolutionising the approach of both students and teachers. The implementation of ChatGPT can bring forth both benefits and challenges. ChatGPT has widened the horizon of education. It has facilitated the creation of different aspect of educational content, offered personalised learning experiences, and ways to get around language hurdles that earlier use to significantly affect teaching and learning outcomes. By producing pertinent content on a certain topic, ChatGPT can be utilised as a tool to assist students with their academic work. Additionally, it can be utilised to give students feedback on their work so they can increase their knowledge. This may motivate students and make them prepared to face the right number of challenges. ChatGPT will aid colleges and universities to provide instructional material. According to each student's preferred learning method and current ability level, ChatGPT will design tasks just for them. By using ChatGPT, teachers can now create questions, tests, projects, and interactive educational materials, such games and simulations, that are tailored to the learning preferences of their students. Additionally, they can assist in the analysis of student data, the discovery of learning trends, and the generation of insightful information that can be used to improve instruction. Generative AI models like ChatGPT can offer students interactive and interesting learning opportunities that are tailored to their requirements and learning preferences. Additionally, they can help students with impairments by improving accessibility and creating more inclusive teaching methods. This can be achieved by ChatGPT's text-to-speech or speech-to-text feature that may aid students with hearing or visual impairments.

The fundamental constraint of this review is the small amount of literature that is available. AI chatbot such as ChatGPT, in educational landscape is a constantly expanding and evolving topic. Hence, its adoption in present day education may be relatively new and hence less investigated when compared to other sectors. Despite this drawback, the review gives a general picture of how ChatGPT has affected education sector. To fully comprehend ChatGPT's impact and advantages, it will be helpful to broaden the study as more research is completed and information about its application in education sector grows.

4.2.3 Challenges generated by ChatGPT for teachers

Although there are multiple benefits, but there are rising concerns regarding the impact of generative AI tools like ChatGPT on education system, educators and students. While the meteoric ascent of ChatGPT has captivated the enthusiasm of students and teachers, irresponsible usage of the technology has been a source for concern. It has been highlighted by many researchers that reliance on ChatGPT may critically affect the creative power and problem-solving skills of students. It may discourage writing skills and thinking ability of pupils. It might adversely affect interaction between students and create bias. It may raise concerns regarding legitimacy of assigned home assignments as students can completely copy and paste instead of completing their assignment themselves. As students began using this technology to produce their assignments several universities and institutions have prohibited the usage of ChatGPT for homework purposes. Moreover, student plagiarism is one of the rising concerns in the field of Education. The plagiarism in the assignments of the students is now being detected with the help of plagiarism detection apps like Turnitin [34]. But ChatGPT has the ability to bypass such plagiarism tools by producing seemingly original content [35]. Students that create high-calibre work using ChatGPT have an unfair edge over their colleagues who aren't aware of it and don't know how to use it. More importantly, when using ChatGPT, teachers are unable to evaluate student performance effectively, making it challenging to track down students who are having difficulties with their learning. In summary ChatGPT is great complementary AI tool with a promising future. If used wisely then it can be foster development and take the educational sector to greater heights.

This review supports the fact that while AI has enormous potential to improve education, it is important to strike a balance between utilising AI's advantages and maintaining the importance of human educators in fostering students' academic and personal development. Therefore, rather than replacing human connection in the educational process, AI should be seen as a tool that supplements and enriches educational process.

4.3 RESEARCH GUIDING QUESTIONS

Keeping in mind that ChatGPT has increased the automation in substituting or assisting teachers, few questions need to be answered:

1. What is the collective perception of the educational system comprising teachers and students regarding the potential of ChatGPT in automation to advance learning while protecting teaching job?
2. What is the overall state of ChatGPT in educational system?
3. What are the challenges generated by ChatGPT for teaching fraternity in educational system?

4.4 RESEARCH METHODOLOGY

4.4.1 Research criteria and sampling technique

The research method used in the analysis is a mixed method approach comprising both qualitative and quantitative techniques of data collection and interpretation. Under the qualitative analysis, the study makes an effort to penetrate deep into the subject matter to examine the role of ChatGPT in substituting or assisting teachers in educational platforms. Further, for quantitative analysis, data is collected from 250 teachers and 250 students from different schools of Odisha to uncover their opinion in regard to the subject matter. However, the sample was collected based on the purposive sampling technique to particularly understand the perception of the two target groups of teachers and students regarding ChatGPT use.

4.4.2 Data collection and analysis

Semi-structured interviews were employed with the targeted participants to understand their perception regarding the subject matter. Further, online questionnaires were circulated among them regarding four major criteria including supportive assistance of the ChatGPT to students and teachers (questionnaire and sub-criteria explained in 5.1), critical challenges of the ChatGPT for students and teachers (questionnaire and sub-criteria explained in 5.2), over-powering of the ChatGPT over classroom teaching (questionnaire and sub-criteria explained in 5.3), and potential of ChatGPT in substituting teachers (questionnaire and sub-criteria explained in 5.4). The answers obtained under these criteria were analysed and presented under results and discussion. However, two Focused Group Discussions were organised before and after the data collection and interpretation to get new insights into the subject matter as well as stimulating an informal discussion in more detail than what is possible through a survey for the participants. Once the data is collected, it is then construed by taking into consideration mean and standard deviation of the same and graphical representation of the interpreted data. The analysis has also made use of Likert type scale ranging from 1 (strongly disagree) to 5 (strongly agree) for data interpretation.

4.5 RESULTS AND DISCUSSION

4.5.1 Supportive assistance by ChatGPT to educational system

With the help of AI, cloud computing, machine learning, and big data approach, software are designed in such a way that it can match the human intelligence [16]. ChatGPT making use of these technologies is able to understand, remark, identify, learn, respond, and resolve complications as human brain rather more effectively [36]. Inevitably, such technology has not just brought a renaissance in the workplace or corporate but has also brought a significant change in the educational sector. Although AI can be helpful for human beings to perform more skilfully and effectively at every stage in every domain, but at the same time it can bring disruptive changes altogether. The current study has made an effort to analyse the perception of both students and teachers in understanding the supportive assistance provided by ChatGPT to Educational system. For the analysis purpose, five queries were raised and were answered by the students and teachers comprising assistance provided by ChatGPT in Classroom Teaching, assistance provided by ChatGPT in doubt clearing, assistance provided by ChatGPT in better understanding or conceptualisation, assistance provided by ChatGPT in terms of advanced learning, and assistance provided by ChatGPT in providing practical examples to existing theories. To these queries raised by the investigator, average student and teacher perception is marked as 3.5 and 2.4 in case of assistance provided by ChatGPT in Classroom Teaching, average student and teacher perception is marked as 3.1 and 2.3 in case of assistance provided by ChatGPT in doubt clearing, average student and teacher perception is marked as 3.5 and 1.4 in case of assistance provided by ChatGPT in better understanding or conceptualisation, average student and teacher perception is marked as 4.6 and 3.4 in case of assistance provided by ChatGPT in terms of advanced learning, and average student and teacher perception is marked as 1.8 and 1.2 in case of assistance provided by ChatGPT in providing practical examples to existing theories. It can be interpreted that maximum strength of students and teachers have opined that ChatGPT assists them in advanced learning and have also stated that ChatGPT is failing in terms of assisting in providing practical examples to existing theories. However, the maximum difference in the opinion exists in the perception that ChatGPT is assisting in better understanding or conceptualisation with a difference of 2.1. However, the graphical representation of the same is illustrated in the Figure 4.1.

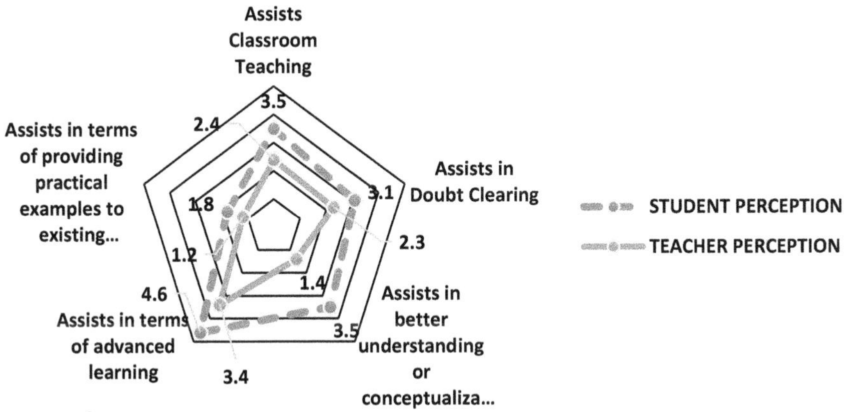

Figure 4.1 Supportive assistance by ChatGPT.

4.5.2 Critical challenges imposed by ChatGPT for students and teachers

The application of AI in educational system has a humungous impact which is reflected from the gradual improvement in the efficiency of the smart educational system via promotion and personalisation of global learning and formation of creative content affecting education process effectively [30]. By optimisation of teaching-learning process, education is no more limited to four walls of classroom rather has turned into a system that goes beyond knowledge acquisition. In this approach, ChatGPT has emerged as a new technology with greater potential fostering of personalised learning considering the needs, interest and requirements of students at any point of time [37]. However, at the same time it is equally important to look into the ethical consideration of ChatGPT usage in educational system comprising data privacy, adverse impact on knowledge-providers, brain storming capability of students and many other [38]. The current research work has tried to identify the challenges of ChatGPT in a responsible manner to understand the importance of maintaining a balance between upgrading technology and role of educators to prevail the holistic growth of the education system in the ever-changing world [39]. For the analysis purpose, five queries were raised and were answered by the students and teachers as presented in Figure 4.2. The students and teachers were questioned about the (i) preference over easy-going method instead of brain storming method

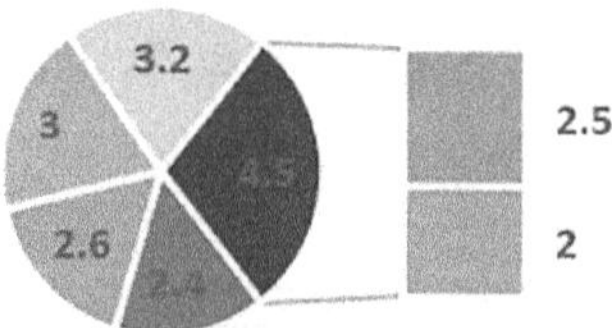

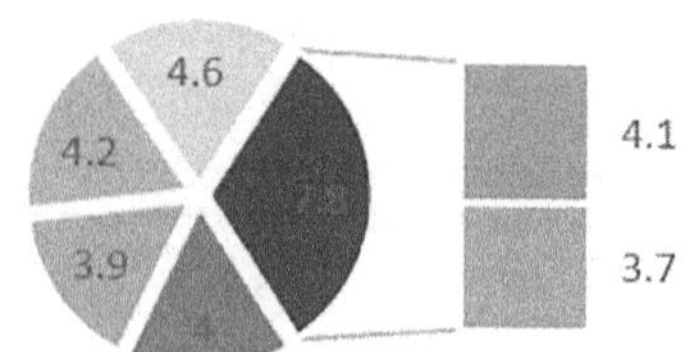

Figure 4.2 Students and teachers perception towards critical challenges.

via ChatGPT usage, (ii) to what extent ChatGPT has Obstructed creativity and critical thinking, (iii) to what extent ChatGPT has limited human interaction, (iv) to what extent ChatGPT has misguided by misleading or incomplete responses, (v) to what extent ChatGPT has clogged problem solving attitude, and (vi) to what extent ChatGPT has risked ethical consideration of self-assessment. To these queries raised by the investigator, average student and teacher perception is marked as 2.4 and 4 under preference over easy-going method instead of brain storming method via ChatGPT usage, 2.6 and 3.9 under to what extent ChatGPT has Obstructed creativity and critical thinking, 3 and 4.2 under to what extent ChatGPT has limited human interaction, 3.2 and 4.6 under to what extent ChatGPT has misguided by misleading or incomplete responses, 2.5 and 4.1 under to what extent ChatGPT has clogged problem solving attitude and 2 and 3.7 under to what extent ChatGPT has risked ethical consideration of self-assessment. It can be interpreted that maximum strength of students and teachers have opined that ChatGPT has misguided by misleading. Both students and teachers scored these statements with least score but with a difference of 1.4 in average score. However, the maximum difference in the opinion exists in the perception that ChatGPT is risking ethical consideration of self-assessment with a difference of 1.7 where student's perception is quite different from teachers stating that ChatGPT is not impacting the ethical consideration of self-assessment adversely.

4.5.3 Role of ChatGPT in over-powering classroom teaching

Enormous research work has been accomplished related to the potential impact of ChatGPT and use of AI tools such as personalised learning experiences, adaptive testing, machine learning, predictive analytics, cloud computing, and chatbots for development of students in educational system [30,37]. These tools have played a significant role in improvising the learning efficacy of the students and have also provided personalised educational support to educators. But, at the same time we need to reconsider the risks as well as the limitations associated with these technologies, one of which is identified as overpowering classroom teaching by ChatGPT [33]. Few research has come up with the allegation that ChatGPT and other related AI tools have revolutionised the educational system for fact, but has also disrupted it to a significant extent in terms of teaching-learning process, and dismantling the academic scenario. Therefore, it is matter of concern that how classroom teaching can continue in balance with the ChatGPT use such that it does not replace the conventional and cognitive process of learning. The current study has made an effort to understand the student and teacher perception of the extent of ChatGPT in over-powering Classroom teaching. For the analysis purpose, four queries were raised and were answered by the students and teachers comprising (i) underestimation of classroom teaching due to the use of ChatGPT, (ii) challenging teacher's knowledge in class due to the use of ChatGPT, (iii) reduction of student strength in class due to lack of confidence on teachers, and (iv) bossy attitude over teachers due to the use of ChatGPT. To these queries raised by the investigator, average student, and teacher perception is marked as 3.6 and 4.2 under underestimation of classroom teaching due to the use of ChatGPT, respectively, 2.2 and 4.6 under challenging teacher's knowledge in class due to the use of ChatGPT, respectively, 3 and 4.1 under reduction of student strength in class due to lack of confidence on teachers, respectively, and 3.8 and 3.7 under bossy attitude over teachers due to the use of ChatGPT, respectively. It can be interpreted that most of the students have opined that due to use of ChatGPT, students maintain a bossy attitude over teachers in classroom by testing educator's intelligence by use of ChatGPT. Further, it is observed that most of the teachers have opined that students keep challenging teacher's knowledge in class by studying beforehand via ChatGPT to underestimate teachers in open forum, which is a very unhealthy practice. This practice will encourage students not to clarify their doubts in order to learn and grow, rather to challenge teachers and insult them in classroom. However, the maximum score of difference is noted to be 2.4 under the statement challenging teacher's knowledge in class due to the use of ChatGPT where perception of students and teachers stands against themselves. The least difference is observed in the perception of students and teachers under bossy attitude over teachers due to the use of ChatGPT which shows that both students and teachers believe that most of students develop bossy attitude in the class due to the use of ChatGPT and consider their intelligence over and above teacher's knowledge and experience. However, the detail graphical representation is portrayed in Figure 4.3.

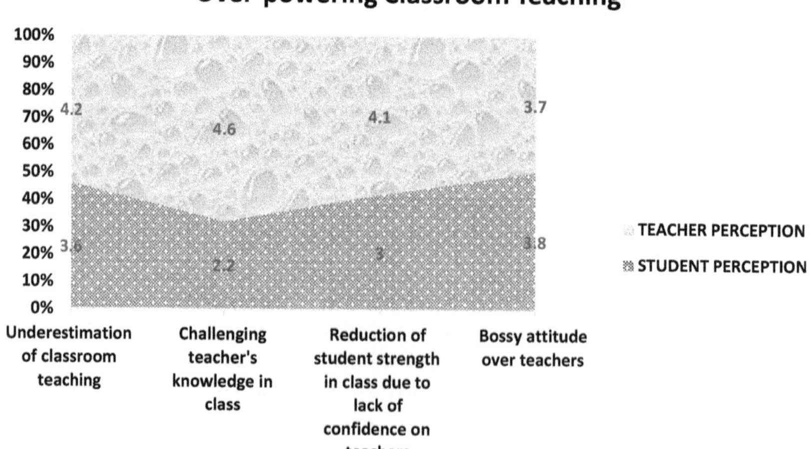

Figure 4.3 ChatGPT overpowering classroom teaching.

4.5.4 Can ChatGPT replace educators in educational institutes?

Around developing motivation and commitment of students towards learning process, ChatGPT has also come to be a competing substitute for educators or teachers in educational institutes in terms of providing educational support at any given point of time without any hassle. Educational support comprises framing research topic and its development, emerging teamwork skills, quick integration and fast-forward understanding and leveraging advanced learning before classroom teaching [40]. However, even after providing up-to-date content and activities for smoothening teaching-learning process, what it lacks behind is the personal interaction, individual touch and personalised feedback. The major concern of use of ChatGPT is to maintain a responsible classroom environment whose use must be carefully monitored to ensure that it should not replace classroom teaching with the passage of time [41]. The academic misuse of ChatGPT is immense in terms of dishonest or unethical purposes in academia. Further, the answers provided by ChatGPT are sometimes not accurate which requires attention of educators for reliability. Even it is evidential that few answers provided by ChatGPT can seem to be well-written and coherent but are actually erroneous or unsubstantiated information. In these cases, it can be settled with the perception that in academia even after these days students are very much relying on ChatGPT to get quick answers to their queries but are not reliable to certain extent, thus it is important that the educators need to rectify their confusion such that it would no more negatively affect their learning and understanding of the topics [25]. For the analysis purpose, four queries were raised and were answered by the students and teachers comprising the use of ChatGPT in (i) doubt clearing, (ii) better understanding, (iii) maintaining classroom connectivity, (iv) maintaining moral standards, and (v) providing emotional touch. To these queries raised by the investigator, average student and teacher perception is marked as 2.5 and 1 under use of ChatGPT in doubt clearing, 2.6 and 1.5 under use of ChatGPT in Better Understanding, 1 and 1 under use of ChatGPT in maintaining Class-room connectivity, 2.6 and 1 under use of ChatGPT in maintaining moral Standards and ultimately, 1.5 and 1 under use of ChatGPT in providing emotional touch. It can be interpreted that maximum strength of students have opined that ChatGPT is successfully providing them better understanding without deviating any moral standards but is definitely failing in terms of providing classroom connectivity. Average number of teachers also agreed that ChatGPT is very useful in providing a better understanding but is not that impactful in remaining four statements which is clearly evidential in the Figure 4.4. However, the maximum score of difference is noted to be 1.6 under the statement of use of ChatGPT in maintaining moral standards where perception of students and teachers stands against themselves. The least difference is observed to be 0 in the perception of students and

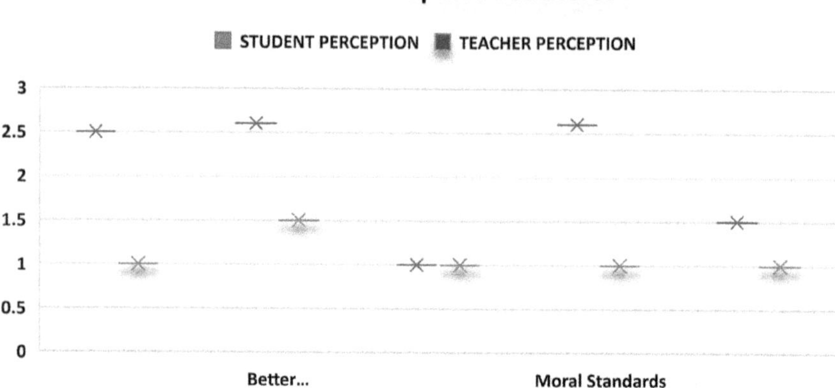

Figure 4.4 Possibility of ChatGPT replacing classroom teaching.

teachers under classroom connectivity provided by ChatGPT which shows that both students and teachers believe that ChatGPT is failing to maintain classroom connectivity.

4.6 CONCLUSION

In the fast growing and techno-driven society, scientific advancement has revolutionised the world and has taken on a fundamental role in our existence which has an incredible mark in our current education system. No wonder, the use of AI chatbot such as ChatGPT, in educational landscape is constantly expanding and dramatically changing the approach of students and teacher's approach of teaching-learning process in classroom. However, ChatGPT has its own benefits as well as challenges for the teaching-learning process. In one hand, ChatGPT interacts with users in a human-like manner while offering them directions and apt information and is capable of performing a wide array of tasks such as organising data for research, translating software and papers, inscribing mails and essays, debug and annotate code, generate poetry, and respond to questions of any genre. Further, it is believed that it might hold tremendous potential to make the process of teaching and learning easier. On the other hand, despite the advantages ChatGPT has also instigated new challenges and threats into the educational landscape but has also raised concerns that ChatGPT can replace teachers and augment automation [25]. Further, there are concerns about AI-assisted cheating because it may be used to produce written assignments and exams on behalf of students and can answer specifically to user requirements. Moreover, many teachers have raised apprehensions regarding the outsourcing of schoolwork by students by using ChatGPT. Hence, ethical and responsible approach is desired to minimise the challenges and maximise the benefits of ChatGPT. And in order to achieve this, a deeper understanding is prerequisite. It's true that the use of ChatGPT can create wonders in teaching-learning process as a promising tool by offering multiple benefits but is still not potent enough to replace educators or teachers. Therefore, in order to give students a more inclusive, accessible, and useful educational experience, educational institutions should keep exploring and creating AI applications while keeping in mind that such promising tool as ChatGPT can never replace the pedagogical methodologies, guidance, and expertise of the educators.

REFERENCES

1. Ng, Wan, and Wan Ng. "Change and continuity in educational uses of new digital technologies." *New Digital Technology in Education: Conceptualizing Professional Learning for Educators*, Wan Ng (ed.), (2015): 3–23. Cham, Switzerland: Springer.
2. Rigby, Darrell. "The future of shopping." *Harvard Business Review* 89.12 (2011): 65–76.
3. Ward, Jeremy P. T., et al. "Communication and information technology in medical education." *The Lancet* 357.9258 (2001): 792–796.
4. Anderson, Monica. "More Americans using smartphones for getting directions, streaming TV." (2016).
5. Poushter, Jacob. "Smartphone ownership and internet usage continues to climb in emerging economies." *Pew Research Center* 22.1 (2016): 1–44.
6. Schindler, Laura A., et al. "Computer-based technology and student engagement: a critical review of the literature." *International Journal of Educational Technology in Higher Education* 14 (2017): 1–28.
7. *Changes in learning process caused by the implementation of ICT in education in Estonian in-service and pre-service teachers perceptions.*
8. Ilomäki, Lig, and Minna Lakkala. "Digital technology and practices for school improvement: innovative digital school model." *Research and Practice in Technology Enhanced Learning* 13.1 (2018): 25.
9. Haleem, Abid, et al. "Understanding the role of digital technologies in education: a review." *Sustainable Operations and Computers* 3 (2022): 275–285.
10. Aggarwal, Karan, et al. "Has the future started? The current growth of artificial intelligence, machine learning, and deep learning." *Iraqi Journal for Computer Science and Mathematics* 3.1 (2022): 115–123.
11. Calo, Ryan. "Artificial intelligence policy: a primer and roadmap." *UCDL Review* 51 (2017): 399.
12. Dignum, Virginia. "Responsible artificial intelligence: designing AI for human values." (2017).
13. Kurian, N., et al. "AI is now everywhere." *British Dental Journal* 234.2 (2023): 72–72.
14. Chatterjee, Sheshadri, et al. "Adoption of artificial intelligence and cutting-edge technologies for production system sustainability: a moderator-mediation analysis." *Information Systems Frontiers* 25.5 (2023): 1779–1794.

15. Devedžić, Vladan. "Web intelligence and artificial intelligence in education." *Journal of Educational Technology & Society* 7.4 (2004): 29–39.

16. Richter, Olaf Zawacki, et al. "Systematic review of research on artificial intelligence applications in higher education: where are the educators?" *International Journal of Educational Technology in Higher Education* 16 (2019): 6.

17. Božić, Velibor, and Indrasen Poola. "Chat GPT and education." Preprint (2023).

18. Brown, Tom, et al. "Language models are few-shot learners." *Advances in Neural Information Processing Systems* 33 (2020): 1877–1901.

19. Sok, Sarin, and Kimkong Heng. "ChatGPT for education and research: a review of benefits and risks." *SSRN 4378735* (2023).

20. Dilmegani, Cem. "ChatGPT education use cases, benefits & challenges in 2023." *AI Multiple* (2023).

21. Rasul, Tareq, et al. "The role of ChatGPT in higher education: benefits, challenges, and future research directions." *Journal of Applied Learning and Teaching* 6.1 (2023): 41–56.

22. Zhai, Xiaoming. "ChatGPT user experience: implications for education." *Available at SSRN 4312418* (2022).

23. Mollick, Ethan R., and Lilach Mollick. "New modes of learning enabled by ai chatbots: three methods and assignments." *SSRN 4300783* (2022).

24. Lo, Chung Kwan. "What is the impact of ChatGPT on education? A rapid review of the literature." *Education Sciences* 13.4 (2023): 410.

25. Ausat, Abu Muna Almaududi, et al. "Can Chat GPT replace the role of the teacher in the classroom: a fundamental analysis." *Journal on Education* 5.4 (2023): 16100–16106.

26. Abd-Elaal, El-Sayed, S. H. Gamage, and Julie E. Mills. "Artificial intelligence is a tool for cheating academic integrity." *30th Annual Conference for the Australasian Association for Engineering Education (AAEE 2019): Educators Becoming Agents of Change: Innovate, Integrate, Motivate* Oct 11, 2019. 2019.

27. Mhlanga, David. "Open AI in education, the responsible and ethical use of ChatGPT towards lifelong learning." In: David Mhlanga (ed.) *FinTech and Artificial Intelligence for Sustainable Development: The Role of Smart Technologies in Achieving Development Goals*, 2023. 387–409. Cham, Switzerland: Springer Nature, University of Johanneburg.

28. Rahman, Md Mostafizer, and Yutaka Watanobe. "ChatGPT for education and research: opportunities, threats, and strategies." *Applied Sciences* 13.9 (2023): 5783.

29. Khaddage, Ferial, and Kim Flintoff. "Say goodbye to structured learning ChatGPT in education, is it a threat or an opportunity?" *Society for Information Technology & Teacher Education International Conference*. Association for the Advancement of Computing in Education (AACE) Mar 13, 2023, 2023. Waynesville, NC USA.

30. Wang, Ting, et al. "Exploring the potential impact of artificial intelligence (AI) on international students in higher education: generative AI, chatbots, analytics, and international student success." *Applied Sciences* 13.11 (2023): 6716.

31. Dwivedi, Yogesh K., et al. "'So what if ChatGPT wrote it?' Multidisciplinary perspectives on opportunities, challenges and implications of generative conversational AI for research, practice and policy." *International Journal of Information Management* 71 (2023): 102642.

32. Gilson, Aidan, et al. "How does ChatGPT perform on the United States Medical Licensing Examination (USMLE)? The implications of large language models for medical education and knowledge assessment." *JMIR Medical Education* 9.1 (2023): e45312.

33. Megahed, Fadel M., et al. "How generative AI models such as ChatGPT can be (mis) used in SPC practice, education, and research? An exploratory study." *Quality Engineering* 36.2 (2024): 287–315.

34. Meo, Sultan A., and Muhammad Talha. "Turnitin: is it a text matching or plagiarism detection tool?" *Saudi Journal of Anaesthesia* 13.Suppl 1 (2019): S48–S51.

35. Khalil, Mohammad, and Erkan Er. "Will ChatGPT get you caught? Rethinking of plagiarism detection." *International Conference on Human-Computer Interaction,* 2023. (pp. 475–487). Cham, Switzerland: Springer Nature.

36. Jain, Suvrat, and Roshita Jain. "Role of artificial intelligence in higher education—an empirical investigation." *IJRAR-International Journal of Research and Analytical Reviews* 6.2 (2019): 144–150.

37. Montenegro-Rueda, Marta, et al. "Impact of the implementation of ChatGPT in education: a systematic review." *Computers* 12.8 (2023): 153.

38. Neumann, Michael, Maria Rauschenberger, and Eva-Maria Schön. "'We need to talk about ChatGPT': the future of AI and higher education." *2023 IEEE/ACM 5th International Workshop on Software Engineering Education for the Next Generation (SEENG)* May 16 2023. IEEE, 2023. Melbourne, Australia.

39. Qadir, Junaid. "Engineering education in the era of ChatGPT: Promise and pitfalls of generative AI for education." *2023 IEEE Global Engineering Education Conference (EDUCON)* 1-4 May 2023. IEEE, 2023. Kuwait.

40. Fuchs, Kevin. "Exploring the opportunities and challenges of NLP models in higher education: is Chat GPT a blessing or a curse?" *Frontiers in Education* 8 (2023): 1166682.

41. Halaweh, Mohanad. "ChatGPT in education: strategies for responsible implementation." *Contemporary Educational Technology,* 15.2 (2023): ep421.

Chapter 5

Defence of DDoS attacks using targets in motion

Kummari Vikas, Vijay Kumar Reddy Julakanti,
Abdulsattar Abdullah Hamad, and
Suresh K. Peddoju

5.1 INTRODUCTION

The Arbor Architectures recently proclaimed a fundamental increase in the impossibility of vastly improved version Distributed Denial of Service (DDoS) assaults [1]. The biggest announced data exchange limit attained by a flood-based DDoS assault managed to come up at 100 Gbps in 2010. Nonetheless, the cost of enacting a DDoS assault has shown to be very inexpensive. Various statistical tools have been recommended in prior to preventing or mitigating DDoS assaults. Isolating-based methodologies usually employ passed channels to restrict undesired traffic sent off the guarded center points. Limit-based insurance devices attempt to keep transporters' resource consumption within the boundaries set by the receivers. Reputable visibility courses of action mediate an overlay association to shady packs between clients and the defended center points, implying that they acclimate and filter. In any case, current rigid assurance strategies generally strain on the whole association of auxiliary characteristics over online switches and otherwise permit the tremendous, fiery virtual association to get through to the consistently heightening attacks.

Furthermore, some of them might be currently unprotected against refined tactics which include clearing as well as adaptable inundation attempts. Throughout this work, we postulate MOTAG, which is a robust DDoS insurance instrument that employs as moving target protect technique to protect distributed digital organizations [2]. It provides DDoS flexibility for endorsed and confirmed shoppers of health-related organizations, for example, electronic banking and e-finance. It uses a veneer of mysterious motion go-betweens to moderate all correspondence among patrons as well as the defended database server.

The channels of the network enveloping the database server simply permit pedestrians from significant delegate center points to show up at the shielded servers.

MOTAG go-between centers have two critical properties. In any case, all go-between centers are 'hidden,' as their IPs are either camouflaged from most user orexclusively known by accredited consumers after proper

DOI: 10.1201/9781003565529-5

verification [3]. To avoid unnecessary information spillage, each genuine user is assigned the Internet address of one working middle person at random. To help secure the buyer's communication channel, we utilize conventional Proof of Work schemes. Additionally, the go-between facilities are 'dynamic' or continuously changing. When a working middle person center closes, it is replaced by another center point in a different region, and the connected users are drifted to elective intermediate participants. We demonstrate that these properties enable us not only to mitigate high-powered but also to locate and disengage noxious insiders who reveal the location of secret middle people to outside aggressors. We accomplish this by relocating clients' appointments to novel delegate centers when their exceptional middle people are already under attack [4].

We recommend computations to definitively assess the multitude of bigwigs and change contractor tasks as well as to protect the most trustworthy clients after each blend. We recommend algorithms to accurately evaluate key individuals and adjust contractor tasks, as well as to safeguard the most trustworthy clients after each interaction. Our solution does not depend on global coordination via Internet routers or collaboration across multiple ISPs [5]. Additionally, we do not rely on resource-heavy intermediary networks prone to high data transfer vulnerabilities, and we ensure resilience against non-critical failures. Taking everything into consideration, we employ our go-between secret and versatile abilities to repel solid aggressors. This includes lower alliance costs and achieves significant combative dexterity, leading to a strong DDoS affirmation [6].

5.2 LITERATURE SURVEY

Critical infrastructure and Internet services are both still significantly at risk from DDoS assaults. In this research, we offer MOTAG, a moving target defense system that protects authenticated clients' service access from DDoS flood attacks. To relay data traffic between authorized clients and the secured servers, MOTAG uses a group of dynamic packet indirection proxies. The attempts of external attackers to directly assault the network infrastructure can be successfully thwarted by our architecture. As a result, in order to find hidden proxies and launch attacks, attackers will need to work along with nefarious insiders.

As part of the literature survey, various case studies have been studied before getting into this study. A study made by Xiaoyong Yuan conveys DeepDefense, a deep-learning–based method for detecting DDoS attacks that can automatically separate high-level characteristics from low-level ones and produce robust representations and inferences. To identify patterns in sequences of network traffic and track network assault operations, they created a recurrent deep neural network [7]. Another study done by Mallikarjunan presents an overview of the state of art in DDoS-based attack detection methods. The identification and rectification of false positives involve predicting potential attackers' behavior based on historical data, including factors such as larger bandwidth data, spoofed IP addresses, bottleneck bandwidth, data flow size, and the type of congestion window they use.

By analyzing differences between low-rate and high-rate traffic flows in the network, effective mitigation strategies can be implemented to prevent DDoS attacks and safeguard the system from their detrimental effects. In the proposed approach, incoming packets undergo a series of checks. Initially, the system examines the packet arrival interval, followed by a comparison of the number of received packets to the number sent. If the received packets outnumber the sent ones, the packets are promptly dropped. If the initial condition is not met, the packet proceeds to the predictive profiling method. Here, attacker prediction is conducted, checking for conditions such as those mentioned earlier. If any of these conditions are met during the prediction, the entity is identified as a potential attacker [8].

According to the study done by Chao, the three methods proposed to prevent DoS (DDoS) attacks are using a router DoS attack prevention, increasing the trusted platform module, and system defences. This paper analyzes the principles of DoS (DDoS) attack prevention and provides a thorough analysis of existing prevention techniques [9]. Another study conducted by Falowo et al. [10] was prompted by notable observations outlined in a recent IEEE-published study titled "Threat Actors' Tenacity to Disrupt: Examination of Major Cybersecurity Incidents." The study delves into the utilization of DDoS attacks, along with other attack methods, by threat actors to disrupt both public and private enterprises across various instances. Specifically focusing on DDoS attacks, this research explores the application of diverse machine learning techniques to identify and mitigate such incidents. Real-world examples of significant DDoS attacks, publicly reported or announced, were scrutinized, revealing an increasing trend over the last 7 years.

Within the dataset encompassing more than 700 major cybersecurity incidents analyzed from 2015 to 2022, evidence of DDoS attacks was consistently present each year. Notably, in the subset examined, the instances of major DDoS attacks surged from 8 to 31 between 2021 and 2022, representing an estimated 288% increase. Recognizing the challenge posed by the substantial volume of network traffic analysis in manually or efficiently detecting DDoS events, this study underscores the significance of incorporating AI technologies, particularly machine learning algorithms, for effective DDoS detection and mitigation. The research also identifies specific machine learning algorithms suitable for detecting and mitigating DDoS attacks. Acknowledging the absence of a universal solution for DDoS mitigation, the study recommends leveraging CIS Benchmarks and encourages organizations to reference an NIST framework titled "AI Risk Management Framework," released to the public on January 26, 2023. This framework provides guidelines for implementing multiple layers of controls to effectively manage DDoS attacks.

Another study [11] emphasizes the utilization of Artificial Neural Networks for the detection of diverse types of DDoS attacks, including UDP-Flood, Smurf, HTTP-Flood, and SiDDoS, with a primary focus on network and transport layer DDoS attacks. Furthermore, the evaluation includes the assessment of time and space complexity to enhance the efficiency of the implemented model and address limitations identified in the existing research gap. Our analysis of the dataset reveals that the methods proposed in this study exhibit improved capabilities in detecting DDoS attacks.

5.3 PROPOSED METHODOLOGY

Throughout this mission, we advocate MOTAG, a mobile impartial warning system that aims to protect validated patrons from inundating attacks (Figure 5.1). It engages connectivity of evolving parcel conduits to transfer network data for both instantly recognizable clients and

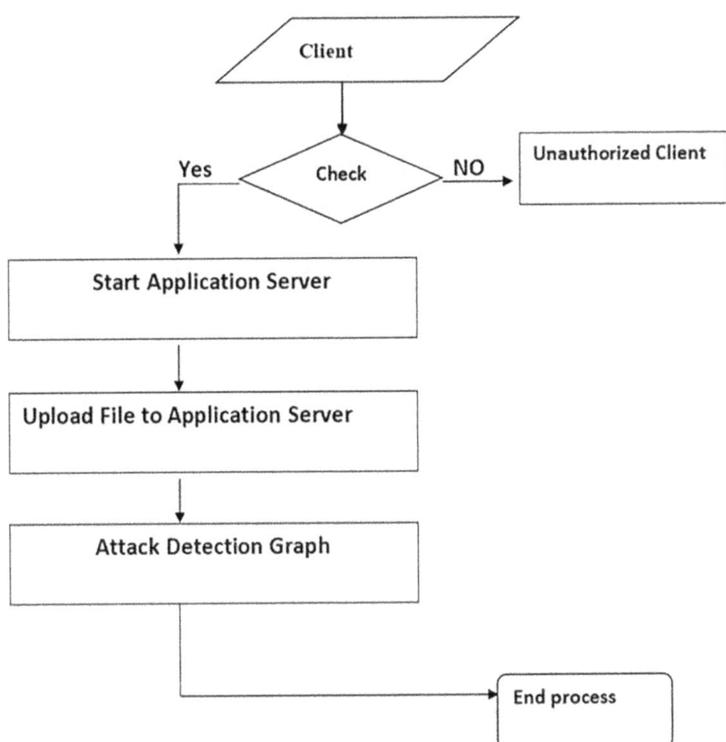

Figure 5.1 Flow chart.

retention. This plan will really restrain outer assailants' endeavors to barrage the organization foundation straightforwardly. Subsequently, assailants should intrigue with malignant insiders in finding secret intermediaries and afterward starting assaults. MOTAG, on the other side, can clearly differentiate system vulnerabilities from business clients by habitually "moving" private center individuals to the latest organize locales while continuing to expand service user initiatives [11].

To reduce the number of ways proxy reallocation, we here used a greedy shuffling algorithm that also maximizes attack isolation.

5.4 IMPLEMENTATION

5.4.1 Authentication server

Initially, the client request is received by the authentication server which internally performs the Random Shuffling Algorithm based on the greedy manner this server selects one of the proxy servers out of the available proxy servers thereafter it assigns the selected proxy to the client. To detect if any malicious user places any insider attack, we are here using the random shuffling method by which the attack could not find which intermediary is currently being under the usage of the server and the port which indeed prevents stealing several applications [12].

Since here there is a continuous shuffling process with randomness the insider attack gets deviated and can't attack the application/proxy server.

5.4.2 Intermediary proxy server

Intermediary servers receive requests from clients and then evaluate them. If the size of the request data is within the processing capacity of the application server, the intermediary will allow the client request to reach the secure application server [13]. In the event that solicitation capacity is going past the defined limit, it will resolve the DDoS on the JAM database server, causing the intermediary servers to withdraw the request and notify the client or confirmation server that the attack is over.

5.4.3 Application server

That component identifies client requests and, after a little delay, cycles the request and returns a reply to the client. Throughout this program, we help the client to send textual data to the database server, which will receive the txt archive and store it a short time later. On the off chance that the size txt record goes past the limit or the client sends a tremendous size of the archive then the middle person won't transport it off the application server [14].

5.4.4 Client

The mill or threatening to clients who move requesting to the server for taking care of malevolent client attacks can be perceived and dropped by proxies are handled by client server [15].

5.4.5 Output screens

Open application server folder and run.bat file from there. Click on "Start Application Server" to begin server we can see application server getting started. Then click on the run file in the folder of authentication server. Likewise click to start server of proxy1. Proxy 2 folder's run bat file has to be opened to get the button to start server of proxy 2.

Open run file present in client folder and click "Upload File to Application Server" button to upload file to the required server where the communication is going to happen between application server, random proxy, and authentication server as shown in the below screen.

Here we are going to upload a text file which is a large file as shown above. Here proxy 2 has been assigned to the client using authentication server where the intermediary server 2 will be sending request to application server. Proxy 2 and the authentication server can provide the requested details. On the screen, we can see that the first authentication server assigns Proxy 2 to the client. Then, Proxy 2 becomes involved and forwards the request to the application server. The application server receives the data and stores it in the 'ReceiveData' folder (Figures 5.2–5.4).

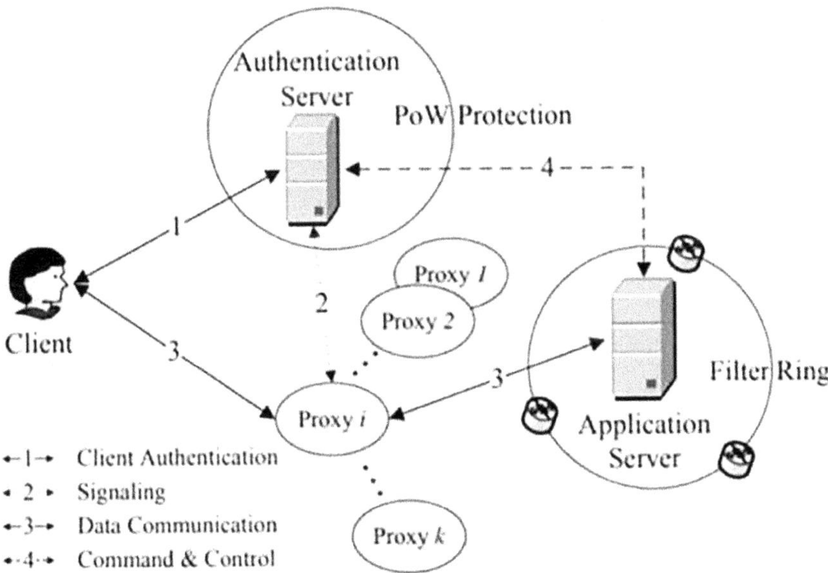

Figure 5.2 Architecture diagram.

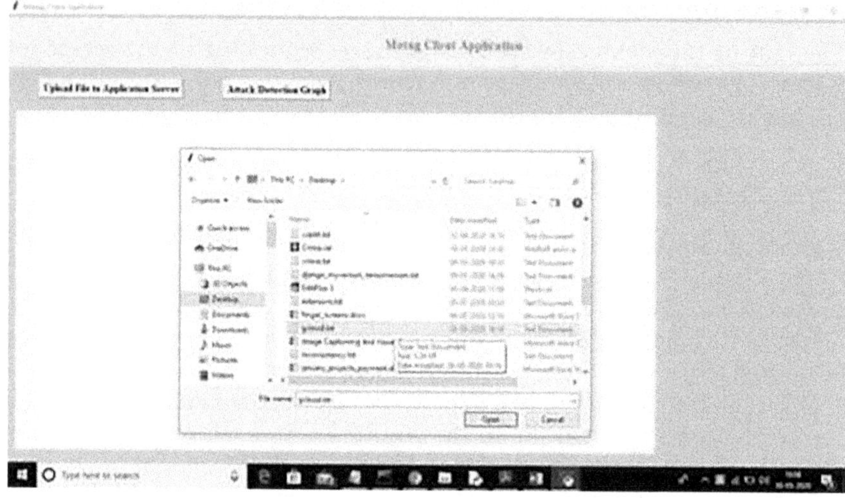

Figure 5.3 File being uploaded to application server.

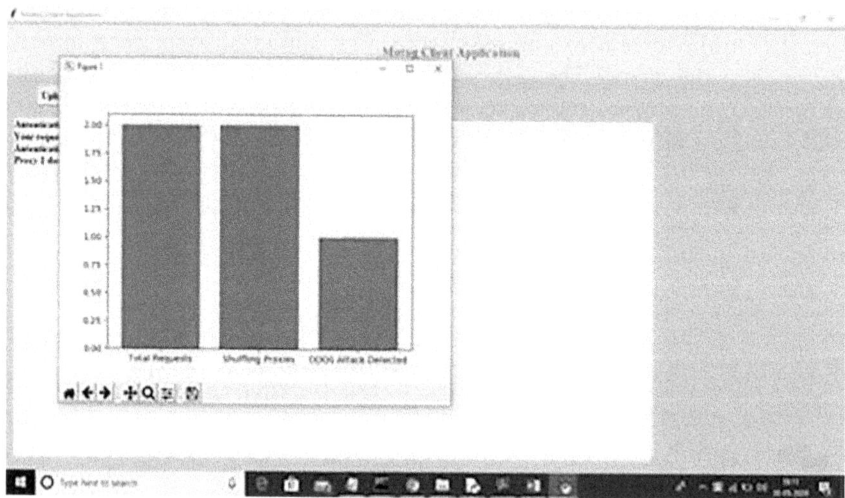

Figure 5.4 Graph showing number of attacks detected.

We could observe the Google Cloud file getting saved in the receive folder. We are uploading one single file of txt format which is of huge size. This time the proxy 1 being assigned by the authentication server where the size of the file is huge hence the proxy 1 has sent output as DDOS attack. There is DDOS attack, whereas in client side we can see the attack detection graph as shown in Figure 5.4.

5.5 CONCLUSION AND FUTURE SCOPE

We will demonstrate a DDOS attack avoidance solution entitled MOTAG, which employs adaptive, obscured servers with mobile concentrations for mitigate channel flooding DDoS. Confirmed clients are assigned to intermediate user center points that perform package sending and meeting policing in order to show up at the secured organization. When a DDoS is undertaken even against intermediaries, the confirmed users associated only with pursued go-betweens are dispersed onto alternative go-betweens throughout execution, allowing them to be able to dodge the persistent attack and continue to serve the protected business with permission. We can truly conceal the protected essential entities from outside aggressors with it. Insiders can be used by complex aggressors to track down and assault our middle person center points. It disengages insider-assisted attacks through an innovative, effective modifying structure.

Our calculations demonstrate how it can protect a higher number of untouchable clients from DDoS attacks aided by several insiders within a few blends. Furthermore, the outcomes of our probing methodology can be used to integrate the execution and course of action of MOTAG-based malware assurance systems. Future Scope: "Expanding the scalability and availability of the MOTAG framework. Increasing the number of clients that can be served by a single MOTAG structure to benefit as many users as possible. Reducing the latency in delivering responses to clients.

REFERENCES

1. N. K. S. Vanitha, S. V. Uma and S. K. Mahidhar, "Distributed denial of service: Attack techniques and mitigation," *2017 International Conference on Circuits, Controls, and Communications (CCUBE)*, 2017, pp. 226–231, doi: 10.1109/CCUBE.2017.8394146.
2. C. Douligeris and A. Mitrokotsa, "DDoS attacks and defense mechanisms: a classification," *Proceedings of the 3rd IEEE International Symposium on Signal Processing and Information Technology (IEEE Cat. No.03EX795)*, 2003, pp. 190–193, doi: 10.1109/ISSPIT.2003.1341092.
3. Q. Jia, K. Sun and A. Stavrou, "MOTAG: moving target defense against internet denial of service attacks," *2013 22nd International Conference on Computer Communication and Networks (ICCCN)*, 2013, pp. 1–9, doi: 10.1109/ICCCN.2013.6614155.
4. R. Sharma, K. Kumar, K. Singh and R. C. Joshi, "Shared based rate limiting: an ISP level solution to deal DDoS attacks," *2006 Annual IEEE India Conference*, 2006, pp. 1–6, doi: 10.1109/INDCON.2006.302831.
5. S. Potluri, M. Mangla, S. Satpathy and S. N. Mohanty, "Detection and prevention mechanisms for DDoS attack in cloud computing environment," *2020 11th International Conference on Computing, Communication and Networking Technologies (ICCCNT)*, 2020, pp. 1–6, doi: 10.1109/ICCCNT49239.2020.9225396.

6. S. Potluri, K. Subba Rao and S. Nandan Mohanty. *Cloud Security: Techniques and Applications*. Berlin, Boston: De Gruyter, 2021. doi: 10.1515/9783110732573.

7. X. Yuan, C. Li and X. Li, "DeepDefense: identifying DDoS attack via deep learning," *2017 IEEE International Conference on Smart Computing (SMARTCOMP)*, 2017, pp. 1–8, doi: 10.1109/SMARTCOMP.2017.7946998.

8. K. N. Mallikarjunan, K. Muthupriya and S. M. Shalinie, "A survey of distributed denial of service attack," *2016 10th International Conference on Intelligent Systems and Control (ISCO)*, 2016, pp. 1–6, doi: 10.1109/ISCO.2016.7727096.

9. Z. Chao-yang, "DOS attack analysis and study of new measures to prevent," *2011 International Conference on Intelligence Science and Information Engineering*, 2011, pp. 426–429, doi: 10.1109/ISIE.2011.66.

10. O. I. Falowo, I. Okpala, E. Kojo, S. Azumah and C. Li, "Exploration of various machine learning techniques for identifying and mitigating DDoS attacks," *2023 20th Annual International Conference on Privacy, Security and Trust (PST)*, Copenhagen, Denmark, 2023, pp. 1–7, doi: 10.1109/PST58708.2023.10320151.

11. Y.-K. Lee, H.-I. Ju, J.-H. Park and J.-W. Han, "User authentication mechanism using authentication server in home network," *2006 8th International Conference Advanced Communication Technology*, 2006, pp. 4 pp.-506, doi: 10.1109/ICACT.2006.206017.

12. O. O. Abiona, T. Anjali, C. E. Onime and L. O. Kehinde, "Proxy server experiment and the changing nature of the web," *2008 IEEE International Conference on Electro/Information Technology*, 2008, pp. 242–245, doi: 10.1109/EIT.2008.4554305.

13. A. Sugiura and C. Dermawan, "In traffic jam IVC-RVC system for ITS using bluetooth," *IEEE Transactions on Intelligent Transportation Systems 6.3* (2005): 302–313. doi: 10.1109/TITS.2005.853704.

14. K. Vlaeminck, S. V. Van Hoecke, F. De Turck, B. Dhoedt and P. Demeester, "Design and implementation of an application server load balancing architecture supporting the end-to-end provisioning of value-added services," *11th International Telecommunications Network Strategy and Planning Symposium. NETWORKS 2004*, Vienna, Austria, 2004, pp. 345–350, doi: 10.1109/NETWKS.2004.241085.

15. S. Simpson, A. T. Lindsay and D. Hutchison, "Identifying legitimate clients under distributed denial-of-service attacks," *2010 Fourth International Conference on Network and System Security*, 2010, pp. 365–370, doi: 10.1109/NSS.2010.77.

Chapter 6

Unlocking sign language interpretation

Leveraging transfer learning in deep learning models

*G. Sucharitha, D. Celesty Bliss Rufus,
C. Mahananda Reddy, and G. Chandra Sekhar*

6.1 INTRODUCTION

Sign Language is a visual communication method, primarily utilized by individuals who are deaf or mute. Traditional methods of Sign Language recognition (SLR) are encumbered by constraints such as low accuracy and restricted vocabulary [1–3]. The primary objective of this chapter is to highlight the utilization of deep learning techniques in SLR. Over the past few years, this methodology has exhibited significant potential in varied computer vision tasks. In line with this trend, the present study suggests a methodology for SLR using convolutional neural networks (CNNs). The proposed model has undergone training on a small-scale dataset of Indian Sign Language, aiming to capture the intricate spatio-temporal patterns inherent in sign gestures. The model's performance was evaluated using standard metrics, yielding satisfactory results. This study will bring the efficacy of deep learning techniques in SLR into limelight. The ultimate goal of this research article is to establish an accurate SLR system leveraging deep learning techniques. Such a system holds the potential to empower the hearing-impaired community by facilitating more efficient and effective communication with non-signers in real time.

6.2 RELATED WORK

6.2.1 Overview of existing research

SLR utilizing deep learning has emerged as a burgeoning field of research in recent years. This area focuses on constructing machine learning models capable of comprehending and deciphering Sign Language gestures. Such models are instrumental in creating tools and applications aimed at enhancing communication between the deaf and mute communities and those who communicate verbally.

DOI: 10.1201/9781003565529-6

The current approaches to SLR can be broadly classified as:

a. Sensor approaches
b. Vision approaches

6.2.1.1 Sensor approaches

Most of the sensor-based approaches consist of three units: the first one being an input or receptor unit, the second one being a processing unit and the third one being an output unit. The input unit consists of sensors which aid in measuring orientation of the hand and its movement, bending of fingers, separation between fingers, etc. A microcontroller is the processing unit which uses the data produced by the sensors for processing and recognition.

Lokhande et al. [4] were successful in achieving 99% accuracy in recognizing Indian Sign Language up to eight common words using three-axis accelerometer and flex sensors. Han et al. [5] were successful in achieving 99.4% accuracy in recognizing up to five static and eight dynamic common gestures using several IMU sensors. Manware et al. [6] explored the creation of smart gloves designed as a communicating aid for individuals who are speech-disabled or hearing-disabled. These gloves incorporate sensors and a microcontroller capable of detecting the wearer's hand signs and converting them to speech or text. The study demonstrated the efficacy of the smart gloves in enhancing communication for users, achieving a high level of accuracy in both detecting and translating hand gestures.

There are a lot of other endeavors in this field, but the above are worth mentioning. Although high accuracies have been obtained, the equipment is quite costly and may take better planning for everyone to have access to these devices. Sruthi and Lijiya [7] argue that vision-based approaches may offer a superior solution to the aforementioned problem.

6.2.1.2 Vision-based approaches

A big share of research efforts in recognizing Sign Language have predominantly concentrated on vision based methodologies, instead of sensor-based approaches. This is because of the fact that vision-based techniques diminish reliance on sensor devices, thus reducing costs associated with specialized sensors. While 2-D vision-based systems exhibit effectiveness with appropriate image or video processing methods, they might lack required depth information. However, the accessibility of 3-D or depth cameras at affordable prices has significantly enhanced research in 3-D vision-based methodologies.

Although 3-D SLR systems entail higher computational costs, their superior recognition rates compensate for this drawback. The integration of the third dimension furnishes additional depth information, effectively addressing numerous challenges encountered in 2-D SLR. Kishore et al. [8] were successful in achieving 98.9% accuracy in recognizing up to 500 Indian Standard Language (ISL) gestures in a 3-D-vision–based system with the assistance of 3-D motion lets and adaptive kernel matching. Sharma et al. [9] were 97.10% accurate in recognizing all the ten digits of the decimal number system in ISL, using their own dataset. At the preprocessing stage, the RGB images were first converted to grayscale images using MATLAB®. They have used the first derivative Sobel edge detector method for identifying the borders. For feature extraction, the researchers employed direct pixel values and hierarchical centroid methods. They then utilized kNN and neural network classifiers for further analysis and classification. Patil et al. [3] in their work have focused on optimizing the time factor, neglecting the accuracy and were successful in recognizing 26 alphabets of the ISL through a 2-D-vision–based system. Scale-Invariant Feature Transform (SIFT) and Key point extraction were primarily used.

The study by Nandy et al., [10] employs SIFT for recognition, emphasizing the significance of static and dynamic gestures produced using both hands. The use of direction histograms as features for classification, considering illumination and orientation invariance, is a notable contribution. The paper discusses the challenges in gesture recognition, including background removal, hand tracking, and illumination variation. The proposed methodology entails several key steps: first, the creation of a video database; next, feature extraction utilizing SIFT; and finally, classification using K-nearest neighbor and Euclidean distance metrics. Experimental results demonstrate the efficacy of this approach, achieving recognition accuracies of up to 100% for certain gestures. The authors highlight the need for additional features and the integration of Hidden Markov Models for real-time recognition of continuous gestures. The paper concludes by suggesting future work involving humanoid robot training and integration with Natural Language to enhance communication further.

Shivashankara and Srinath [11] introduce an optimal approach aimed at translating 24 static Sign Language alphabets and numbers from ASL to a decipherable English manuscript for both humanoid and machine comprehension. The paper not only elucidates a detailed technical approach but also includes a statistical evaluation of results. Moreover, it offers comparative graphical depictions between existing techniques and the proposed method.

Garcia, Brandon, and Sigberto Alarcon Viesca [12] leveraged transfer learning with GoogLeNet architecture on the ILSVRC2012 dataset from Surrey University and ASL dataset from Massey University. The system aims to classify ASL letters in real time, addressing challenges such as environmental concerns, occlusion, sign boundary detection, and co-articulation. The authors conducted experiments with different classifier development techniques, such as transfer learning and reinitialization, achieving promising results for letters a-e and a-k. Their study acknowledges the need for more diverse datasets and proposes future work on exploring additional neural network architectures, image preprocessing techniques, and language model enhancements for sentence-level translation. The work conducted by Krizhevsky et al. [13] introduces a groundbreaking approach to object recognition using deep CNNs. The authors trained a sizable CNN on the ImageNet dataset, which comprises up to 1.2 million high-resolution images belonging to 1,000 classes. Remarkably, the CNN outsmarted the previous state-of-the-art results, achieving low error rates. The architecture consists of five convolution layers, followed by quite a many max-pooling layers and up to three fully connected layers culminating in a 1,000-way softmax layer. To mitigate overfitting, the authors employed dropout regularization and data augmentation techniques. The paper underscores the significance of large datasets and the formidable learning capacity of CNNs in tackling intricate object recognition tasks. The authors also discuss the efficiency of GPU implementations, non-saturating neurons, and the use of ReLU nonlinearity, contributing to the success of their model.

Tang et al. [14] introduced a two-stage Hand Pose Recognition (HPR) system, harnessing the capabilities of a Kinect sensor. Notably, this approach exhibited versatility by accommodating diverse background conditions and addressing scenarios where hands were closely positioned to other body parts or are not the closest objects to the camera, even in the presence of occlusion. Deep Neural Networks were employed to autonomously derive features from hand gesture images, showcasing resilience to scaling, movement and rotation. The system showcased rapid and accurate performance, achieving a commendable recognition accuracy of 98.12%.

The research paper authored by Murakami and Taguchi [15] outlines a methodology for recognizing gestures in Japanese Sign Language. Initially, the authors devised a gesture recognition system employing neural networks to identify 42 symbols. To address the more challenging task of gesture recognition, which involves dynamic processes, they introduced a recurrent neural network. The paper discusses the development of a continuous gesture recognition method and the results of their research. The experiments involved training the system to recognize Japanese Sign Language words and evaluating recognition rates. The authors also explored the use of neural networks for recognizing finger alphabet patterns and introduced methods to enhance recognition rates, such as augmented positional data and filtering in time-space. The effectiveness of the proposed neural network approach in gesture recognition was evident in the results. Enhancements were observed through the utilization of encoding methods and recurrent neural network architectures. The research contributes valuable insights to the field of artificial reality and natural interaction between humans and virtual environments.

Bantupalli and Xie [16] utilized Inception, a CNN, for capturing spatial features, alongside an RNN for temporal feature training. This reflects the integration of state-of-the-art technologies in their approach. The research drew attention to the challenges in cross-modal communication and demonstrated a practical solution using proficient deep learning techniques. Their choice of the American Sign Language Dataset as the training dataset adds credibility to the model, ensuring relevance and applicability to real-world situations involving communication through Sign Language.

Wadhawan and Kumar [17] designed a system for recognizing static signs in Sign Language employing CNNs. They amassed a dataset comprising 35,000 images of static signs encompassing 100 signs from diverse users. The efficacy of their system was assessed across nearly 50 CNN models and various optimizers. Their approach attained the highest training accuracy, reaching 99.90% for grayscale images and 99.72% for color images. Performance evaluation was conducted based on precision, recall, and F-score metrics.

Overall, SLR using deep learning has shown promising results in recent research, but there are still challenges to overcome such as dealing with limited vocabulary, lack of completeness, variations in Sign Language dialects, lighting and camera angles, and incorporating context and facial expressions in recognition models.

6.3 CNN MODEL ARCHITECTURE

The CNN is extensively recognized for its effectiveness in computer vision tasks. CNNs utilize kernels or filters to derive features from input images via the convolution operation. This process involves taking the input image, defining a weight matrix, and performing convolution to extract specific features while preserving spatial arrangement details. This is in contrast to traditional Artificial Neural Networks, which tend to lose spatial features of input images.

In CNNs, the concept of parameter sharing is integral. This principle is achieved by employing a single filter across different sections of the input image, resulting in a feature map. As the weight matrix traverses across the image, the pixel values are reused, facilitating effective parameter sharing.

Architecture of the CNN typically consists of three broad layers:

- The Convolution Layer
- The Pooling Layer
- The Output Layer

6.3.1 The convolutional layer

The fundamental component of a CNN is the convolutional layer. It plays a vital role in deriving useful features from the input images through convolutional operations with learnable filters or kernels. For extracting the features, a filter which is a weight matrix is traversed across the image in strides. It helps in extracting specific information from the original matrix. Weight matrices can perform multiple tasks. While one weight combination extracts edges, the other might extract a specific color, while all that another might do is blur the image.

The weights would be learnt in a way such that the loss function gets minimized. In the case of multiple convolutional layers, generic feature extraction is done by the initial layers. As the network gets deeper, the weight matrices extract more and more complex features suitable to our problem. By initializing a hyperparameter called stride, we can decide the movement of the weight matrix across the image. To rectify the problem of reduction of image size with increase in the stride value, zero padding layer can be added. Same padding ensures that the spatial size of the feature maps remains consistent throughout the convolutional layers in a CNN architecture. In this way, multiple filters of the same dimension are applied together and their resultant is a stacked output of the filters also known as an activation map.

6.3.2 The pooling layer

The pooling layer is typically utilized when the size of the input image is relatively large. It helps in reducing the trainable parameters. This is especially required to decrease the spatial dimensions of the image. The pooling layers are required to be introduced periodically between the convolutional layers. The dimensions of the image remain unchanged since the pooling is done individually on each dimension. The most commonly used is the max-pooling layer. Other dominant pooling layers used are L2 norm and average pooling.

6.3.3 The output layer

The output is expected to be a class. The convolution layer and the pooling layer can only reduce the number of parameters and extract features. The output is expected to be a class. The convolutional and pooling layers primarily reduce the number of parameters and extract features. Applying a fully connected layer (FCL) ensures that we generate an output corresponding to the number of required classes. The FCL is limited to processing one-dimensional (1-D) data. After the convolutional and pooling layers, the output is typically flattened to create a 1-D vector. This vector is then fed into one or more fully connected layers, which perform classification or regression tasks. These fully connected layers transform the flattened feature representation to match the number of classes specified by the network architecture. The output layer utilizes a loss function, such as categorical cross-entropy or mean square loss, to assess the prediction error. Following the completion of the forward pass, the process of backpropagation commences to adjust the weights and biases, aiming to minimize the error and loss. A single forward and backward pass is called a single training cycle.

6.4 PROPOSED METHODOLOGY

6.4.1 Proposed framework

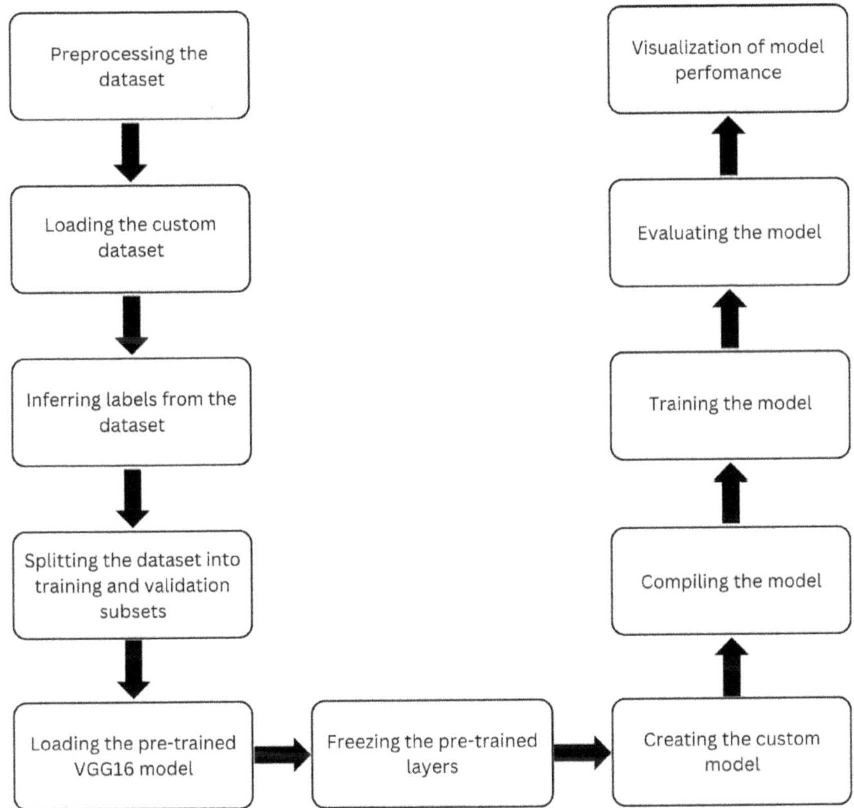

Figure 6.1 Methodology flow diagram.

6.4.2 Model training

6.4.2.1 Transfer learning for deep neural network training

In transfer learning, a model that has been previously trained on a large dataset for a particular task is used as a base to address a different yet related task. These pre-trained models undergo training on extensive datasets using massive computational resources and may require days or even weeks to complete training. Pre-trained weights play a vital role in transfer learning within deep learning. Integrating pre-trained models, such as the VGG16 model with weights that underwent training on the

ImageNet dataset, is a common practice in model training. By utilizing these pre-trained weights, the trainable parameters are reduced, facilitating computational efficiency and requiring less training data.

6.4.2.2 The VGG16 model

The VGG16 CNN is the top-performing model on the ImageNet dataset, out of all the setups [18]. The VGG16 architecture is composed of 16 layers, which can be categorized as 13 convolutional layers and 3 are fully connected layers. It adheres to a straightforward and consistent structure, utilizing (3×3) convolution filters and max-pooling layers throughout. Each convolutional layer is succeeded by ReLU activation functions, while the fully connected layers conclude with a Softmax activation function for classification purposes. This uniform architecture enables the VGG16 model to effectively extract hierarchical features from input images and perform accurate classification tasks.

It takes a fixed 224 by 224 image with RGB channels as input and normalizes each pixel's RGB values. The input image undergoes processing through multiple stacks of convolution layers interspersed with ReLU activations and a few max-pooling layers in the architecture. This process gradually reduces the spatial dimensionality of the feature maps parallelly incrementing the number of filters to extract hierarchical features from the input image. The network architecture comprises several stacks of convolutional layers, each with different numbers of filters, followed by fully connected layers. The final output layer consists of 1,000 neurons, corresponding to the classes of the ImageNet dataset, and employs a softmax activation function for classification.

6.4.3 Dataset used

We utilized a pre-acquired Indian Sign Language dataset sourced from Kaggle [19] and augmented it by appending diversified images to enhance the model's performance. RGB images were employed for both training and testing to ensure compatibility with real-time scenarios. The dataset comprises 36 classes of static signs, including 26 English alphabets and 10 digits from the decimal number system. The dataset has been divided into train and test datasets, with each class containing 800 and 200 images, respectively.

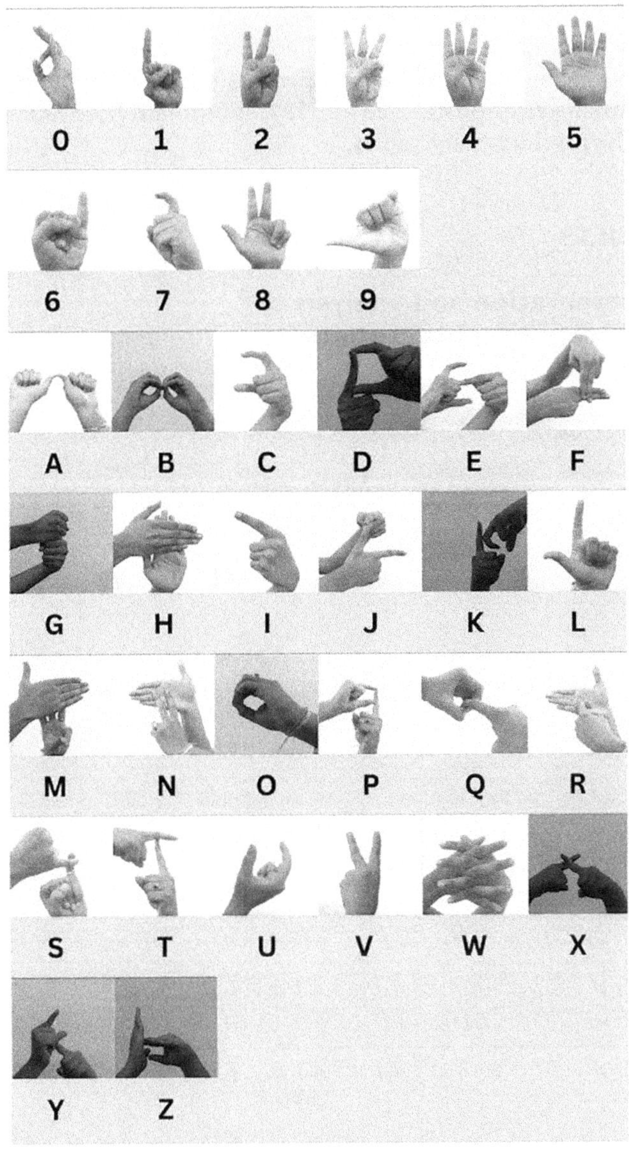

Figure 6.2 Images from the dataset used.

6.4.4 Data preprocessing

During the data preprocessing phase, noise was removed from the images by applying various configurational operations. The images were resized to form square matrices of size 225×225, and normalization was performed to adjust the pixel intensity values.

6.5 RESULTS

6.5.1 Presentation and analysis of the experimental results

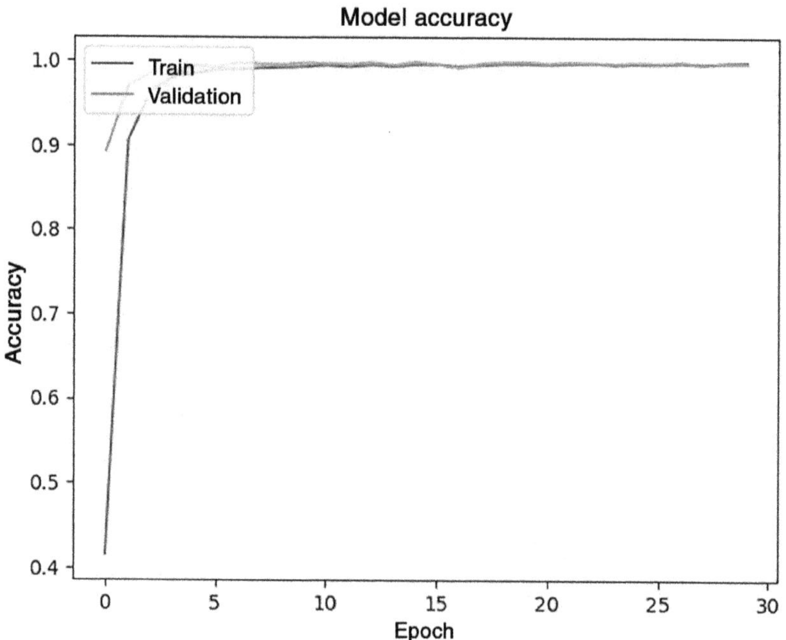

Figure 6.3 Plot of model accuracy (epochs vs. accuracy).

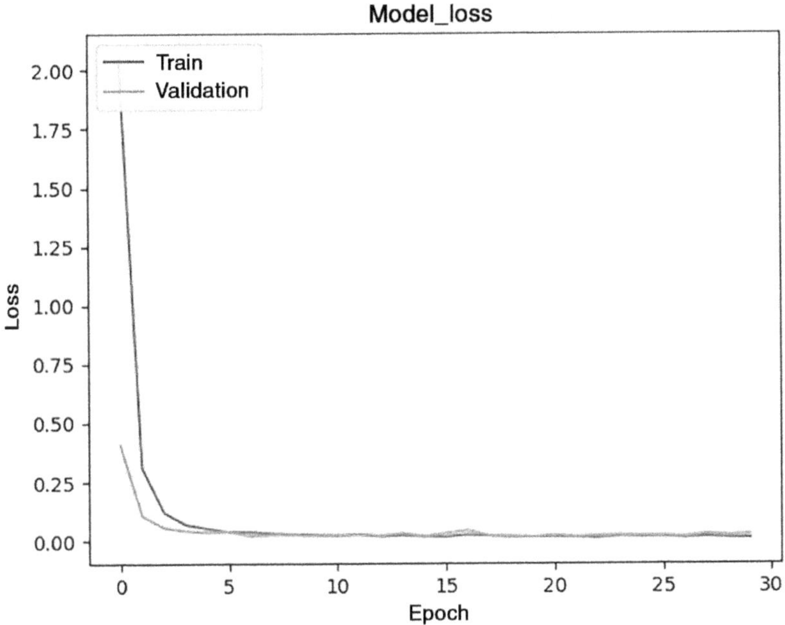

Figure 6.4 Plot of model loss (epochs vs. loss).

Table 6.1 Table representing loss and accuracy values during training and validation phases

Epoch	Train loss	Train accuracy	Validation loss	Validation accuracy
1	2.0514	0.4158	0.4065	0.8925
2	0.3104	0.9057	0.104	0.9692
3	0.1177	0.9634	0.0533	0.9854
4	0.0658	0.9794	0.0392	0.9904
5	0.0492	0.9843	0.0332	0.9935
6	0.0351	0.989	0.0356	0.9921
7	0.0341	0.9894	0.0179	0.9972
8	0.0282	0.9911	0.0215	0.9957
9	0.0246	0.9921	0.022	0.9954
10	0.0224	0.993	0.0187	0.9975
11	0.0185	0.9943	0.0199	0.9962
12	0.0239	0.9949	0.0215	0.996
13	0.0146	0.9928	0.0184	0.9978

(Continued)

Table 6.1 (Continued) Table representing loss and accuracy values during training and validation phases

Epoch	Train loss	Train accuracy	Validation loss	Validation accuracy
14	0.0199	0.9934	0.0282	0.9947
15	0.0145	0.9954	0.0149	0.9987
16	0.0131	0.9958	0.0285	0.9957
17	0.0214	0.9936	0.0395	0.9917
18	0.0164	0.9949	0.0171	0.9971
19	0.0114	0.9961	0.0161	0.9981
20	0.0114	0.9964	0.0121	0.9987
21	0.0123	0.9958	0.0187	0.9969
22	0.0102	0.9969	0.0135	0.9987
23	0.075	0.9975	0.0172	0.9972
24	0.0153	0.9956	0.0199	0.9968
25	0.0132	0.9967	0.0162	0.9979
26	0.0116	0.9962	0.0159	0.9969
27	0.0077	0.9976	0.0135	0.9985
28	0.0122	0.9961	0.024	0.9953
29	0.0082	0.9977	0.0185	0.9978
30	0.0061	0.9986	0.0219	0.9974

6.6 CONCLUSION

6.6.1 Summary

This chapter introduced an effective approach for recognizing alphabets, digits, and commonly used words in Indian Sign Language. Leveraging transfer learning, the study utilizes the VGG16 model to extract valuable features from the images. Subsequently, the model is fine-tuned with a custom architecture tailored for the specific image classification task defined by the custom dataset. This methodology enables the model to achieve accurate recognition of Indian Sign Language gestures.

6.6.2 Implications in real-world applications

SLR technology plays a pivotal role in fostering smooth communication between communities that use sign gestures and those who do not. This capability facilitates interactions across various domains, including education, customer service, healthcare, and public spaces. In educational settings, SLR can provide real-time transcription of lectures, enabling them to follow along more effectively. It can also aid in creating interactive learning tools and resources in Sign Language, making education much more accessible and inclusive. Indeed, SLR can be seamlessly adjoined with varied assistive devices and applications, such as smartphones, tablets, and wearable technology. By doing so, deaf and hard of hearing individuals gain enhanced access to information and communication tools in their daily lives.

SLR can support the legal and civic participation of weak hearing and mute individuals by providing access to legal proceedings, government services, and public announcements in Sign Language.

6.6.3 Future directions and recommendations for further research

The coming work may be integrating the proposed model with Natural Language Processing to generate real time textual or spoken language that can facilitate communication between the verbally weak and verbally sound communities. However, challenges persist, including the need for robust datasets, addressing variations in Sign Language gestures, and ensuring real-time accuracy and reliability of the developed systems.

REFERENCES

1. Hussain, Imran, Anjan Kumar Talukdar, and Kandarpa Kumar Sarma. "Hand gesture recognition system with real-time palm tracking." *2014 Annual IEEE India Conference (INDICON)*, December 12 to December 14, 2014 in Delhi. IEEE, 2014.
2. Sharma, Kalpana, Garima Joshi, and Maitreyee Dutta. "Analysis of shape and orientation recognition capability of complex Zernike moments for signed gestures." *2015 2nd International Conference on Signal Processing and Integrated Networks (SPIN)*, Noida, India, from February 26 to February 27, 2015. IEEE, 2015.
3. Patil, Sandeep Baburao, and G. R. Sinha. "Distinctive feature extraction for Indian Sign Language (ISL) gesture using scale invariant feature Transform (SIFT)." *Journal of the Institution of Engineers (India): Series B 98.1* (2017):19–26.
4. Lokhande, Priyanka, Riya Prajapati, and Sandeep Pansare. "Data gloves for sign language recognition system." *International Journal of Computer Applications 975* (2015): 8887.
5. Han, Rui, et al. "A data glove-based KEM dynamic gesture recognition algorithm." *International Journal of Performability Engineering 14.11* (2018): 2590.
6. Manware, Ajit, et al. "Smart gloves as a communication tool for the speech impaired and hearing impaired." *International Journal of Emerging Technologies and Innovative Research 4* (2017): 78–82.
7. Sruthi, C. J., and A. Lijiya. "Signet: A deep learning based Indian sign language recognition system." In *2019 International Conference on Communication and Signal Processing (ICCSP)*, April 4 to April 6, 2019. It was held in India at Adhiparasakthi Engineering College, Melmaruvathur, Tamil Nadu. (pp. 0596–0600). IEEE, 2019.

8. Kishore, P. V. V., et al. "Motionlets matching with adaptive kernels for 3-D Indian sign language recognition." *IEEE Sensors Journal* 18.8 (2018): 3327–3337.

9. Sharma, Madhuri, Ranjna Pal, and Ashok Kumar Sahoo. "Indian sign language recognition using neural networks and KNN classifiers." *ARPN Journal of Engineering and Applied Sciences* 9.8 (2014): 1255–1259.

10. Nandy, Anup, et al. "Recognition of isolated Indian sign language gesture in real time." *Information Processing and Management: International Conference on Recent Trends in Business Administration and Information Processing, BAIP 2010,* Trivandrum, Kerala, India, March 26–27, 2010. *Proceedings.* Springer Berlin Heidelberg, 2010.

11. Shivashankara, S., and S. Srinath. "American sign language recognition system: An optimal approach." *International Journal of Image, Graphics and Signal Processing* 11.8 (2018): 18.

12. Garcia, Brandon, and Sigberto Alarcon Viesca. "Real-time American sign language recognition with convolutional neural networks." *Convolutional Neural Networks for Visual Recognition* 2.225–232 (2016): 8.

13. Krizhevsky, Alex, Ilya Sutskever, and Geoffrey E. Hinton. "Imagenet classification with deep convolutional neural networks." *Advances in Neural Information Processing Systems* 25 (2012): pp. 1–9.

14. Tang, Ao, et al. "A real-time hand posture recognition system using deep neural networks." *ACM Transactions on Intelligent Systems and Technology (TIST)* 6.2 (2015): 1–23.

15. Murakami, Kouichi, and Hitomi Taguchi. "Gesture recognition using recurrent neural networks." *Proceedings of the SIGCHI Conference on Human Factors in Computing Systems,* New Orleans, USA, 1991.

16. Bantupalli, Kshitij, and Ying Xie. "American sign language recognition using deep learning and computer vision." *2018 IEEE International Conference on Big Data (Big Data),* Seattle, Washington, USA, from December 10 to December 13, 2018. IEEE, 2018.

17. Wadhawan, Ankita, and Parteek Kumar. "Deep learning-based sign language recognition system for static signs." *Neural Computing and Applications* 32 (2020): 7957–7968.

18. Abu-Jamie, Tanseem N., and Samy S. Abu-Naser. "Classification of sign-language using vgg16." (2022).

19. **Dataset:** https://www.kaggle.com/datasets/atharvadumbre/indian-sign-language-islrtc-referred.

Chapter 7

AI engineering for the best sustainable practices towards social applications

A. Krishna Chaitanya, N. Jaya Krishna, M. Manideep, Y. Manohar Reddy, and G. Sucharitha

7.1 INTRODUCTION

Artificial intelligence (AI) [1] is an example of modern technology which takes major role in industry development, waste reduction, and environment impact. This chapter explores existing literature to find key trends and insights. Furthermore, the chapter examines the role of AI in order to promote sustainable practice within the organization for intelligent work planning and risk management to adaptive workforce optimization. AI enhances the work through operational efficiency and reduces environmental impact. AI is sustainable in many industries, including agriculture, food production, and industrial work.

It also emphasizes the scalability of product which increases the revenue of a company and also it finds for replicable solutions. The ethical consideration and challenges associated with AI in sustainable engineering and management are also being discussed in this chapter. The ethical considerations and challenges associated with AI in sustainable engineering and management [2] are also discussed.

Applications of AI in sustainable engineering and management are not an exception. AI has been a disruptive factor in many sectors. The goal of sustainable engineering is to satisfy current demands without endangering the capacity of future generations to satisfy their own. The integration of AI into sustainable engineering methods may help firms decrease their environmental footprint, increase efficiency, and make more informed decisions. The combination of AI with sustainable engineering not only solves immediate problems but also advances long-term economic and environmental sustainability.

By providing creative solutions and insights, AI technologies such as machine learning, data analytics, and optimization algorithms [3] are vital to sustainable engineering and management. These applications fall into broad categories, such as resource management, monitoring of the environment, energy efficiency [4], and the development of sustainable infrastructure [5].

Resource management: Sustainable engineering calls attention to the prudent use of the environment's resources. AI makes a contribution by using advanced analysis of information to optimize resource efficiency. AI-powered systems, for instance, may enhance accuracy in agriculture through assessing waste and increasing crop output.

Environmental monitoring: Real-time gathering and analysing information for environmental conditions is made possible by AI-driven sensors and monitoring systems. This supports monitoring the health of ecosystems, following animal patterns, and determining pollution levels. Timely action taken due to the early detection of environmental issues promotes long-term conservation efforts.

Development of sustainable infrastructure: AI helps in the design and construction of green infrastructure. Algorithms can assess a variety of data sources to improve the effectiveness of building procedures, cut down on wasted material use, and increase the longevity of constructions. AI (AI)-enabled smart cities can advance sustainable urban development through improving waste management, energy conservation, and transportation systems.

Decision support systems: By evaluating huge amounts of data, AI gives decision-makers insightful information. In engineering and management, this makes decision-making more sustainable and well-informed. AI helps to connect choices with over time ecological and economic goals, whether it is through the selection of eco-friendly products, optimizing the supply chain, or sustainable business practices.

7.2 LITERATURE STUDY

The literature on AI applications in engineering for sustainable management [2,6] and engineering shows a growing recognition of the potential benefits of AI in resolving environmental issues and advancing sustainable practices. Researchers and academics have looked into numerous aspects of this confluence, including the ways in which technology based on AI might be used to improve efficiency, lessen environmental effect, and support sustainable development. An outline of the main concepts and conclusions from the literature is given below

Scholars have looked into the utilization of machine learning algorithms to optimize energy usage across many industries. This covers research on manufacturing procedures, smart electrical networks, and energy-efficient structures [7,8]. Energy distribution networks have been optimized, renewable energy systems have been made more efficient, and patterns of energy use were foretold thanks to machine learning algorithms.

The literature emphasizes how AI can reduce waste and maximize resource use to support the ideas of the circular economy, including applications in manufacturing processes, waste reduction strategies, and supply chain management. Research looks at how AI may help in material monitoring, recycling, and repurposing to make goods that are less harmful to the environment and sustainable [9].

Environmental surveillance and preservation:

Researchers have looked at the use of AI in environmental monitoring, including sensor networks, satellite image processing, and remote sensing. The creation of AI models for tracking biodiversity, forecasting environmental changes, and assisting with conservation efforts is the main emphasis of this field of study.

Urban planning and sustainable infrastructure:

The literature looks at how AI may be included into the planning and building of environment-friendly infrastructure. Studies on streamlining the building process and making structures using green building technique are included in this.

The use of AI in making smart city is being investigated as a way to enhance urban sustainability by optimizing resource allocation, waste management, and mobility. Academics underscore the significance of AI-powered decision support systems in steering sustainable decision-making in diverse industries. This entails incorporating AI into supply chain management, policy creation, and business procedures. Studies examine how AI tools help businesses run sustainably and make well-informed decisions that take the economy, society, and environment into account.

Difficulties and ethical issues:

The literature recognizes the difficulties and moral dilemmas that come with using AI in sustainable engineering. This includes worries about algorithmic unfairness, data privacy, and the environmental effects of training massive AI models.

Case studies and real-world application:

Numerous studies include case studies from real-world situations as well as useful applications of AI in sustainable engineering and management. These examples highlight effective uses, things that may be learnt, and possible directions for further study.

7.3 METHODOLOGY

Applications of AI engineering for sustainable engineering and management must be developed and implemented using a methodical approach to guarantee efficacy, efficiency, and a favourable environmental impact. The following are proposed in the theoretical research as shown in Figure 7.1.

Describe particular sustainability objectives and problems in the fields of engineering and management.

Determine the domains, such resource management, environmental monitoring, or energy efficiency, where AI can support sustainability and

Figure 7.1 Green sustainability with AI-based water maintenance system.

assemble pertinent data sources, such as engineering characteristics, environmental statistics, and historical data.

To guarantee precision and dependability, clean up and preprocess the data. Examine the possibility of using Internet of Things (IoT) devices to get data in real time.

Select the right AI algorithms (e.g., optimization algorithms for resource management and machine learning for predictive analysis) based on the particular sustainability objectives. Create models with data analysis capabilities that can offer insights into sustainable engineering methods.

Make sure that existing engineering and management systems and AI apps integrate seamlessly. Create Application Programming Interfaces (APIs) to facilitate communication between hardware and software components.

Create key performance indicators (KPIs) to gauge how AI applications affect environmental objectives. To guarantee the precision and dependability of the AI models, carry out extensive testing and validation.

Incorporate a feedback loop to iteratively update and enhance AI models according on their performance in the real world and keep an eye on how well AI applications are working and tweak as necessary to improve sustainability results.

Encourage cooperation between engineers, environmental scientists, management specialists, and AI specialists. Make sure the development and implementation of AI application follow sustainable development guidelines and address moral issues with AI applications, such as algorithmic prejudice or unforeseen environmental effects. Establish accountability and transparency policies to foster stakeholder confidence. Consider scalability while designing AI systems to handle expansion and shifting needs.

To enable replication in other settings or industries, document the approach and execution methods. The creation of environment-friendly infrastructure is another area where AI and sustainable engineering work well together. Here, algorithms help to optimize building procedures, reduce waste, and encourage the use of sustainable materials. AI-driven smart cities are prime examples of how technology can be used to develop urban settings that put an emphasis on waste minimization, energy efficiency, and sustainability all around.

AI is a useful tool for firms pursuing sustainable practices. It gives decision-makers the ability to negotiate intricate situations and match plans with both economic and environmental objectives. Applications of AI engineering for sustainable practices have a good influence on ecological responsibility and human growth, which is a step in the right direction. Incorporating AI into sustainable engineering and management not only solves the environmental issues of today but also paves the way for a time in the future when technology is used to balance human advancement with environmental preservation. Sustainability is a dynamic journey, and AI is a major driver in creating a resilient and eco-friendly global ecosystem.

Developing AI solutions for sustainable engineering and management necessitates a methodical approach to guarantee successful execution. A thorough process for creating AI engineering applications for sustainable engineering and management is provided in Figure 7.2.

Stakeholder evaluation plays the key role in sustainable development.

Figure 7.2 shows the major stakeholders in that process, including, managers, engineers, end users, and environmental specialists, and ensures that the AI application meets their demands and is aware of their expectations, worries, and requirements. Appropriate gathering of required data is

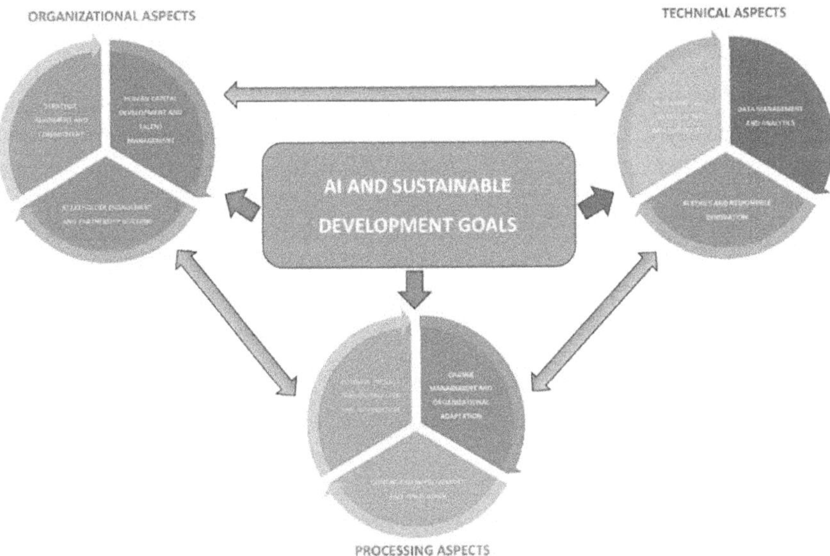

Figure 7.2 Process for the sustainability development.

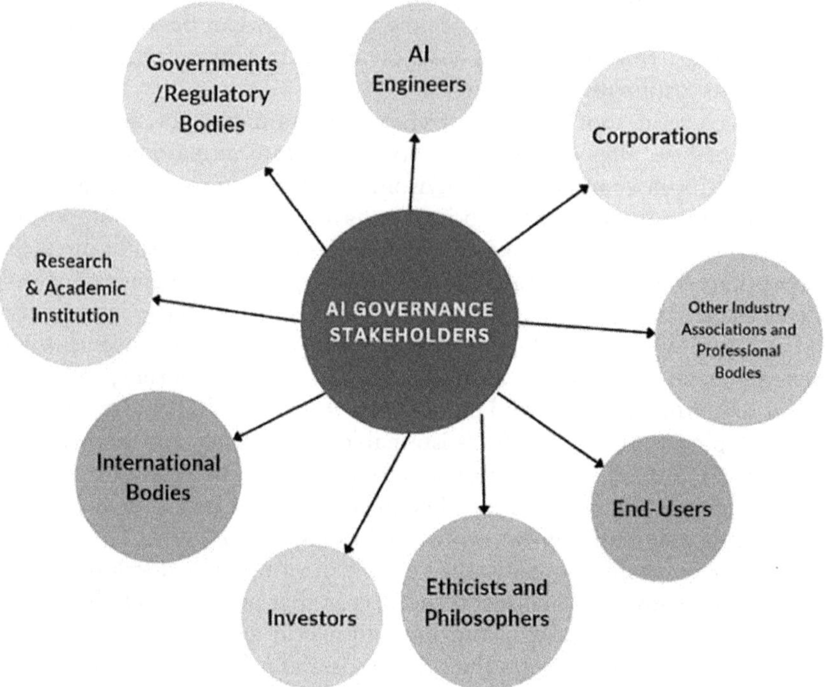

Figure 7.3 Stakeholders recognition in the process of sustainability development.

processed in further development module to perform the analysis task. This process is shown in Figure 7.3.

Collect pertinent information on the sustainable engineering and management issues and examining the past data to find trends, patterns are majorly focused on important factors in the sustainability (Figure 7.4).

Engineering features gives refinements in addition of the gathered data. It is useful for the efficient extraction and providing of pertinent characteristics that may be utilized to train AI models for the target users. It encompasses the AI-designed components of sustainable engineering and management methodologies. Figures 7.5 and 7.6 show this emerged architecture and its implementations.

Implementation ideas

Install the AI program in a live setting, making sure it integrates with the current systems.

Create a maintenance schedule that takes updates, security flaws, and changing sustainability standards into account.

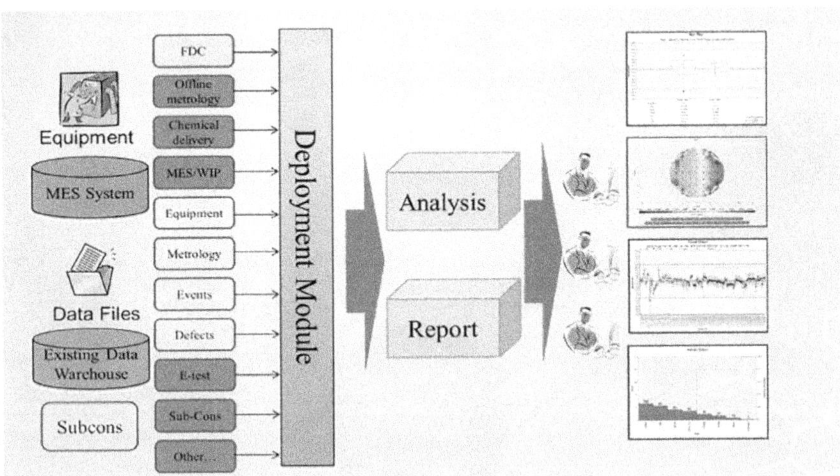

Figure 7.4 Analysis of results of sustainability.

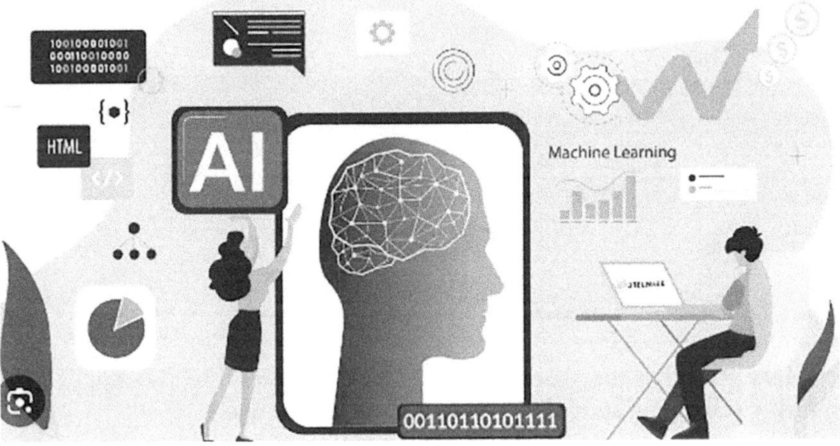

Figure 7.5 Component of engineering features for AI-based results.

7.3.1 Discussion of implemented models and benefits

7.3.1.1 Flexibility and scalability

Feature of scalability is increased while developing the AI application so that it can handle an increase in data and user numbers for the social sustainability applications

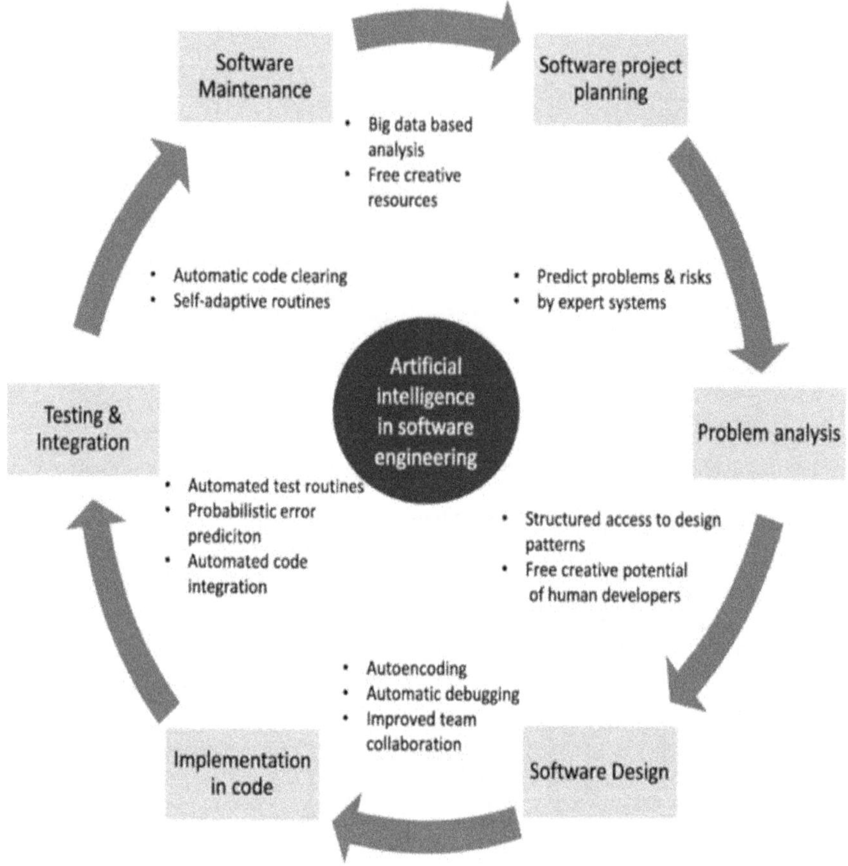

Figure 7.6 Implementation ideas.

Make sure that the adaptability is smoothly done wherever the cases to adjust to changing stainability laws and requirements.

7.3.1.2 *Integration of quantitative and qualitative data*

It provides the AI models with data that is both quantitative and qualitative, so that useful for giving more outcomes that take into account subjective elements, expert opinions, and qualitative evaluations.

7.3.1.3 *Effective evaluation of risk and resilience*

Incorporate risk assessment techniques to pinpoint possible hazards linked to sustainable decision-making with the created robust AI models that are able to adjust to unanticipated difficulties and uncertainties.

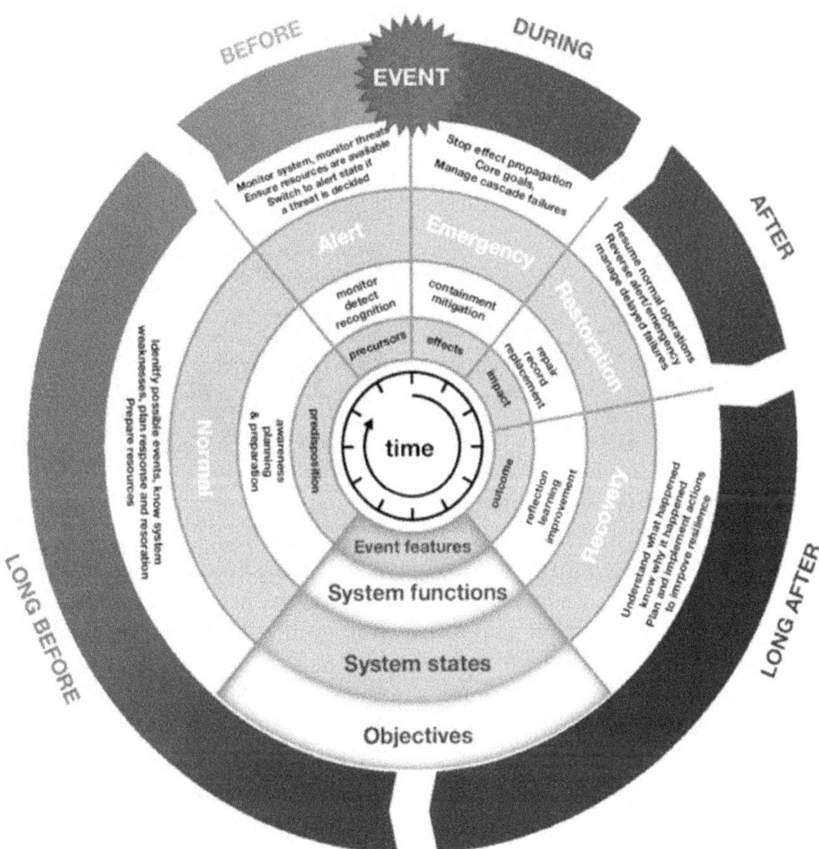

Figure 7.7 Evaluation of risk and resilience.

7.3.1.4 Explainability and interpretability

Make sure the AI models can be understood and that they offer justifications for the forecasts or suggestions they make.

Increase transparency to foster greater decision acceptability and confidence among stakeholders (Figure 7.8).

7.4 CONCLUSION AND FUTURE SCOPE

In summary, AI in sustainable engineering and management is a critical turning point toward a future that is resource- and environmentally-aware. Combining AI with sustainable practices provides a multimodal way to tackle today's issues with energy use, resource depletion, environmental

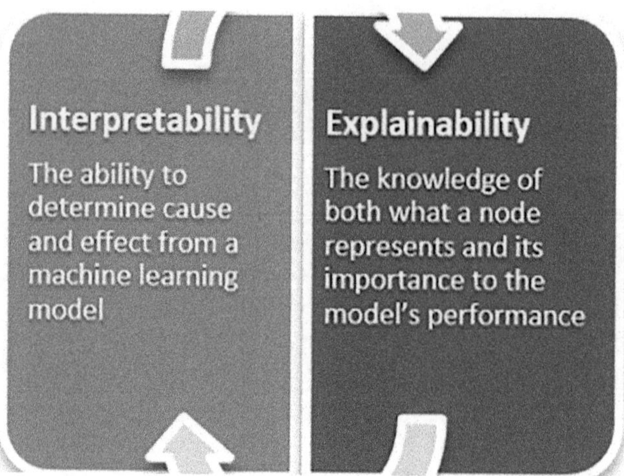

Figure 7.8 Explainability and interpretability.

monitoring, infrastructure development, and decision-making. As AI can evaluate large and complicated information, it helps companies make data-driven decisions that support sustainability over the long run. AI applications are essential for reducing the negative environmental effects of human activity, from improving resource management in agriculture to optimizing energy use in a variety of industries.

Furthermore, AI-driven systems' real-time monitoring features offer a pro-active strategy for environmental preservation. Resilience in the face of ecological difficulties and climate change is fostered by early issue detection and timely actions, which allow more effective responses.

REFERENCES

1. J. Liu, et al., "Artificial intelligence in the 21st century," *IEEE Access*, vol. 6, 2018, pp. 34403–34421, doi: 10.1109/ACCESS.2018.2819688.
2. K. Rajendra Prasad, K. Narasimhulu, Ch. N. Santhosh Kumar, and N. Ramanjaneya Reddy. "Sampling-based fuzzy speech clustering systems for faster communication with virtual robotics toward social applications," *Journal of Soft Computing*, 2023, pp. 1–9, doi: 10.1007/s00500-023-08273-y.
3. V. Sitharamulu, K. Rajendra Prasad, K.S. Reddy, A.V.K. Prasad, and M. V. Dass. "Hybrid classifier model for big data by leveraging map reduce framework," *International Journal of Data Mining, Modelling and Management*, vol. 16, Issue 1, 2024, pp. 23–48.
4. W. L. S. Mahmoudi, G. Pellegrino, and E. Armando, "Efficiency maps of electrical machines," *2015 IEEE Energy Conversion Congress and Exposition (ECCE)*, Montreal, QC, Canada, 2015, pp. 2791–2799, doi: 10.1109/ ECCE.2015.7310051.

5. S. M. Monjurul Hasan and A. Trianni, "Energy management: sustainable approach towards industry 4.0," *2020 IEEE International Conference on Industrial Engineering and Engineering Management (IEEM)*, Singapore, 2020, pp. 537–541, doi: 10.1109/IEEM45057.2020.9309939.

6. G. Sucharitha, M. Sirisha, K. Pravalika, and K. N. Gayathri, "A study on the performance of deep learning models for leaf disease detection," *EAI Endorsed Transactions on IoT*, vol. 10, 2023, pp. 1–9.

7. G. Sucharitha, V. Sitharamulu, S. N. Mohanty, A. Matta, and D. Jose, "Enhancing secure communication in the cloud through blockchain assisted-CP-DABE," *IEEE Access*, vol. 11, 2023, pp. 99005–99015, doi: 10.1109/ACCESS.2023.3312609.

8. S. Rahul and V. Bhardwaj, "Optimization of resource scheduling and allocation algorithms," *2022 Second International Conference on Interdisciplinary Cyber Physical Systems (ICPS)*, Chennai, India, 2022, pp. 141–145, doi: 10.1109/ICPS55917.2022.00034.

9. Q. R. Xu, Z. Y. Wu, S. P. Zhang, and S. Y. Liu, "Total innovation management paradigm for firm innovation system," *2014 IEEE International Conference on Management of Innovation and Technology*, Singapore, 2014, pp. 359–364, doi: 10.1109/ICMIT.2014.6942453.

Chapter 8

Healthcare robots enabled with IOT and artificial intelligence in healthcare applications

Rahul Joshi, Krishna Pandey, and Suman Kumari

8.1 INTRODUCTION

The need for automated assistance has emerged due to the need to address complex challenges using solutions surpassing human capabilities. Incorporating expertly encoded practical expertise and concepts into a computer system was essential, streamlining the problem-solving process. Expert systems and knowledge-based systems are endeavours aimed at achieving this goal. The advancement of methods such as artificial neural networks, fuzzy logic, and evolutionary algorithms has significantly improved the problem-solving capabilities of these machines. Artificial intelligence (AI) emerged with the introduction of the first modern computer in the mid-20th century. In essence, a computer emulates the collective cognition of several individuals to handle large amounts of data efficiently and with rapidity and precision. Currently, we are seeing a reduction in both the physical dimensions and financial expenses associated with computers while simultaneously experiencing an increase in their overall functionality. The increasing capabilities of computers indicate that they are getting more proficient at surpassing human talents. These improved skills are then used to boost the effectiveness of corporate operations. The objective of AI scientists has always been to create a cognitive machine capable of problem-solving in a manner that would be deemed "intelligent" if performed by a person. This serves as the foundation for our notion of AI.

According to the Oxford definition, "artificial" refers to anything that is unnatural or does not correspond to nature, while "intelligence" is the aptitude to comprehend information. The term AI was first used by McCarthy, who defined it as "the computational aspect of the capacity to accomplish objectives in the physical realm" [1]. AI is a discipline that develops methods to empower computer systems to execute tasks quickly and surpass human intellect. According to Patterson, AI is a field within computer science that focuses on studying and developing computer systems capable of demonstrating Intelligence. These systems can learn new concepts and tasks, reason and make meaningful conclusions about the

DOI: 10.1201/9781003565529-8

world, understand natural language, or interpret visual scenes, and perform other complex tasks that typically require human-like intelligence [2]. Konar defines AI as replicating human intellect on a computer, enabling the machine to find and use appropriate information during problem-solving effectively [3]. The emergence of AI was driven by the need to address little and inconsequential issues. It was deemed essential to incorporate meticulously encoded practical knowledge and principles into a computer system to address this issue, streamlining the problem-solving process. AI is not intended to replace humans. It enhances our capabilities and improves our performance. AI amplifies the velocity, accuracy, and efficacy of human endeavours. In financial institutions, AI approaches may detect potentially fraudulent transactions, implement quick and precise credit scoring, and automate labour-intensive data administration operations. AI is used in many sectors, including industries, hospitals and healthcare, education, agriculture, legal services, research, and development (R&D), retail, banking, households, etc.

AI is already making its mark in several areas of healthcare, and it is only now starting to have a profound impact, from the development of treatment methods to the automation of repetitive chores in medicine delivery and drug research, among other areas. Now that everything is digitised, a deluge of data is ready to be analysed. Additionally, healthcare generates massive amounts of data daily. Current algorithms can analyse patient data, including their dietary habits, geographic location, medical history, etc. Many authorities and institutions are mining massive databases, including medical information, to predict potential illnesses that could increase in a particular region. Studying animal populations and their environments may help predict when epidemics will strike, and studying outbreaks can reveal where they started, where the clusters formed, and how the mutations occurred. People may be promptly alerted of their forthcoming medical appointments and checkups. Vaccination courses might be recommended to individuals depending on their location and other variables. An individual's genetic makeup may shed light on their vulnerability to certain diseases and illnesses and any faulty genes present via genetic testing. Another way to discover whether problems cause other diseases to emerge is to look at people's lifestyles. Unlike 70 years ago, doctors nowadays do not need to commit nearly as much material to memory. AI is ready to level up this world. Getting into a position to ponder, choose, or investigate anything is what "thinking" time is all about. Finding or acquiring information took a lot more time than analysing it. After several hours of computation, the data was finally transformed into a similar format. A few seconds were all it needed to decide when they were in similar shape. AI is poised to revolutionise three critical areas of human thought: sophisticated computing, statistical analysis, and hypothesis creation [4]. These three domains represent three separate waves of AI.

8.2 HISTORICAL BACKGROUND OF ROBOTS IN MEDICINE

In the early 1900s, robots had yet to become a prominent element in popular science fiction. The notion of robotics entered everyday awareness with the publication of Joseph Capek's short novel "Opilec" in 1917, which described automats, and Karel Capek's play Rossum's Universal Robots in 1921 [5,6]. Attributing the word "robot" to its original brother is a contention within the Czech literary sphere. The name "robot" originates from the Czech word "robot," which translates to "serf" or "labourer." Karl Capek's play, RUR, was designed as a protest against the fast advancement of contemporary technology. In the play, Capek portrays a progression of robots with escalating skills, leading to a rebellion by these robots against their human creators [7]. Unintentionally, the Capek brothers popularised the word "robot" in contemporary parlance and ignited public intrigue with their inventions.

Robots in medicine are increasing due to advancements in AI, miniaturisation, and computer processing capabilities. Medical robots began around 40 years ago when an industrial robot and computed tomography (CT) navigation were used to surgically implant a probe into the brain to extract a biopsy specimen [8]. Subsequently, a series of robots were developed with the ability to do specific urological treatments, including complete hip arthroplasty. However, these autonomous robots were not well-received by surgeons, and as a result, the following robots were built to be subservient to the control of surgeons. Figure 8.1 depicts the chronological progression of Robotics in the healthcare sector.

Medical robots are recognised for their involvement in surgical operations, particularly in precisely manipulating surgical instruments using robots, computers, and software via one or more tiny incisions [9]. A three-dimensional (3D) high-definition magnified picture of the surgical area allows the surgeon to perform operations with exceptional accuracy and mastery. The Food & Drug Administration approved system, da Vinci, has been used globally for over 6 million operations since its introduction in 2000. The advantages for patients undergoing robot-assisted surgery mainly stem from the laparoscopic technique, which includes smaller incisions, less blood loss, and quicker recovery. There is no significant difference in the long-term surgical results compared to conventional surgery, and the system sometimes malfunctions. Unlike older methods, surgeons experience enhanced ergonomics and dexterity using advanced laparoscopic techniques. Significant disadvantages include the substantial expense and the need to train doctors and the surgical crew. The minimum cost of a da Vinci system is at least $1 million. Several businesses are now in the process of creating surgical robots that are specifically tailored for performing specific procedures, such as knee or hip replacement surgeries. Several

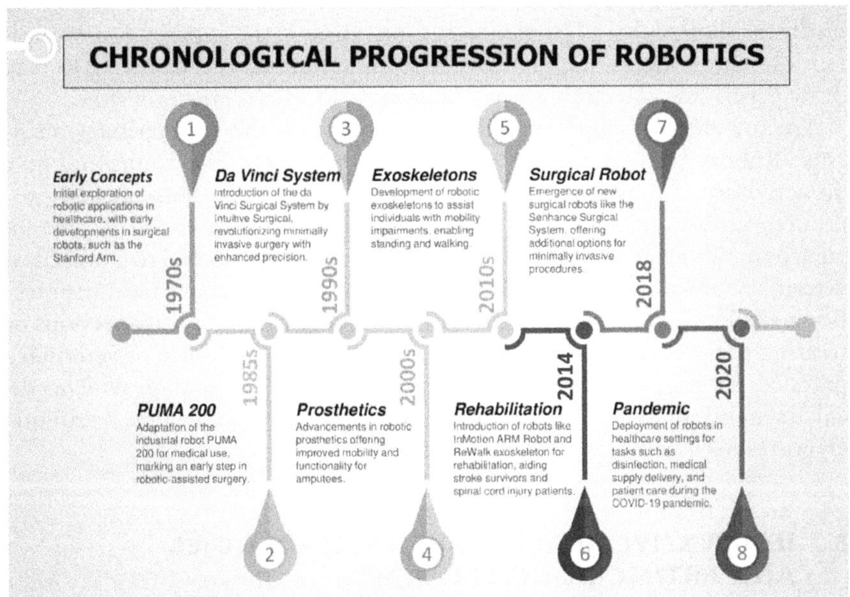

Figure 8.1 Timeline of Robotics in Healthcare.

organisations are endeavouring to develop systems that integrate AI to aid surgical decision-making [10].

The Modus V is an automated robotic arm and digital microscope used in neurosurgery. A Toronto-based business develops it and utilises the Canadarm technology from the space shuttle. The robotic arm precisely follows the movement of surgical tools, autonomously relocates to the specific region where the surgeon is operating, and displays an enlarged, high-definition picture on a screen [11]. New structures and control mechanisms significantly enhance the functionality of prostheses. Integrating bionic skin and brain systems with robotic limbs enables users to exert surprising control. Robotic exoskeletons, or orthoses, are used in rehabilitation to aid individuals with paralysis in walking and addressing anatomical abnormalities. Hospitals use robots to maintain cleanliness by employing high-intensity UV light to disinfect hospital rooms. Conventional endoscopy will likely be replaced by miniature robots that may be directed to sites to perform diverse functions, such as extracting a tissue sample or cauterising a leaking blood artery. Microrobots may be used to navigate through blood arteries and provide targeted treatment, such as radiation or medicine, to a precise location. Robotic endoscopic capsules are ingested to monitor the digestive system, collect data, and transmit diagnostic information to the operator. Additionally, robotic nurses are specifically engineered to aid or replace overburdened nurses in performing duties such as digital data input, tracking patients, plasma extraction, and cart transportation. An area of medical robotics that is particularly captivating is the substitution of antibiotics. Nanorobots equipped with receptors that bacteria can cling to may draw germs into the circulatory system or at localised infection sites.

Are any of these significant advancements applicable to veterinary medicine? Robots are now used in simulations to teach veterinarians and may do activities such as animal lifting. Unless the cost of robot-assisted procedures significantly decreases and its benefits in comparison to present laparoscopic techniques are clearly shown, it is improbable to be used in veterinary practice. Robot assistants, robotic prostheses, hospital disinfection machines, and microrobots capable of doing endoscopic inspections or treating patients are all potential advancements in the future of veterinary practice. Shortly, robotics veterinarians will likely emerge to provide medical assistance to animals with prosthetic limbs, implanted chips, or robotic creatures used in diverse environments.

8.3 INNOVATIVE HEALTHCARE TECHNOLOGIES AND ARTIFICIAL INTELLIGENCE

Health systems worldwide are now facing a critical situation as they struggle with escalating healthcare expenses far exceeding GDP growth rates. This presents problems to the long-term sustainability of these systems. The problem was simplified and unambiguous due to the emergence of the 2019 coronavirus (COVID-19) outbreak and the crisis in Ukraine. The confluence of constrained financial resources, an ageing population, the increasing frequency of chronic diseases, and the strain on healthcare systems have traditionally posed difficulties in providing the escalating demand for affordable and readily accessible treatments. Moreover, the COVID-19 pandemic is causing the rapid deterioration of healthcare systems in other countries, including India, Brazil, and Indonesia. In healthcare, AI uses mathematical algorithms and data science derived from the human body to provide diagnoses that surpass clinicians. This enables specialists to intervene promptly in illnesses that may otherwise progress significantly. Healthcare systems are a diverse and widespread network that varies in kind, location, and prevalence. These systems operate using different protocols, include various medical equipment, and are tailored to the needs of different entities. These entities are established by individuals living in diverse circumstances and pursuing distinct objectives.

Consequently, architecture has been developed to facilitate medical applications through an organised system for integrating, distributing, and storing medical data and electronic medical records. This system is a web spider within an intelligent information-processing system consisting of various subsystems with specific functions. It also governs the flow of information and control among these subsystems, allowing for adaptable autonomy. The implementation of web-based simulation systems will enhance the quality of service. AI is being used in several capacities within the healthcare industry. The most prominent use of AI in healthcare is the monitoring and organisation of medical records and data management. This process includes consolidating, storing, standardising, and tracing the origins of the data. This is the first stage in revolutionising the existing healthcare systems. Google DeepMind Health, the AI research division of Google, has just introduced the "Google DeepMind Health project." This initiative utilises medical data analysis to provide highly accurate and prompt healthcare services [12].

8.3.1 Recent advancements in AI within the healthcare sector

1. **Rising Prevalence of Telemedicine Services:** Telemedicine facilitates the delivery of medical consultation at distant locations and has several benefits, especially for those living in underserved and rural regions. Utilising the metaverse enables people to participate in telemedicine consultations with doctors and experts, eliminating the need for physical transportation and lowering healthcare expenses. The treatment offers several advantages to patients, such as convenience, effectiveness, and cost-efficiency. This alternative to conventional in-person consultations minimises appointment waiting times, simplifies the recovery process, and provides a cost-efficient option [13]. Moreover, it streamlines patient care management by providing continuous assistance and direction from healthcare professionals, improving treatment outcomes.

2. **Spread of Technologies for Remote Monitoring:** In India, remote monitoring technology is becoming more widespread as more people look for ways to improve their general health and well-being. Patients may actively participate in their care and get timely treatments, when necessary, by using these tools. Many popular remote monitoring devices, such as glucometers, fitness trackers, blood pressure monitors, and heart rate monitors, are on the market. The devices allow for regular data transfer to the healthcare provider, and the patient may wear or keep them in their home [14].

3. **Advanced Medical Evaluations:** Transformational diagnostics are vital when managing a population's health since they help find individuals prone to certain diseases and conditions. Genomic sequencing, next-generation sequencing, whole-genome sequencing, and RNA-based testing are among the revolutionary diagnostic approaches that have arisen as powerful instruments in the fight against the pandemic [15]. There is encouraging evidence that these tests may provide valuable

diagnostic data. In addition, they will provide faster and more accurate diagnoses of medical issues, potentially altering the healthcare industry completely.

4. **The Rise of Medical Technology as a Service:** Healthcare facilities increasingly use software (SaaS) to improve operational efficiency and treatment quality. Healthcare providers want pragmatic solutions for storing electronic health records and data management for population health [16]. SaaS solutions are vital in assisting healthcare workers in managing medical records for patients, appointments, and billing data.

5. **Enhancing Healthcare with the Use of a Digital Twin:** Research on digital twin (DT) has significantly grown due to rapid advancements in communication technology, sensor technology, vast data analysis, the Internet of Things (IoT), and simulation technology. The concept of DT was expanded to include digital representations of both animate and inanimate physical entities, enabling the use of DT in domains such as healthcare and social well-being. DT is a dynamic notion that closely imitates human organs, tissues, cells, or microenvironments. It constantly adjusts to fluctuations in online data and has the potential to predict the future of its equivalent counterpart. The DT is a virtual duplicate and a sophisticated entity that may be connected to a natural counterpart using advanced technology. The closed-loop optimisation between DT and its environment enables it to function as a living, intelligent, and evolving model [17]. It can optimise processes and consistently predict future conditions, including defects, damages, and failures. The technologies necessary for digital transformation may be categorised into two broad groups: mechanical models, including multi-scale knowledge and data, and data-driven statistical models. After training using samples and numerical data, the AI model obtains real-time structural performance information by analysing real-time sensor data. DT technology significantly transforms the healthcare sector, enabling several prominent enterprises to enhance efficiency and detect problems.

6. **Robotic Surgery Gain Acceptance:** Like other new technology, robotic surgery was first met with resistance. Moreover, given the inherent difficulties in adjusting to change, several surgeons wanted additional autonomy in the operating room (OR) beyond what robotic surgery treatments might provide. Initially, using automated technology in complex surgical operations was more of a curiosity than a radical change in healthcare practices. Robotic surgery has become a significant hit in medical procedures due to its restricted geographical limits, complex settings, and sometimes inaccessible locations [18]. This surge in popularity is likely explained by the robotic surgical system's integration of cutting-edge technology elements, including a 3D video

system and wrist equipment. Robotic surgery has grown in popularity in several medical specialities, including general surgery, especially for treating abdominal wall hernias. With the aid of this cutting-edge technology, surgeons may now perform minimally invasive treatments with greater accuracy and speed.

7. **Emergence of Rehabilitation:** AI has pioneering uses in the realm of rehabilitation. The concept encompasses both the physical aspect, which involves robotics, and the virtual aspect, which requires informatics. Furthermore, within AI, a specific branch, machine learning (ML), focuses on developing algorithms that enhance their performance via iterative practice. ML is used in rehabilitation for several applications, such as perioperative medicine, brain-computer interface technologies, myoelectric control, and symbiotic neuroprosthetics. ML techniques have been used in the musculoskeletal domain, such as assessing patient data, aiding clinical decision-making, and analysing diagnostic imaging. A treatment session used an artificial cognition application to assess rehabilitation activities by analysing the signals obtained from the machine [19]. Technological advancements have led to the transformation of rehabilitation research and practice by integrating AI and robots. Smart homes can aid inhabitants in everyday tasks and promptly notify carers when necessary. Furthermore, intelligent mobile and wearable gadgets may gather data and provide consumers with information to evaluate health enhancement and track advancements towards individualised rehabilitation objectives [20]. Furthermore, wearable technology utilises inertial sensors to verify whether folks are correctly executing and sticking to training routines [21]. The study evaluated the level of compliance with a rotator-cuff workout programme among healthy persons who wore an Apple iWatch. Various supervised learning techniques were used to effectively classify workout accuracy across all algorithms [22].

8.4 IOT & HEALTHCARE SCENARIO

The healthcare sector is being transformed by the Internet of Medical Things (IoMT) as the IoT technology advances. This transformation is achieved by enhancing patient health and facilitating virtual connections between physicians and patients. The IoMT refers to a network of interconnected devices that gather and transmit data in real time. IoT has emerged as a crucial tool in healthcare, significantly benefiting the sector and offering possible solutions to medical problems [23]. The IoMT can proactively avoid future ailments and effectively treat existing ones. Wearables have used technology to provide convenient monitoring of patient health.

The IoMT is categorised into several divisions that address the healthcare business's specific medical challenges.

1. **On Body Segment:** This pertains to wearables designed for monitoring consumer health and other wearables that meet medical-grade standards. Commonly used items in this sector include Fitbit, Apple Watch, Amazfit Smartwatches, and other similar options.
2. **In-Home Segment:** This pertains to customised health monitoring systems for people. This area includes customised response systems for emergencies (PERS), telehealth virtual visits, and remote patient monitoring (RPM) [24].
3. **Community Segment:** The system has five elements: Kiosks that distribute items and provide services, mobility services for patients' automobiles, emergency response intelligence, point-of-care devices resembling medical camps, and logistics for tracking medical equipment and supplies.
4. **In-Clinic Segment:** This sector comprises IoMT devices specifically designed to provide healthcare services within a clinical setting. In addition to point-of-care devices, the medical IoT solutions in the in-clinic market also provide other healthcare solutions.
5. **In-Hospital Segment:** This encompasses several IoT healthcare solutions designed for many administration areas. Asset management, staff management, patient flow management, inventory management, and power and environmental monitoring are some of the devices included in this section.
6. **For Patients:** Wearable intelligent gadgets in healthcare, such as glucometers, heart rate cuffs, and fitness bands, allow patients to receive individualised treatment. The continuous monitoring capability of these gadgets might significantly affect those who live alone since they can transmit notifications to their relatives.
7. **For Physicians:** Physicians may enhance their ability to monitor patients' health using medical IoT wearables and home surveillance devices equipped with IoT technology. The data collected by IoT healthcare devices may assist clinicians in determining the optimal patient treatment and monitoring protocol. This convergence combines the IoT with healthcare to see improved outcomes.
8. **For Hospitals:** Hospitals use IoT technology by integrating sensors to monitor the live whereabouts of medical devices and equipment, including defibrillators, wheelchairs, oxygen pumps, nebulisers, and other similar monitoring devices. Furthermore, this technology allows for the examination of the distribution of medical personnel across several sites. Figure 8.2 shows that the Medical IoT addresses many challenges in the healthcare ecosystem by actively engaging multiple stakeholders.

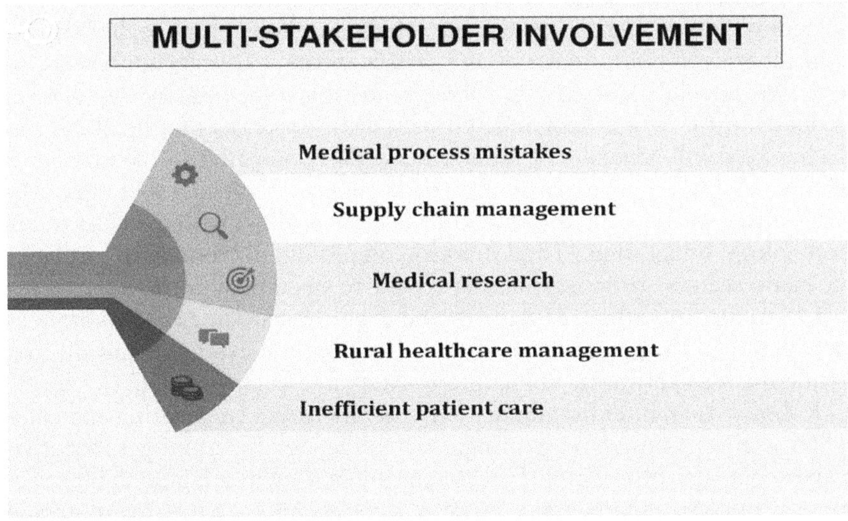

Figure 8.2 IoMT involving different stakeholders.

8.4.1 Functioning mechanism of the internet of healthcare things

Medical IoT firms pragmatically use a four-stage method for functioning IoT solutions [25]. Every step is intricately interconnected in terms of data acquisition and processing.

1. **Deployment of Interconnected Devices:** This encompasses a range of devices, such as sensors, actuators, monitors, detectors, and video systems, which are responsible for gathering data.
2. **Data Digitisation:** The sensory data obtained from various devices is then consolidated and transformed into a digital format.
3. **Cloud Migration:** The data is moved to a cloud data centre following the pre-processing and standardisation procedure.
4. **Insight Extraction:** After the data is managed and analysed, advanced analytics provide actionable insights.

8.5 INTRODUCTION TO AI APPLICATIONS IN ROBOTICS

The convergence of robots and AI is rapidly emerging as a catalyst for developing new industries, state-of-the-art technology, and enhanced productivity and efficiency in established sectors. The uses of AI in robotics are becoming more evident as the technology continues to advance.

AI significantly impacts several industries and enhances everyday life, including sectors such as self-driving vehicles, customer service, healthcare, and robotics in both industrial and service settings. Despite apprehensions over the possibility of AI and robots rendering some human employment redundant, the World Economic Forum (WEF) forecasts that by 2025, this technology will generate 12 million more jobs compared to the number it eliminates [26]. This expansion offers a chance to educate and equip the staff with new skills and invest in knowledge that aligns with the most recent technology. Integrating AI and robotics can profoundly transform job duties in many sectors, from automating repetitive operations in manufacturing plants to incorporating adaptability and cognitive capacities in monotonous applications. The applications of AI in robotics are extensive and diverse, rendering it a captivating domain to research and comprehend [27].

Robotics is an interdisciplinary field encompassing engineering and computer science, focusing on creating, assembling, and functioning robots that can execute pre-programmed tasks autonomously without human intervention. Robotics fundamentally involves using technology to streamline and enhance the efficiency and safety of jobs via automation. Traditionally, robots have been used for jobs that are too challenging or hazardous for people to execute, such as the lifting of bulky machinery, or for activities that involve monotonous repetition, such as the assembly of vehicles. Robotics solutions may increase productivity and boost safety by automating these operations, allowing human workers to allocate their time and effort towards more intricate and innovative pursuits. It is essential to mention that the same constraints as humans do not bind robots. For instance, although a person may experience fatigue, boredom, or disinterest from repeatedly doing a job, a robot will persist in executing the same activity with consistent efficiency and accuracy. Robotics solutions significantly influence several sectors, from precise agriculture harvesting to transportation and automobile assembly.

8.5.1 Are AI and robotics the same thing?

Although AI and robotics are commonly used interchangeably, they are different although connected areas. Although AI and robots can profoundly influence numerous businesses and elements of life, they have distinct purposes and work in specific ways. AI neural network models resemble biological brain networks, while robots may be likened to the human body. AI encompasses creating systems capable of executing activities that traditionally need human intellect, including learning, problem-solving, and decision-making. These systems can operate autonomously, without requiring continuous instructions, since they are designed to acquire knowledge and adjust independently. Robotics, in contrast, pertains to the advancement of robots capable of executing precise physical tasks. These robots may be trained to do rudimentary, repetitive tasks, such as categorising objects or assembling tiny components. While AI may be implemented into robotics to increase the robot's capabilities and improve decision-making, it is not always essential. Specific robotics applications need robots to perform predetermined tasks without requiring extra cognitive capacities [28]. Although AI and robotics are distinct concepts, they synergistically enhance each other and collaborate to provide various advantages and progress in numerous domains.

8.5.2 How AI is used in robotics

AI has made significant advancements in recent years, and its seamless integration with robots has been a logical and expected development. Although the use of AI in robotics is not yet prevalent, it is quickly gaining traction as AI systems continue to progress. Integrating AI with robots presents significant opportunities, resulting in heightened productivity and effectiveness, enhanced safety, and strengthened adaptability for professionals across several fields. *Machine learning* is a fundamental method by which AI is used in robotics. This methodology facilitates the acquisition and execution of specific tasks by robots via monitoring and imitating human activities. AI provides robots with computer vision capabilities, allowing them to move, identify objects, and respond appropriately. This enables them to transcend the execution of monotonous duties and evolve into genuine *cognitive collaborators*. AI is also used in robotics via the utilisation of *edge computing*. AI in robotics necessitates the real-time analysis of vast data collected by sensors integrated into robots [29]. Consequently, this data is examined near the robot rather than sent to the cloud for processing. This methodology gives machines instantaneous consciousness, enabling robots to execute judgements far faster than human skills permit.

8.6 MEDICAL ROBOTS TRANSFORMING HEALTHCARE

Medical robots have lately emerged as significant tools to assist healthcare practitioners in treating medical disorders and optimising hospital procedures. Healthcare professionals started the exploration of this technology in the 1980s, and robots became a prominent fixture in the surgical domain in the 1990s. Subsequently, medical robots have assumed many functions within the healthcare sector [30]. Telepresence robots facilitate remote patient check-ins for doctors; rehabilitation robots assist patients in executing specific post-surgery movements; surgical robots aid surgeons in performing diverse surgeries and operations; and companion robots alleviate anxiety and depression in older adults. Healthcare companies use robots for tasks such as transporting hospital supplies, dispensing prescription medications, and assisting with cleanliness and clinical administration. Medical robots alleviate healthcare staff's workload by assuming additional duties, enabling them to dedicate more attention to patients. This prospect has generated enthusiasm among industry experts over the future of robots in healthcare. The following are the primary categories of medical robots that incorporate cutting-edge technology into the healthcare industry and improve the interactions between healthcare practitioners and patients.

8.6.1 Types of medical robots

Hospitals and healthcare organisations are incorporating medical robots into their operations, while technology businesses are expediting automation in the healthcare sector.

8.6.1.1 Surgical robots

Surgical robots collaborate with surgeons to provide less invasive operations. These robots include several instruments, machinery, sensors, and algorithms that interact with each other to create a linked ecosystem capable of providing valuable information and guiding a surgeon's choices. These robots are used in operations requiring exceptional precision and little invasiveness, such as chemical and radiofrequency ablation. Leading companies are increasing their focus on research and development in robotic surgical equipment. Intuitive surgical has a substantial portion of the total business; however, the market dynamics are changing fast [31]. The MedTech surgical robotics sector is strengthened by the participation of well-established manufacturers such as Johnson & Johnson and Medtronic. Each business has distinct product portfolios concentrating on various therapeutic areas for minimally invasive robotic surgery. The following are examples of surgical robots.

1. **Vicarious Surgical's Robotic System:** Vicarious Surgical's robotic system provides surgeons with a console and a robot equipped with miniature human-like arms, enabling them to do more precise procedures. The company's robot is equipped with 28 sensors on each arm, allowing it to replicate the exact motions of a surgeon. Additionally, it has a remarkable turning ability. Vicarious Surgical's robot is equipped with a camera, enabling surgeons to get a more detailed view of a patient's anatomy while performing incisions as tiny as one centimetre for minimally invasive treatments.
2. **Intuitive's Da Vinci Platform:** The Da Vinci platform from Intuitive is a prominent medical robot consisting of networked systems, software, and medical devices. The platform aggregates statistics for teams to

analyse, enabling them to optimise their processes and provide a superior patient experience. Furthermore, the platform enables teams to conduct simulations, allowing surgeons to practise with authentic situations before engaging in the OR.

3. **Asensus Surgical's Senhance Surgical System:** The Senhance Surgical System from Asensus Surgical provides surgeons with enhanced control via a comprehensive range of technologies and equipment. The device enables surgeons to navigate more accurately using an eye-controlled camera while providing alerts when excessive pressure or tension is applied during an operation. Due to the use of reusable equipment, several healthcare organisations consider the Senhance system a cost-effective choice.

4. **Energid Technologies' Actin SDK Software:** Given the valuable assistance that surgical robots provide to doctors, it is crucial to maintain their cost-effectiveness and optimal condition. Energid technology tackles this challenge by advancing and perfecting medical robot technology using its Actin SDK software. Actin SDK is specifically designed to regulate the motions of several robots, ensuring that various components of medical robots do not obstruct each other. This approach allows surgeons to make precise adjustments to their systems without requiring complete replacement.

8.6.1.2 Service robots

Service robots carry out fundamental activities that do not involve direct interaction with patients to optimise the administrative aspects of healthcare institutions. These robots use AI and sensors to acquire the ability to traverse their environment and engage with patients and healthcare staff [32]. They can execute basic tasks such as transporting goods, providing personal protective equipment (PPE) to teams, and delivering supplies and drugs to patients. The following are example of Service robots.

1. **Diligent Robotics' Moxi Robot:** It utilises AI-driven automation and social intelligence to accomplish fundamental healthcare duties. In addition to dispensing drugs and transferring medical supplies, the robot can adjust to workflows via repetition while manoeuvring around people and accommodating requests for selfies. Moxi may alleviate hospital and nursing personnel from monotonous tasks, mitigating nursing shortages and diminishing burnout among existing nurses.

8.6.1.3 Exoskeleton robots

These robots are wearable devices with sensors, levers, motors, and other components to facilitate human movements. These healthcare robots use sensors that detect electrical impulses inside a patient's body, which may initiate movement or adapt to the patient's motions. Certain robots allow individuals to create and manipulate the exoskeleton via buttons and other mechanisms [32]. The following are examples of Exoskeleton robots.

1. **ReWalk Robotics' Personal Exoskeleton:** The personal exoskeleton developed by ReWalk Robotics aids individuals in reacquiring the ability to walk after spinal cord injury. Made primarily of hip, knee and ankle levers, the exoskeleton helps users maintain a natural gait and can be adjusted to meet the walking patterns of each person. Users can modify the robot's reactions by using wrist buttons and changing their torso motions. This allows them to navigate through various surroundings, including stairs and obstacles.

2. **Harvard Biodesign Lab's Soft Exosuits:** This Lab has significantly advanced exoskeletons by developing its soft exosuits. Unlike exoskeletons, these suits snugly envelop the user's body using pliable materials, eliminating the unwieldiness of prior exoskeletons that may impede normal human movement. This most recent iteration of exoskeleton robots aims to provide users with a smoother and more effortless experience while assisting those with less muscular strength in their movements.

8.6.1.4 Rehabilitation robots

Rehabilitation robots, such as exoskeleton robots, rely on sensors to detect electrical impulses on a patient's skin. These sensors then activate motors to aid mobility or carry out the intended actions for the patient. These robots focus on assisting patients in recovering lost skills and regaining body control and autonomy [33]. Here are some instances of rehabilitation robots.

1. **Barrett Technology's Burt Robot:** The Burt robot developed by Barrett Technology may collaborate with therapists to assist patients in overcoming weakness or paralysis in their hands and arms. Burt is affixed to a patient's forearm and helps with the mobility of the upper limbs as patients participate in games. This kind of treatment, which involves the use of robots, is well-suited for patients in the process of recuperating from various ailments such as strokes, brain injuries, spinal cord injuries, Parkinson's disease, and multiple sclerosis.

2. **Myomo's MyoPro Robot:** This orthotic device securely fastens to the arm and hand and is specifically designed to rehabilitate and enhance the functionality of the upper limbs. The brace recognises the nerve impulses inside a patient while the patient strives to accomplish a desired motion. Patients retain complete control over their limbs throughout the duration since the MyoPro robot only amplifies

muscle impulses to direct movements. Consequently, patients may achieve recovery from traumas and neurological illnesses using non-invasive methods.

8.6.1.5 Social robots

Social robots are specifically engineered to engage in intricate relationships with people, often displaying human-like characteristics to transmit and elicit appropriate emotional reactions effectively. These robots are equipped with AI, ML capabilities, and sensor cameras, enabling them to analyse human speech and actions and provide context-based replies [34]. Here are some instances of social robots.

1. **Moxie Robot:** The robot developed by Embodied assists children with learning impairments in developing their social and emotional intelligence. The robot is an AI companion for children, enhancing its plush body and lifelike movements with conversational AI that enables engaging interactions. Moxie is equipped with a wide array of games and a screen that effectively screens out improper information, providing children with activities suitable for their age and promoting their socio-emotional development [35].
2. **Pepper Robot:** The Pepper robot developed by Softbank Robotics is a companion for elderly patients and anyone undergoing rehabilitation. Pepper has a user-friendly design and a humanoid form that is recognisable to patients, characterised by its curvaceous characteristics. Furthermore, Pepper can identify human faces and discern basic emotions, all while being equipped with voice recognition capabilities in 15 different languages. These elements facilitate more personal contact between Pepper and patients from diverse backgrounds [36].

8.6.2 Successful stories of robotic surgeries in India

Indian doctors and institutions have embraced robotic-assisted procedures in response to the need to maintain physical distance during and after the COVID-19 outbreak. Proficient surgeons specialising in robotic operations in India are gaining recognition internationally, as they have executed several successful treatments. Although the expenses associated with robotic-assisted operations remain relatively expensive, the prices in India are often affordable due to excellent medical facilities and a high percentage of success. Many Indian robotic surgery facilities boast exceptional clinical coordination and patient care teams. The convergence of these elements has prompted several patients from across the globe to seek robotic surgery in India. Based on the statistics provided by NCBI, the success rate of robotic surgery in India ranges from 94% to 100%. Robotic surgery has

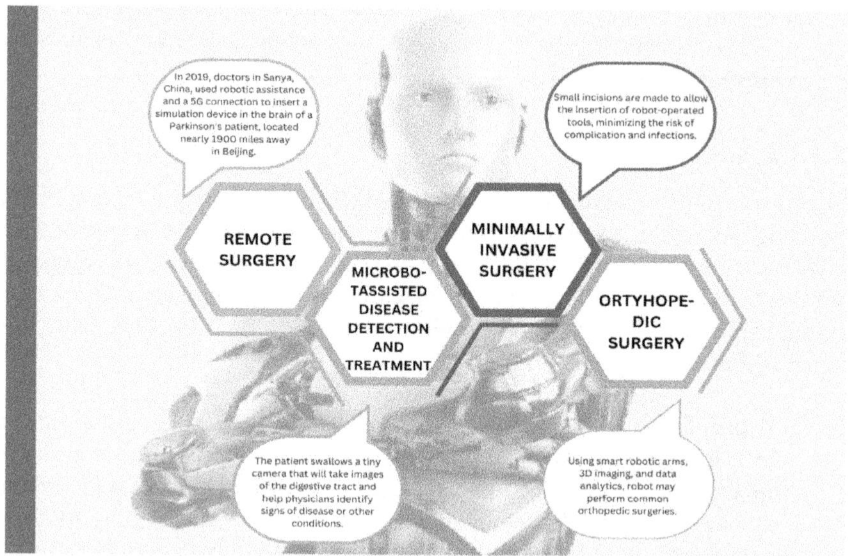

Figure 8.3 Use cases of various robotic surgeries.

fewer problems than traditional techniques. Additionally, the time of the surgery has lowered as a consequence. Robotic surgery further guarantees a substantial reduction in blood loss, increasing the success rates even more. Figure 8.3 illustrates the use cases of various robotic surgeries conducted worldwide.

Robotic surgical operations must be covered by health insurance in India as per the recommendations released by IRDAI. Health insurance companies often provide coverage for many aspects of robotic surgery, including expenses related to hospitalisation, surgical fees, physician fees, nursing facility costs, intensive care unit expenses, pre- and post-hospitalisation expenses, and sometimes even ambulance charges. Robotic surgery offers advantages in the proficient and accurate execution of intricate treatments. Robotic surgery benefits surgeons because it maintains flexibility and control, which is particularly valuable when dealing with illnesses with complex procedures. Consequently, in present times, several surgeons in India rely on robotic-assisted procedures to get optimal outcomes. The effectiveness of a surgical programme relies on the presence of a substantial number of patients, the uniformity of all surgical procedures, and repeated exposure to the same illness, which reinforces strict management guidelines and ensures the efficient and cost-efficient use of resources. The rapidly increasing population of India, although a significant societal issue, leads to a high volume of patients being attended to by a solitary surgeon. This situation is beneficial since it facilitates the learning process for most robotic surgeons and offers them many possibilities for learning. Several institutes in India provide training programmes in robotic operations under the guidance of experienced consultants from government institutions and commercial hospitals.

1. **Stryker Mako Robotic System in Kokilaben Dhirubhai Ambani Hospital:** Kokilaben Dhirubhai Ambani Hospital in Navi Mumbai has recently introduced the Stryker Mako robotic-arm-aided surgical system, which is considered one of the most modern ortho-robotic technologies worldwide for performing joint replacement surgeries. Dr Subhash Dhiware, an Orthopaedics Consultant at Kokilaben Dhirubhai Ambani Hospital in Navi Mumbai, successfully conducted the first two cases with his colleagues using the new robotic technology. The hospital's objective with this clinical launch is to use the surgical system for more than 80% of knee and joint procedures. The US FDA has authorised the Mako robotic system for complete hip, total knee, and partial knee replacements. Advancements in technology have revolutionised orthopaedic surgery, allowing for very accurate and error-free joint replacements, surpassing the capabilities of traditional manual surgery. Mako's CT-based planning generates a 3D representation of the affected joint, providing the surgeon with a comprehensive visualisation of the patient's joint and the specific abnormality. This enables the surgeon to design a customised strategy for every unique patient [37]. It is crucial to consider that each patient has a unique joint structure, and conditions such as arthritis may cause further changes to the joint.

2. **Apollo Hospitals, Kolkata Launched Eastern India's First Da Vinci Xi Robot:** Apollo Multispeciality Hospitals in Kolkata has obtained the Da Vinci Xi Robot, marking the first instance of its presence in Eastern India. The Xi Robot can aid surgeons in many medical procedures, including gynaecology, urology, oncology, cardiology, gastroenterology, etc. The Robot is very beneficial for reaching the organs located in the lower abdomen that are difficult to achieve manually. The da Vinci Xi robot enables surgeons to do minimally invasive procedures with excellent vision and skill, resulting in precise motions and enhanced patient safety. It achieves high accuracy while minimising the time the docking equipment is attached to the patient's body before the surgical operation. Patients have advantages such as less scarring, shorter hospital stays, little blood loss, and speedier recovery. Robotic surgery entails the doctor occupying the console, where a monitor presents the surgical region, while robotic arms aid the doctor in the operation. One of the arms of the surgical device has a compelling 3D camera that is introduced into the belly via a small incision. This camera offers a clear and detailed image of the operating area, with a magnification of 12 times and a high-definition 3D

display [38]. The remaining artificial arms have surgical tools replicating human hands and wrists' dexterity and motion.

3. **Manipal Hospital Performed a Record 50 Robotic Spinal Procedures in a Month:** Manipal Hospital has achieved a significant milestone by becoming a leader in spinal surgery, with a remarkable 50 successful robotic spinal procedures performed within the first month of its introduction. The hospital, known for its dedication to healthcare quality, has used precise automated techniques that have transformed spinal surgery. The introduction of robotic surgery in treating intricate spinal conditions provides a promising prospect for individuals searching for accurate and secure remedies. Manipal Spine Care Centre, a leading institution in spine treatment for over 15 years, has performed over 15,000 successful spine procedures. Each year, the centre has a success rate of 1,000 procedures out of 25,000 outpatient department. Significantly, 40% of the patients receiving advantages from this service are from West Bengal and the northern region. Manipal Hospital demonstrated its dedication to advanced medical techniques by using state-of-the-art equipment, such as a 3D imaging operating table, O-arm, and a radiolucent Purba Majorx robot, during operations conducted on patients ranging from 15 to 85 years old [39].

8.7 LIMITATIONS OF ROBOTICS IN HEALTHCARE SECTOR

Although robotic surgery is often regarded as a notable breakthrough in contemporary technology, it does include several disadvantages. Now, let us analyse the disadvantages linked to this advanced surgical technique:

1. **Abnormal Operation of Robots:** The several facets of a robotic surgical method are thoroughly examined and closely scrutinised. Thoroughly inspecting all surgical components, such as the equipment and the robotic system used in robotic surgery, is crucial. Nevertheless, it is essential to acknowledge that machines are synthetic creatures and, hence, are prone to encountering failures. Thus, this surgical approach is marred by a significant shortcoming, namely mechanical dysfunction [40]. Moreover, the existence of faulty devices has the risk of causing nerve palsies, tissue damage, and other related issues. These challenges hinder the efficient implementation of surgical techniques [41].

2. **Data Input Error:** Integrating robotic technology into surgical operations does not mean a surgeon is wholly replaced. The system serves as an auxiliary instrument for a surgeon during the implementation of a surgical operation. Medical practitioners must precisely input the

pertinent data into the robotic device to perform the surgical process. Inaccurate information might result in unanticipated complications. There is a potential for the robotic system to have challenges in reaching the desired surgical site, which could lead to unintended injury to the surrounding anatomical areas. Furthermore, it has been shown to lead to death in a limited number of cases. Hence, a physician needs to demonstrate unwavering assurance in such situations.

3. **Decreased Availability of Centres:** Although other hospitals are nearby, it is essential to acknowledge that just a select number provide complete facilities. Consequently, those who need assistance may face obstacles such as the absence of availability to robotic surgery. Hospitals may need to improve site equipment to perform technologically advanced robotic procedures proficiently.

4. **Expense:** Robotic technology in surgical operations incurs further charges and expenses due to its capacity to provide improved manoeuvrability and visualise complex anatomical areas. Hence, the economic viability of robotic surgery is limited by the financial capabilities of individual patients. Furthermore, using advanced equipment and intricate techniques, such as state-of-the-art technology, increases expenses. As a result, many people are forced to re-evaluate their behaviour because of the more significant financial consequences of the process above.

5. **Procedure Latency:** Computers need a specific duration of time to process and execute activities that have been planned. Hence, providing a specific duration for the computer to transfer the obtained data to the robotic arms engaged in the surgical operation is essential. The item or ingredient in the issue may lack significant significance or relevance in the context of routine surgical operations. Nevertheless, the need for prompt medical procedures, such as cardiac surgeries, may cause significant anxiety. Therefore, a medical professional must exercise prudence and accurately foresee the consequences of the machine's delay in such situations.

8.8 FUTURE CHALLENGES AND RECOMMENDATIONS

The future trajectory of robots in medical research is characterised by state-of-the-art developments and promising prospects. Scientists and engineers are now engaged in many critical domains to advance the limits of robots in the healthcare field:

1. **Nanorobotics:** These nanoscale robots can traverse the human body, administer precise medications, conduct minimally invasive operations, and heal cellular-level tissue damage. Nanorobots can

revolutionise medicine delivery by accurately targeting particular cells or organs, enhancing therapeutic effectiveness while minimising adverse reactions [42].

2. **Automated Medication Execution:** Advanced sensors and microprocessors may be used in robots to achieve accurate and specific medication administration. These automated devices can manoeuvre through intricate anatomical structures, guaranteeing the precise delivery of drugs to targeted regions inside the body while optimising dose and timing. This methodology can substantially enhance the effectiveness of drugs and minimise negative consequences [43,44].

3. **Automating Medicine Using Robots:** Robotics is mainly focused on automating mundane chores in healthcare settings. Healthcare providers may free up more time for direct patient care with the help of robotic systems that handle logistics, inventory, and monitoring. In healthcare institutions, automation has the potential to improve efficiency, simplify procedures, and decrease the likelihood of human mistakes [45].

4. **Cooperation between Humans and Robots:** Facilitating cooperation between people and robots is a critical focus for future progress. Efforts are underway to enhance communication and collaboration between people and robots by developing user-friendly interfaces, tactile feedback systems, and sophisticated AI algorithms. Integrating people and robots in surgical operations, rehabilitative treatment, and caring may provide substantial benefits by facilitating seamless interaction and collaborative decision-making, ultimately resulting in enhanced results [46,47].

5. **Adherence to Regulations and Ethical Considerations:** The regulatory frameworks and ethical issues must catch up to the fast innovation progress in healthcare robots. Meeting regulatory criteria and resolving ethical issues about patient privacy and safety poses a substantial barrier. Develop explicit protocols and regulatory frameworks explicitly tailored to healthcare robots [48,49]. In order to tackle the ethical problems, promote cross-disciplinary cooperation among ethicists, policymakers, and technologists.

Solving these difficulties requires the cooperation of scholars, healthcare professionals, legislators, and industry leaders. The future of robots in medical research has the potential to profoundly transform healthcare delivery, enhance patient outcomes, and define the future of medicine via collaborative endeavours. Figure 8.4 displays the primary suggestions for progress.

8.9 CONCLUSION AND IMPLICATIONS OF THE STUDY

To summarise, robotics has become a revolutionary influence in medical research, displaying a wide range of applications and holding great

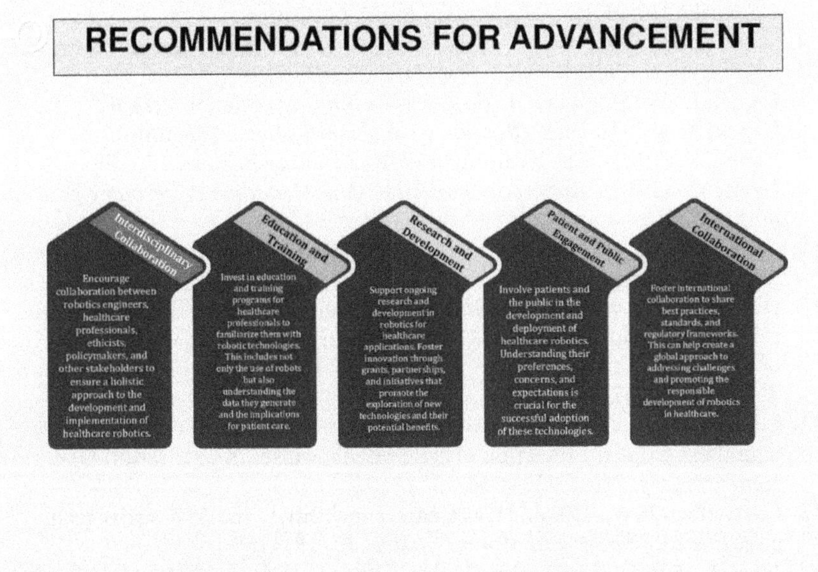

Figure 8.4 Recommendations for advancement in robotics.

potential for future advancements. Progress in robotic surgery, rehabilitation robotics, diagnostics, and medication administration has fundamentally transformed how patients are cared for, surgical results are achieved, and therapeutic methods are used. The fusion of AI and robotics has bolstered the capabilities of medical systems, enabling more significant diagnosis, treatment planning, and personalised medicine. The field of robotics in medical science has excellent promise as it advances in nanorobotics, robotic medication delivery, healthcare automation, and human-robot cooperation. Nevertheless, tackling critical issues such as safety, ethical considerations, incorporation into healthcare systems, and availability is essential. To fully use the potential of robots in medical science, academics, healthcare practitioners, policymakers, and industry leaders must collaborate to address the issues at hand. Due to ongoing progress and innovations, the integration of robots in medical research is positioned to revolutionise healthcare provision, enhance patient results, and influence the future of medicine.

REFERENCES

1. McCarthy, J. (1987). Generality in artificial intelligence. *Communications of the ACM, 30*(12), 1030–1035.
2. Patterson, D. (1990). *Introduction to Artificial Intelligence and Expert Systems*. Prentice-Hall, Inc.

3. Konar, A. (2018). *Artificial Intelligence and Soft Computing: Behavioral and Cognitive Modeling of the Human Brain*. CRC press.

4. Salathé, M., Wiegand, T., & Wenzel, M. (2018). Focus group on artificial intelligence for health. *arXiv preprint arXiv:1809.04797*.

5. Capek, J. (1925). Opilec. *Lelio A Pro Delfina*. Aventinum, Prague.

6. Capek, K. (2004). *RUR (Rossum's Universal Robots)*. Penguin.

7. Capek, K. (1923). The meaning of RUR. *Saturday Review, 136*, 79.

8. Gyles, C. (2019). Robots in medicine. *The Canadian Veterinary Journal, 60*(8), 819.

9. Tomlinson, Z., & Tomlinson, Z. (2018). Medical robots that are changing the world. October 11, 2018.

10. D'Ettorre, C., Mariani, A., Stilli, A., y Baena, F. R., Valdastri, P., Deguet, A., ... & Stoyanov, D. (2021). Accelerating surgical robotics research: A review of 10 years with the da vinci research kit. *IEEE Robotics & Automation Magazine, 28*(4), 56–78.

11. Schupper, A. J., Price, G., & Hadjipanayis, C. G. (2021). Robotic-assisted digital exoscope for resection of cerebral metastases: A case series. *Operative Neurosurgery, 21*(6), 436–444.

12. Coeckelbergh, M. (2010). Health care, capabilities, and AI assistive technologies. *Ethical Theory and Moral Practice, 13*, 181–190.

13. Vyas, S., Gupta, S., Kapoor, M., & Khan, S. (Eds.). (2024). *Handbook on Augmenting Telehealth Services: Using Artificial Intelligence*. CRC Press.

14. Srivastava, M., Siddiqui, A. T., & Srivastava, V. (2024). Application of Artificial Intelligence of Medical Things in Remote Healthcare Delivery. In *Handbook of Security and Privacy of AI-Enabled Healthcare Systems and Internet of Medical Things*, Agbotiname Lucky Imoize, Valentina Emilia Balas, Vijender Kumar Solanki, Cheng-Chi Lee, and Mohammad S. Obaidat (Eds.), (pp. 169–190). CRC Press.

15. Younis, H. A., Eisa, T. A. E., Nasser, M., Sahib, T. M., Noor, A. A., Alyasiri, O. M., ... & Younis, H. A. (2024). A systematic review and meta-analysis of artificial intelligence tools in medicine and healthcare: Applications, considerations, limitations, motivation and challenges. *Diagnostics, 14*(1), 109.

16. Gowda, D., Shashikala, S. V., Manu, Y. M., Kaur, M., & Jha, S. K. (2024). Introduction to Cloud Computing and Healthcare 5.0: Transforming the Future of Healthcare. In *Federated Learning and AI for Healthcare 5.0*, Ahdi Hassan, Vivek Kumar Prasad, Pronaya Bhattacharya, Pushan Dutta, and Robertas Damaševičius (Eds.), (pp. 26–45). IGI Global.

17. Subasi, A., & Subasi, M. E. (2024). Digital Twins in Healthcare and Biomedicine. In *Artificial Intelligence, Big Data, Blockchain and 5G for the Digital Transformation of the Healthcare Industry*, Patricia Ordonez de Pablos and Xi Zhang (Eds.), (pp. 365–401). Academic Press.

18. Silvera-Tawil, D. (2024). Robotics in healthcare: A survey. *SN Computer Science, 5*(1), 189.

19. Anderson, D. (2019). Artificial intelligence and applications in PM&R. *American Journal of Physical Medicine & Rehabilitation, 98*(11), e128–e129.

20. Luxton, D. D., & Riek, L. (2019). Handbook of Rehabilitation Psychology. In *Artificial Intelligence and Robotics for Rehabilitation*. 3rd ed. American Psychological Association Books.

21. Goldzweig, C. L., Orshansky, G., Paige, N. M., Towfigh, A. A., Haggstrom, D. A., Miake-Lye, I., ... & Shekelle, P. G. (2013). Electronic patient portals: evidence on health outcomes, satisfaction, efficiency, and attitudes: A systematic review. *Annals of Internal Medicine, 159*(10), 677–687.

22. Sinsky, C. A., Willard-Grace, R., Schutzbank, A. M., Sinsky, T. A., Margolius, D., & Bodenheimer, T. (2013). In search of joy in practice: A report of 23 high-functioning primary care practices. *The Annals of Family Medicine, 11*(3), 272–278.

23. De Michele, R., & Furini, M. (2019, September). Iot healthcare: Benefits, issues and challenges. In *Proceedings of the 5th EAI International Conference on Smart Objects and Technologies for Social Good,* Barcelona, Spain, from September 18 to September 20, 2019, (pp. 160–164).

24. Malasinghe, L. P., Ramzan, N., & Dahal, K. (2019). Remote patient monitoring: A comprehensive study. *Journal of Ambient Intelligence and Humanized Computing, 10,* 57–76.

25. Ketu, S., & Mishra, P. K. (2021). Internet of healthcare things: A contemporary survey. *Journal of Network and Computer Applications, 192,* 103179.

26. Urwin, M. (2021, July 26). AI taking over jobs: What to know about the future of jobs. *Built In.* https://builtin.com/artificial-intelligence/ai-replacing-jobs-creating-jobs.

27. Murphy, R. R. (2019). *Introduction to AI robotics.* MIT press.

28. Avila-Tomás, J. F., Mayer-Pujadas, M. A., & Quesada-Varela, V. J. (2020). Artificial intelligence and its applications in medicine I: Introductory background to AI and robotics. *Atencion Primaria, 52*(10), 778–784.

29. Van Roy, Vincent, Vertesy, Daniel, & Damioli, Giacomo (2020). AI and Robotics Innovation. In *Handbook of Labor, Human Resources and Population Economics*, Springer, (pp. 1–35).

30. Naik, N., Hameed, B. M., Sooriyaperakasam, N., Vinayahalingam, S., Patil, V., Smriti, K., ... & Somani, B. K. (2022). Transforming healthcare through a digital revolution: A review of digital healthcare technologies and solutions. *Frontiers in Digital Health, 4,* 919985.

31. Tiwari, A., Dubey, A., Yadav, A. K., Bhansali, R., & Bagaria, V. (2024). A review of Smart future of healthcare in the digital age to improve Quality of orthopaedic patient care in metaverse called: The Healthverse!! *Journal of Clinical Orthopaedics and Trauma, 48,* 102340.

32. Smith, R. (2024). Explosion of Robotics in Healthcare. *The Rise of the Intelligent Health System, 22,* 87–111.

33. Kimmatudawage, S. P., Srivastava, R., Kachroo, K., Badhal, S., & Balivada, S. (2024). Toward Global Use of Rehabilitation Robots and Future Considerations. In *Rehabilitation Robots for Neurorehabilitation in High-, Low-, and Middle-Income Countries,* Michelle Jillian Johnson, and Rochelle J. Mendonca (Eds.), (pp. 499–516). Academic Press.

34. da Rocha Ferreira, A. R., De Sousa, M. G. M. M., de Oliveira Correia, M., de Almeida, R. S., Simões-Silva, V., & Trigueiro, M. J. (2024). Use of Socially Assistive Robots in Mental Health: Barriers and Facilitators. In *Handbook of Research on Advances in Digital Technologies to Promote Rehabilitation and Community Participation*, Raquel Simões de Almeida, Vítor Simões-Silva, and Maria João Trigueiro (Eds.), (pp. 269–292). IGI Global.

35. Quinn, A. (2023). *Expanding Autism Treatment Through Social Robotics: Prospective Solutions in Clinical Therapy* (Doctoral dissertation, Azusa Pacific University).

36. Tanioka, T., Yokotani, T., Tanioka, R., Betriana, F., Matsumoto, K., Locsin, R., ... & Schoenhofer, S. (2021). Development issues of healthcare robots: Compassionate communication for older adults with dementia. *International Journal of Environmental Research and Public Health*, 18(9), 4538.

37. Writer, S. (2023, February 2). Kokilaben Hospital deploys Stryker Mako ortho robot. *healthcareradius.in*. https://www.healthcareradius.in/features/surgery/kokilaben-hospital-deploys-stryker-mako-ortho-robot.

38. Apollo Multispeciality hospitals Kolkata brings Da Vinci XI robot for the first time in eastern India. (2023, June 6). *Apollo Multispeciality Hospitals*. https://kolkata.apollohospitals.com/apollo-multispeciality-hospitals-kolkata-brings-da-vinci-xi-robot-for-the-first-time-in-eastern-india/.

39. Manipal hospital sets new record with 50 successful robotic spinal surgeries in a month. (2023, November 17). *Telegraph India*. https://www.telegraphindia.com/business/manipal-hospital-sets-new-record-with-50-successful-robotic-spinal-surgeries-in-a-month/cid/1980673.

40. Abdi, J., Al-Hindawi, A., Ng, T., & Vizcaychipi, M. P. (2018). Scoping review on the use of socially assistive robot technology in elderly care. *BMJ Open*, 8(2), e018815.

41. Ilmudeen, A., & Nayyar, A. (2022). Novel Designs of Smart Healthcare Systems: Technologies, Architecture, and Applications. In *Machine Learning for Critical Internet of Medical Things: Applications and Use Cases*, Fadi Al-Turjman and Anand Nayyar (Eds.), (pp. 125–151). Springer International Publishing.

42. Robertson, A. R., De Boer, N. K. H., Marlicz, W., & Koulaouzidis, A. (2021). An overview of robotic capsules for drug delivery to the gastrointestinal tract. *Journal of Clinical Medicine*, 10(24), 5791–5791.

43. Iacovacci, V., Tamadon, I., Kauffmann, E. F., Pane, S., Simoni, V., Marziale, L., ... & Menciassi, A. (2021). A fully implantable device for intraperitoneal drug delivery refilled by ingestible capsules. *Science Robotics*, 6(57), eabh3328.

44. Ozkil, A. G., Fan, Z., Dawids, S., Aanes, H., Kristensen, J. K., & Christensen, K. H. (2009, August). Service robots for hospitals: A case study of transportation tasks in a hospital. In *2009 IEEE International Conference on Automation and Logistics*, Qingdao, China, from August 5 to August 8, 2009, (pp. 289–294). IEEE.

45. Ajoudani, A., Zanchettin, A. M., Ivaldi, S., Albu-Schäffer, A., Kosuge, K., & Khatib, O. (2018). Progress and prospects of the human–robot collaboration. *Autonomous Robots*, 42, 957–975.

46. Buxbaum, H., Sen, S., & Kremer, L. (2019). An investigation into the implication of human-robot collaboration in the health care sector. *IFAC-PapersOnLine, 52*(19), 217–222.

47. Giallanza, A., La Scalia, G., Micale, R., & La Fata, C. M. (2024). Occupational health and safety issues in human-robot collaboration: State of the art and open challenges. *Safety Science, 169*, 106313.

48. Kaledio, E., Oloyede, J., & Olaoye, F. (2024). The evolution of robotics: Advancements, impacts, and ethical considerations.

49. Naithani, K., & Tiwari, S. (2024). Deep Learning for the Intersection of Ethics and Privacy in Healthcare. In *Machine Learning Algorithms Using Scikit and TensorFlow Environments* (pp. 154–191). IGI Global.

Chapter 9

AI strategies for the realm of sustainable management and practices

*K. Rajendra Prasad, G. Sucharitha,
Mudimanchi Pranay Kumar, and Ganji Adithya*

9.1 INTRODUCTION

The integration of environmental, social, and economic factors into engineering [1,2] and management methods has gained significant attention due to the global quest of sustainable development. Within this framework, artificial intelligence (AI) technologies have encouraging prospects for tackling intricate sustainability issues through creative resolutions, process enhancements, and enhanced decision-making. This section highlights the necessity for cutting-edge technology to accomplish sustainability goals and discusses the use of AI in the context of sustainable engineering and management [3]. It states the goals of the article, which include a survey of the material already in existence, an analysis of AI techniques [4], and a discussion of the implications for sustainability.

Technologies based on AI have shown to be effective instruments for promoting sustainable growth in a number of sectors, including management and engineering. Unprecedented opportunities to maximize resource use, reduce environmental impact, and improve decision-making processes arise from the integration of AI into sustainable practices. This introduction lays the groundwork for examining the complex relationship between AI and sustainable engineering and management, emphasizing both the advantages and disadvantages of doing so.

The international community has been working harder in recent years to solve urgent environmental and socioeconomic issues such social injustice, resource depletion, and climate change. As a result, the necessity of implementing more environmentally friendly engineering and management techniques is becoming increasingly apparent. The goal of sustainable engineering is to maximize long-term economic viability while minimizing detrimental effects on the environment and society through the design, construction, and operation of systems and processes. In a similar vein, sustainable management focuses on maximizing organizational procedures to accomplish financial success while advancing social justice and environmental conservation.

DOI: 10.1201/9781003565529-9

9.2 LITERATURE STUDY

This section draws on a wide range of research articles, reports, and case studies to provide a thorough overview of the body of knowledge regarding AI applications [5] in sustainable engineering and management. We are able to obtain a deeper understanding of the development, patterns, and difficulties related to AI-driven sustainability projects by conducting a methodical literature review.

AI Applications in Sustainable Engineering: The various ways that AI technologies [6], such as machine learning, optimization algorithms, and predictive analytics, are applied to improve the sustainability performance of engineering systems are covered in this subsection. We clarify the ways in which AI might enhance resource utilization, lower emissions, and improve energy efficiency across a range of engineering disciplines by looking at case studies and real data.

AI for Sustainable Management Practices: In this section, we examine how AI may be used to optimize supply chains, decision-making frameworks, and management procedures in order to advance sustainability goals. We identify the possible advantages of utilizing AI tools for improving operational efficiency, cutting waste, and stimulating innovation in sustainable management [7] contexts through a comparative examination of AI-driven management techniques.

Emerging Trends and Future Directions: The combination of blockchain, augmented reality, and Internet of Things (IoT) technologies is one example of an emerging trend and future direction in AI-driven sustainability research that is covered in this area. We offer insights into the possible influence of AI on changing the sustainability environment and bringing about systemic change across industries by projecting future developments.

AI for Sustainable Energy Systems: In this section, we explore how AI might optimize energy production, distribution, and consumption to support grid stability, energy efficiency, and the integration of renewable energy sources. We present novel strategies for integrating distributed energy resources, reducing grid disruptions, and balancing supply and demand in sustainable energy systems through a comparative examination of AI-driven energy management systems.

AI in Sustainable Supply Chains: The use of AI technology to optimize supply chain operations, cut waste, and improve transparency and traceability in sustainable supply chains is the main topic of this subject. We clarify the possible advantages of AI-driven supply chain management, such as enhanced inventory management, demand forecasting, and supplier collaboration, for accomplishing sustainability goals by looking at case studies and industry reports.

9.3 AI STRATEGIES FOR SUSTAINABILITY MANAGEMENT

A thorough summary of the many AI methods and resources relevant to sustainable engineering and management may be found in the methodology section. Through an analysis of the fundamental ideas, working methods, and practical uses of AI technologies, we provide readers with the knowledge they need to fully realize AI's potential for sustainability.

Machine Learning for Sustainability: The use of machine learning algorithms [8], such as supervised learning, unsupervised learning, and reinforcement learning, to analyse data pertaining to sustainability and arrive at well-informed conclusions, is examined in this subsection. We demonstrate how machine learning may be used to tackle sustainability issues in a variety of ways by looking at case studies and best practises.

Optimization Techniques: In this section, we will examine various optimization techniques that are employed in resource allocation, sustainable system design, and operational efficiency enhancement. These techniques include genetic algorithms, linear programming, and swarm intelligence. We clarify the advantages, disadvantages, and suitability for various sustainability scenarios of optimization techniques by a comparative study of them.

Data Analytics and Visualization: The methods covered in this subsection for extracting useful information from data pertaining to sustainability include data mining [9], descriptive analytics, and predictive modelling. We illustrate how data analytics can support evidence-based decision-making and propel ongoing development in sustainable engineering and management processes using practical examples and visualization tools.

Methods of Optimization for Sustainable Decision-Making: In order to improve decision-making processes in sustainable engineering and management, this part examines optimization approaches such as multi-objective optimization, integer programming, and linear programming. Through the formulation of optimization problems, specification of decision variables and constraints, and computational solution of optimization models, we empower practitioners to pinpoint ideal solutions for intricate sustainability issues such as resource allocation, project scheduling, and system design.

Artificial intelligence (AI) has made it possible to implement the following approaches and solutions for sustainable engineering and development:

9.4 ENERGY EFFICIENCY OPTIMIZATION

AI algorithms have the ability to optimize energy consumption in a variety of industries, including manufacturing, buildings, and transportation. Reducing carbon emissions and resource consumption can be achieved by using machine learning models to evaluate data, find patterns, and recommend energy-saving actions. AI algorithms can be used to forecast energy demand, assess trends in energy usage, and optimize building systems (lighting, Heating, Ventilation, and Air Conditioning (HVAC), etc.).

At the vanguard of sustainable engineering and management is Energy Efficiency Optimization [10], driven by AI transformational power of AI. By using AI-driven approaches and solutions to limit energy usage, reduce environmental impact, and improve operational efficiency, this novel approach transforms conventional energy management practices. Fundamentally, energy efficiency optimization uses AI to examine large datasets on equipment performance, weather, building attributes, and patterns of energy use. AI analyses this data using complex machine learning algorithms to find inefficiencies, forecast future energy consumption, and suggest focused solutions. These interventions could involve applying energy-saving techniques in manufacturing processes, modifying lighting schedules, and optimizing HVAC systems.

Renewable Energy Management: AI has the potential to enhance the dependability and efficiency of renewable energy sources, such as wind and solar energy. By forecasting energy generation, predictive analytics can improve grid integration and lessen dependency on fossil fuels, as shown in Figure 9.1.

AI-enabled sustainable engineering and development uses innovative techniques and solutions for renewable energy management that maximize the integration and use of renewable energy sources. Data collection is the first step in the process; information is gathered on weather forecasts, energy consumption patterns, grid status, and the production of renewable

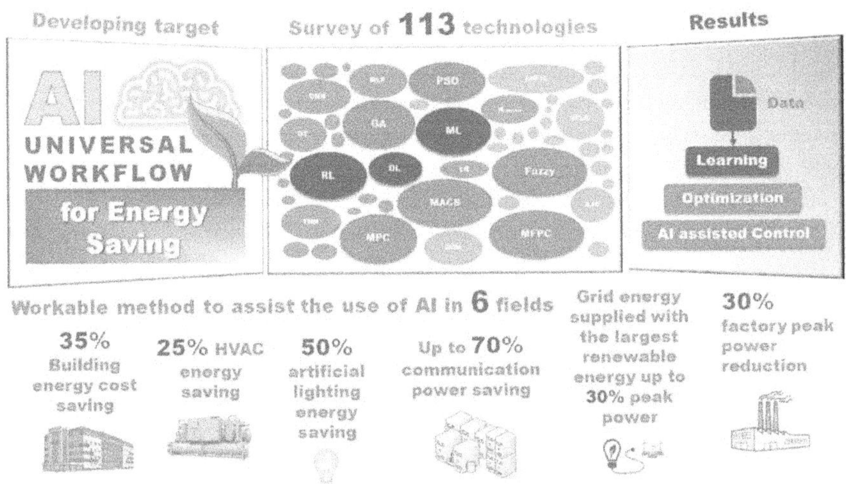

Figure 9.1 Energy efficiency optimization method.

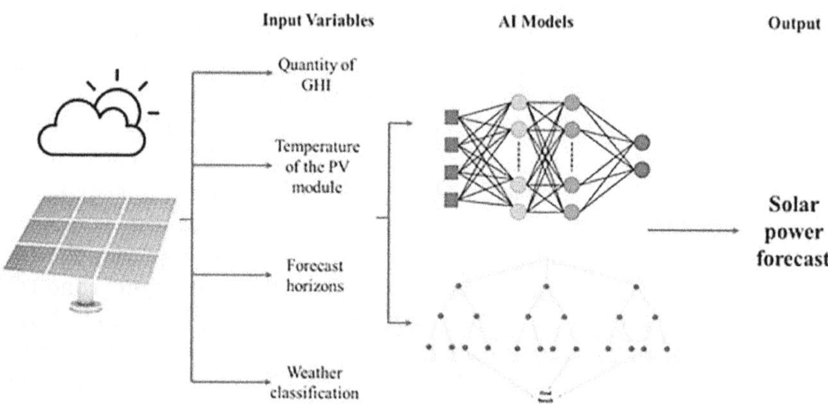

Figure 9.2 Reference sustainable renewable energy architecture.

energy. Next, AI algorithms are used to examine past data, spot trends, and create forecasting models for the production of renewable energy.

These models predict sun irradiance, wind speeds, or hydroelectric output using machine learning approaches such as regression or time-series analysis. See Figures 9.2 and 9.3 for the architectural diagrams.

Intelligent Power Systems (AI): These systems have the potential to improve the efficiency and stability of power distribution networks. In order to optimize energy flow, identify problems, and forecast demand, machine learning algorithms can evaluate data from sensors and metres. This helps create an energy infrastructure that is more sustainable and robust.

Smart grids, which are greatly enhanced by AI technologies, are a cornerstone of sustainable engineering and development. These intelligent grids combine AI-driven technologies to maximize sustainability, dependability, and efficiency, revolutionizing the traditional energy distribution system. In essence, smart grids gather data on energy generation, consumption, and grid status in real time by utilizing sophisticated sensors, metres, and communication networks. This data is examined using AI algorithms in order to spot trends, forecast demand, and enhance grid performance. Predictive models for energy demand forecasting, renewable energy integration, and grid stability evaluation can be created thanks to machine learning techniques. AI-driven optimization algorithms optimize the grid topology, voltage levels, and energy distribution dynamically to minimize losses, minimize traffic, and optimize the use of renewable energy sources.

Precision Agriculture: By offering insights into crop health, soil conditions, and ideal planting practices, AI and machine learning techniques have the potential to completely transform the agricultural sector. AI can decrease chemical inputs, maximize crop yields, improve resource utilization, and lessen environmental impact by utilizing data from sensors, satellites, and drones.

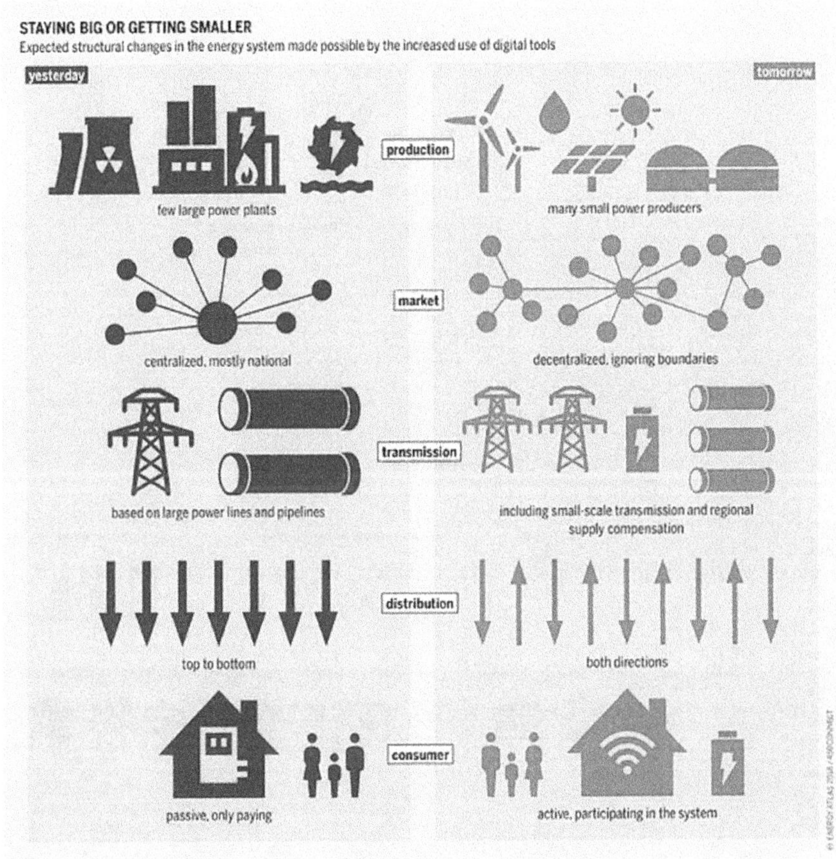

Figure 9.3 Intelligent system for sustainable power method.

Thanks to AI, precision agriculture is a significant development in sustainable engineering and development (AI). By utilizing AI-driven technologies to maximize output, reduce environmental impact, and improve resource allocation, this novel strategy transforms conventional farming practices. Precision agriculture is essentially the integration of a wide range of AI-powered tools and methodologies for the analysis of various datasets, such as crop health, weather patterns, soil composition, and pest infestations. This data is processed using advanced machine learning algorithms to produce recommendations and insights that farmers may use. With the help of these insights, accurate decisions can be made about fertilization methods, crop yield timing, and pest control, all of which maximize crop yields while consuming the fewest resources possible. Additionally, real-time field condition monitoring is made possible by AI-enabled drones and satellite imaging, which makes it easier to identify crop diseases and nutrient deficits early on (Figure 9.4).

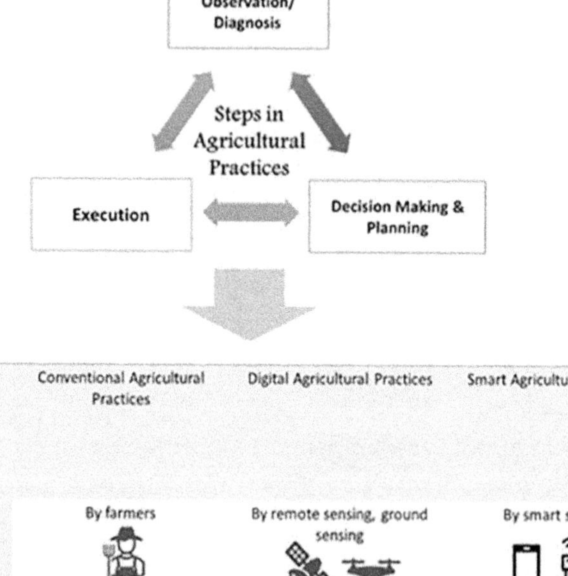

Figure 9.4 Intelligent precision agriculture method.

Management of Natural Resources: AI can help with the observation and administration of natural resources such as forests, water, and wildlife. When paired with AI algorithms, remote sensing technologies can monitor environmental changes, spot illicit activities such as poaching and deforestation, and aid in conservation efforts.

The foundation of sustainable engineering and development, enabled by AI technology, is natural resource management. Through the use of AI-driven solutions, this multidisciplinary approach transforms conventional resource management techniques and promotes long-term sustainability, environmental degradation mitigation, and optimal resource usage. Natural Resource Management, at its core, is the integration of a wide range of AI-powered tools and methodologies for the analysis of complex datasets, such as climate models, geological surveys, biodiversity assessments, and satellite imaging. This data is processed using advanced machine learning algorithms to provide useful insights and support decision-making. Predictive analytics powered by AI makes it possible to spot trends, patterns, and possible dangers related to resource extraction and environmental deterioration. AI-powered optimization algorithms also make it easier to create adaptive management plans that strike a balance between conflicting objectives such as ecological preservation and economic growth. AI technologies improve the efficacy and efficiency of natural resource management initiatives by enabling real-time monitoring, analysis, and decision assistance, guaranteeing the prudent stewardship of valuable resources for future generations. AI-enabled natural resource management fosters resilience, equity, and prosperity for both the present and the future generations by acting as a catalyst for sustainable engineering and development through ongoing innovation and collaboration. AI algorithms are able to examine enormous volumes of climate data in order to forecast future patterns and evaluate the effects of different mitigation techniques. AI-powered models can help with planning for climate adaptation, provide insights into policy choices, and generate creative responses to the problems presented by climate change. An essential component of sustainable engineering and management is climate change mitigation, which is made possible by AI's revolutionary potential.

One approach is the analysis of large datasets of socioeconomic, emissions, and climate variables using AI algorithms. AI can detect trends, patterns, and correlations using advanced machine learning techniques. This allows for a deeper comprehension of the factors contributing to climate change and the efficacy of mitigation measures. Climate models driven by AI have the capacity to model future conditions and forecast the possible effects of climate change on infrastructure, human populations, and ecosystems. The development of focused mitigation and adaptation strategies is influenced by these realizations, as shown in Figure 9.5.

Trash Management: By anticipating patterns in trash generation, streamlining collection routes, and recognizing recyclable materials, AI can enhance waste collection and recycling operations. Sorting operations in recycling plants can be automated using computer vision algorithms, increasing productivity, and lowering contamination.

Using AI algorithms to examine huge databases about trash composition, generation, and disposal trends is one approach. AI is capable of trend identification, trash generation prediction, and process optimization for waste collecting and sorting through sophisticated machine learning algorithms. Predictive models powered by AI help waste management firms and governments predict demand, distribute resources effectively, and keep operating expenses to a minimum.

AI-enabled solutions include intelligent garbage collection systems, robotic sorting technologies, and material recovery facilities, among other

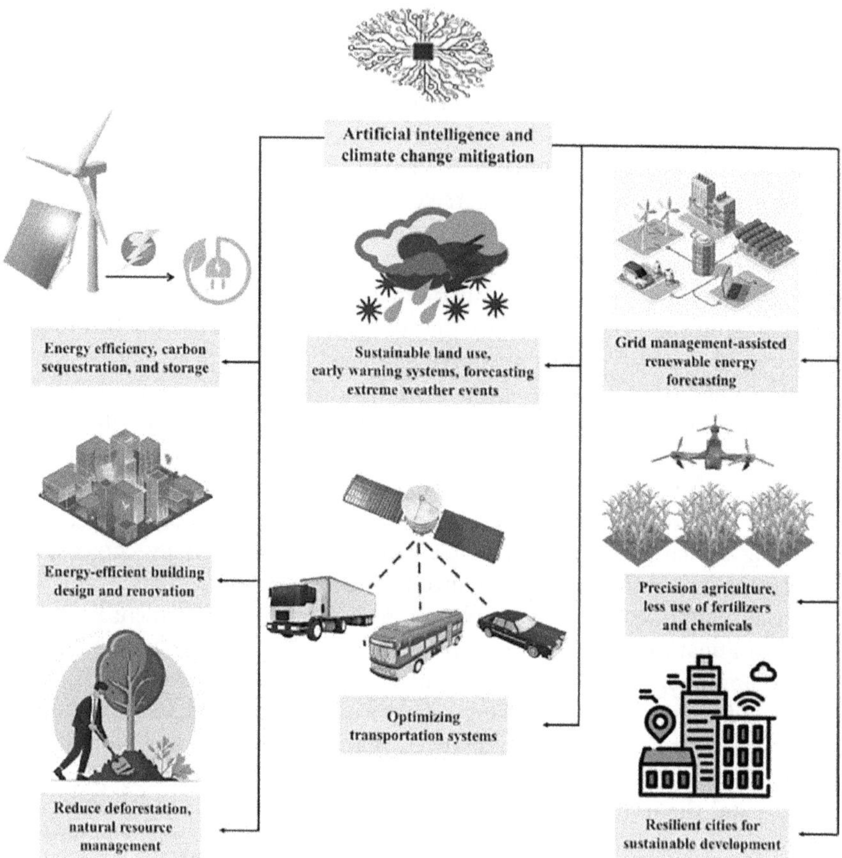

Figure 9.5 AI-based sustainable climate monitoring.

projects. Waste bins can be monitored in real time by AI-powered sensors and IoT devices, which can then be used to optimize collection routes and schedules based on demand patterns and fill levels. Furthermore, recycling materials may be sorted and separated with great accuracy by robotic systems powered by AI, increasing resource recovery and minimizing contamination.

Transportation System Optimization: By using AI-powered algorithms, transportation systems can be made more efficient in order to cut down on pollution, traffic, and fuel use. Mobility may be improved while reducing environmental effect with the help of technologies such as intelligent traffic management systems, driverless cars, and ride-sharing platforms.

Transformative AI powers Transportation Optimization, a fundamental component of sustainable engineering and management. This important field deals with the requirement to improve transportation systems' dependability, efficiency, and environmental sustainability. With the help of intelligent automation, predictive analytics, and data-driven decision-making, AI technologies provide creative answers to these problems. AI-enabled solutions cover a wide range of projects, such as autonomous car technology, dynamic routing algorithms, and intelligent traffic control systems.

Environmental Monitoring and Pollution Control: AI is capable of analysing sensor data to track contamination sources, identify hotspots for pollution, and monitor the quality of the air and water. AI-powered real-time monitoring systems can support early warning systems and timely action methods to reduce environmental dangers.

AI has the potential to revolutionize engineering and management through its revolutionary capability in environmental monitoring and pollution control. The essential need to monitor environmental quality, reduce pollution, and protect ecosystems is addressed in this crucial domain. With the help of intelligent automation, predictive analytics, and data-driven decision-making, AI technologies provide creative answers to these problems.

AI-enabled solutions cover a wide range of projects, such as predictive modelling tools, real-time monitoring systems, and smart sensor networks. Real-time data on soil contamination, water quality, and air quality can be gathered by AI-powered sensor networks, giving fast and reliable information for pollution control initiatives. These methods of AI offer transformative opportunities to address pressing sustainability challenges and accelerate progress towards a more resilient, equitable, and environmentally conscious future. By leveraging AI-driven methodologies and solutions, we can unlock new insights, optimize resource usage, and drive innovation across various sectors. As we continue to harness the potential of AI for sustainability, it's imperative to remain vigilant, adaptable, and committed to the principles of responsible innovation and sustainable development. Together, we can harness the power of AI to build a more sustainable world for current and future generations.

9.5 CONCLUSION

The conclusion of this research highlights the transformative potential of AI technology for sustainable engineering and management. Reiterating the critical role AI plays in fostering resilience against global challenges and realizing sustainability objectives, we summarize the key findings from the sections on methodology and literature evaluation. Furthermore, we emphasize that interdisciplinary collaboration, stakeholder involvement, and capacity building are critical to fully utilizing AI to advance sustainable development. As we move towards a more sustainable future, it is imperative that we make the most of AI in order to positively impact the environment, society, and economy for the benefit of both the present and future generations.

It's crucial to understand, nevertheless, that the ethical, social, and environmental consequences of AI for sustainability must be carefully considered before it can be successfully implemented. Transparency, accountability, and equity must be given top priority as AI technologies advance in order to guarantee that rewards are shared fairly and possible risks are reduced.

REFERENCES

1. M. Gepp, M. Amberg, J. Vollmar, T. Schaeffler and S. Horn, "Economic Considerations of Engineering in the Industrial Plant Business," *2013 IEEE International Conference on Industrial Technology (ICIT)*, Cape Town, South Africa, 2013, pp. 1472–1476, doi: 10.1109/ICIT.2013.6505889.
2. R. C. Creese, Basic economic factors and equations. In: *Strategic Cost Fundamentals. Synthesis Lectures on Engineering*, pp. 93–111. Springer, Cham, 2018.doi: 10.1007/978-3-031-79394-3_8.
3. M. Installe, "Engineering and Technology Management for a More Sustainable Development," *Innovation in Technology Management. The Key to Global Leadership. PICMET '97*, Portland, OR, USA, 1997, pp. 282, doi: 10.1109/PICMET.1997.653361.
4. S. Y and M. Challa, "A Comparative Analysis of Explainable AI Techniques for Enhanced Model Interpretability," *2023 3rd International Conference on Pervasive Computing and Social Networking (ICPCSN)*, Salem, India, June 19–20, 2023, pp. 229–234, doi: 10.1109/ICPCSN58827.2023.00043.
5. S. Zhao, F. Blaabjerg and H. Wang, "An Overview of Artificial Intelligence Applications for Power Electronics," *IEEE Transactions on Power Electronics*, vol. 36, no. 4, pp. 4633–4658, April 2021, doi: 10.1109/TPEL.2020.3024914.
6. C. Huang, Z. Zhang, B. Mao and X. Yao, "An Overview of Artificial Intelligence Ethics," *IEEE Transactions on Artificial Intelligence*, vol. 4, no. 4, pp. 799–819, Aug. 2023, doi: 10.1109/TAI.2022.3194503.
7. R. Imran, M. N. Alraja and B. Khashab, "Sustainable Performance and Green Innovation: Green Human Resources Management and Big Data as Antecedents," *IEEE Transactions on Engineering Management*, vol. 70, no. 12, pp. 4191–4206, Dec. 2023, doi: 10.1109/TEM.2021.3114256.
8. S. Ray, "A Quick Review of Machine Learning Algorithms," *2019 International Conference on Machine Learning, Big Data, Cloud and Parallel Computing (COMITCon)*, Faridabad, India, February 14–16, 2019, pp. 35–39, doi: 10.1109/COMITCon.2019.8862451.
9. S. Agarwal, "Data Mining: Data Mining Concepts and Techniques," *2013 International Conference on Machine Intelligence and Research Advancement*, Katra, India, December 21–23, 2013, pp. 203–207, doi: 10.1109/ICMIRA.2013.45.
10. J. Zou, Y. Cui, Y. Liu and S. Sun, "Energy Efficiency Optimization for Integrated Sensing and Communications Systems," *2022 IEEE Wireless Communications and Networking Conference (WCNC)*, Austin, TX, USA, April 10–13, 2022, pp. 216–221, doi: 10.1109/WCNC51071.2022.9771575.

Chapter 10

Unveiling life's horizon

Predictive insights with automated machine learning

G. Sucharitha, Ravipalli Shreya, and G. Chandra Sekhar

10.1 INTRODUCTION

Life expectancy data analysis (Figure 10.1) is a vital undertaking in public health research, offering profound insights into population longevity and the factors shaping human health outcomes [1]. Life expectancy, defined as the anticipated lifespan of individuals based on prevailing mortality rates, serves as a pivotal indicator of societal well-being and healthcare system effectiveness [2]. Analysing life expectancy data unveils trends, disparities, and predictors crucial for informed decision-making in healthcare policy, resource allocation, and intervention strategies. It plays a crucial role in forecasting future healthcare needs and planning for an aging population [3]. By projecting changes in life expectancy over time, policymakers can anticipate the demand for healthcare services, long-term care facilities, and other support services for older adults. The analysis of life expectancy data enables us to

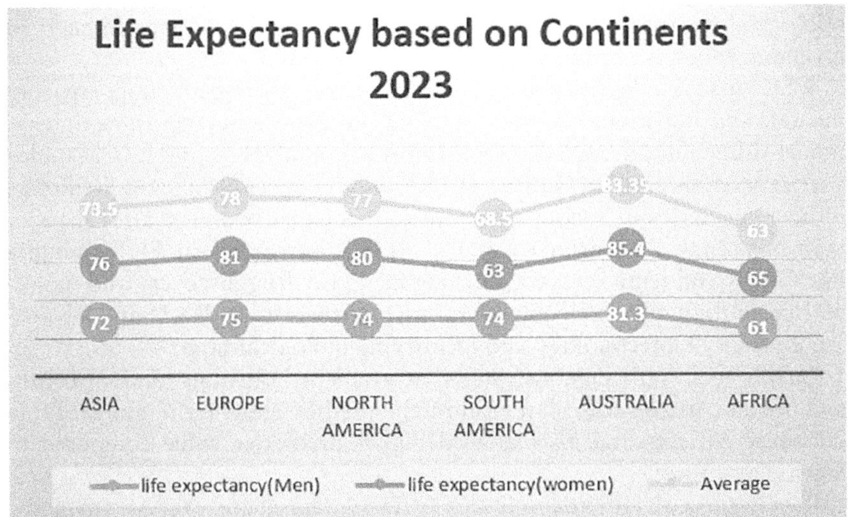

Figure 10.1 Life expectancy data analysis of continents in the year 2023.

DOI: 10.1201/9781003565529-10

identify disparities in health outcomes and inform targeted interventions to improve public health outcomes [4]. By identifying populations with lower life expectancies or higher rates of premature mortality, policymakers can allocate resources more effectively and implement policies and programs aimed at reducing health inequalities and improving overall population health [5]. The synergy between machine learning and tools such as Power BI enhances the visualization and interpretation of life expectancy data. Power BI's intuitive interface enables dynamic visualization of machine-learning–driven insights, facilitating comprehensive understanding and decision-making [5]. Automated machine learning (AutoML) further streamlines the process, automating model selection, feature engineering, and hyperparameter tuning, thereby accelerating analysis and enhancing efficiency [6]. The integration of machine learning techniques amplifies the efficacy of life expectancy analysis by harnessing complex data patterns of e-machine learning algorithms, such as Gradient Boosting Machine (GBM), offering predictive capabilities that delve deep into life expectancy determinants, identifying intricate relationships, and forecasting future trends with unparalleled accuracy. By leveraging GBM and other advanced algorithms, life expectancy analyses can uncover nuanced insights unattainable through traditional statistical methods [7,8]. By amalgamating cutting-edge technologies with rich datasets, we unlock unprecedented insights into life expectancy determinants, paving the way for informed policy decisions, targeted interventions, and improved health outcomes for all.

Li et al. discover that Machine learning can be used in material discovery for tasks like characterization, property prediction, and synthesis [9]. In healthcare industry, it aids in integrating life cycle assessment (LCA) and life cycle cost (LCC), analyzing study categories, institutions, and methodological approaches. It checks environmental and economic impacts of hospitals which is explained by Franca et al. [10].

When machine learning is applied to analyze socio-economic, environmental, and nutritional factors affecting life expectancy across countries, models like multiple regression used by Kabir et al. [11], probit, and multivariate analysis used by Gulis et al. [12] find the most significant predictors of life expectancy like motality trends in this paper by Lee et al. [13].

In this study, Zhavoronkov et al. [14] tells how Modern AI Techniques like GANs and reinforcement learning facilitate drug discovery and molecular generation, potentially accelerating pharmaceutical R&D and advancing longevity biotechnology and identifying biological targets.

Parikh et al. [15] demonstrates that gradient boosting, random forest, and logistic regression, can accurately predict short-term mortality in oncology patients and also showed higher predictive value compared to logistic regression.

Raja Babu et al. [16] analyzes a COVID-19 dataset and proposes machine learning regressors, including linear regression, polynomial regression,

Decision Tree, and Random Forest, to predict future cases. This study finds that the polynomial regression technique provides the best predictions.

Kang et al. [17] explores using emerging technologies like IoT, big data, and mHealth to develop personalized life expectancy (PLE) predictions by integrating health-related data from wearable devices, apps, and cloud computing. The proposed multi-phased approach utilizes machine learning and data analytics to extend current life expectancy models, allowing for lifetime predictions based on individual health data.

Zahedi et al. [18] uses an artificial neural network in MATLAB to classify breast cancer tumors as benign or malignant. The AI-based method outperforms previous approaches in tumor classification from mammography images, offering a more efficient and accurate alternative for early diagnosis of breast cancer.

10.2 LITERATURE SURVEY

In a study conducted by Agarwal et al. [1], Linear Regression models for disease prediction across various continents, including Europe and Asia, were explored. By leveraging multiple linear regression techniques [2], these researchers aimed to enhance the accuracy of disease prediction models, yielding improved results. Furthermore, their investigation extended to employing machine learning algorithms such as Random Forest Classifier [4] and K-Nearest Neighbours (KNN). Upon analysis, they concluded that KNN demonstrated superior performance compared to other algorithms tested in their study. This research underscores the importance of employing diverse methodologies, including traditional statistical models and machine learning algorithms, to address complex healthcare challenges such as disease prediction.

Beeksma et al. [19] used supervised machine learning techniques to employ assess life expectancy using medical records data. Specifically, a Long Short-Term Memory (LSTM) recurrent neural network model was utilized to analyse patient medical records and extract relevant features for life expectancy evaluation. The objective of this study was to automate the processing of medical records, ultimately facilitating improved decision-making within the healthcare industry. By harnessing the power of advanced machine learning algorithms, such as LSTM, researchers aimed to enhance the efficiency and accuracy of life expectancy prediction, thereby contributing to more informed healthcare interventions and outcomes.

In this study, Levantesi et al. [3] have implemented clustering method in LSTM to estimate the levels of longevity at birth which helps in national frameworks to achieve people's health goals and welfare of the nation. It provides forecasting based on the historic data.

Lipesa et al. [5] have utilized the XGBoost algorithm to predict life expectancy by considering health, socioeconomic, and behavioural factors.

It aimed to identify important predictors impacting life expectancy and assess the accuracy of XGBoost's predictions. The researchers likely pre-processed the data, trained the model using techniques such as cross-validation, and evaluated its performance using metrics such as MAE or RMSE. Overall, it demonstrated XGBoost's effectiveness in predicting life expectancy, showcasing its potential for informing public health research and policy decisions.

The study of Brink-Kjaer et al. [20] demonstrates how deep learning, particularly using gradient SHAP, improves age estimation and mortality prediction from PSGs compared to traditional methods. By training AE models on PSG data, we extracted medically relevant information, such as sleep-stage transitions and apnoea, enhancing predictive accuracy. The ensemble model, averaging outputs from various physiological signals, showed promising results in mortality prediction. PSGs, the gold standard for sleep evaluation, offer rich data for deep learning algorithms to uncover vital insights into individuals' health.

Desuky et al. [7] assessed machine learning algorithms, including Naive Bayes, using metrics such as accuracy, F measure, ROC curve, Gmean, TNR, and TPR. It stressed the significance of alternative metrics such as F measure and ROC curve for imbalanced data. Results indicated that attribute ranking methods improved prediction accuracy for certain techniques, diverging from the assumption that boosting always enhances performance. Further research may explore additional attribute selection and machine learning techniques to refine prediction models for lung cancer patients' post-thoracic surgery.

Lakshmanarao et al. [8] proposed a framework where it begins collecting and analysing a Kaggle dataset to identify features influencing life expectancy. Four machine learning algorithms—linear regression, Random Forest, Decision Tree, and SVM—are then applied for prediction. Random Forest emerges as the top performer, achieving a 96% r-squared value. The dataset, sourced from WHO, comprises 2,938 samples from 193 countries, including factors such as immunization, mortality rates, and income. Analysis highlights country-specific features affecting life expectancy, affirming Random Forest's efficacy for prediction.

10.3 METHODOLOGY

10.3.1 Data collection

This methodology takes data from WHO. However, for this study we take from Kaggle [21] and utilize it to analyse health data from 193 countries across the years 2000, 2010, 2015, and 2019. This dataset includes various indicators such as life expectancy rates, mortality rates, disease prevalence, healthcare expenditure, and their relationship with GDP and population.

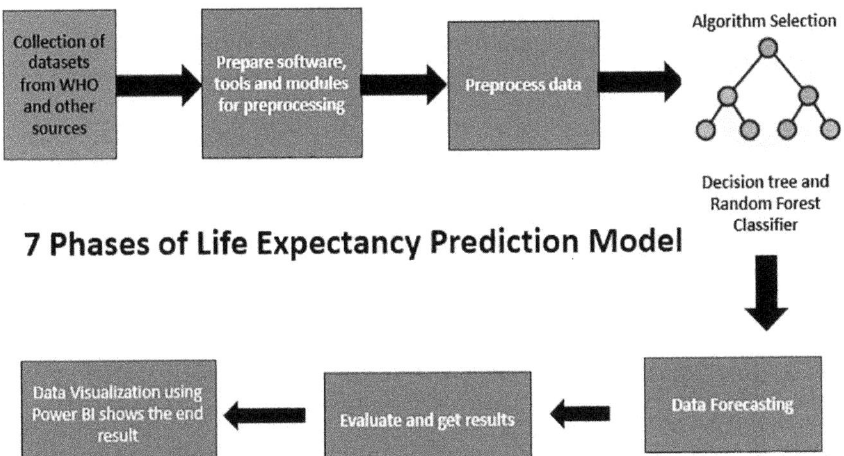

Figure 10.2 Structured approach of life expectancy data analysis.

	Country	Year	Status	Life expectancy	Adult Mortality	infant deaths	Alcohol	percentage expenditure	Hepatitis B	Measles	...	Polio	Total expenditure	Diphtheria	HIV/AIDS	GDP
0	Afghanistan	2015	Developing	65.0	263.0	62	0.01	71.279624	65.0	1154	...	6.0	8.16	65.0	0.1	584.259210
1	Afghanistan	2014	Developing	59.9	271.0	64	0.01	73.523582	62.0	492	...	58.0	8.18	62.0	0.1	612.696514
2	Afghanistan	2013	Developing	59.9	268.0	66	0.01	73.219243	64.0	430	...	62.0	8.13	64.0	0.1	631.744976
3	Afghanistan	2012	Developing	59.5	272.0	69	0.01	78.184215	67.0	2787	...	67.0	8.52	67.0	0.1	669.959000
4	Afghanistan	2011	Developing	59.2	275.0	71	0.01	7.097109	68.0	3013	...	68.0	7.87	68.0	0.1	63.537231

Figure 10.3 Dataset taken for this study.

By analysing these indicators such as life expectancy rates, mortality rates, disease prevalence, and healthcare expenditure, we can understand the overall health status of a population and identify trends, disparities, and areas for improvement. Additionally, examining how these factors relate to GDP and population size offers insights into the socioeconomic determinants of health and helps in formulating targeted interventions to improve public health outcomes (Figure 10.3).

10.3.2 Data preprocessing

In this process, we clean the data by using normalization techniques, removing outliers, and employing feature engineering. By doing this, we ensure that the data is of high quality and suitable for deriving actionable insights. This approach provides a solid foundation for conducting in-depth analysis.

10.3.3 Data analysis

In this study, our analysis provides data-driven insights on the overall health and well-being of populations, allowing policymakers and healthcare professionals to identify trends, disparities, and areas for improvement. By analysing life expectancy data, we understand the impact of various factors such as healthcare access, socioeconomic status, and environmental factors on life expectancy outcomes. This information helps us in developing targeted interventions and allocating resources effectively to improve public health outcomes. It helps in prediction and better decision-making. In machine learning, we took unsupervised learning to perform clustering of the dataset with Random Forest Classifier and Decision Tree algorithms to analyse and get predictive insights from the complex data and make it understandable.

10.3.4 Algorithm selection

We chose Random Forest Classifier and Decision Tree algorithms for the analysis of life expectancy due to their ability to handle multidimensional datasets and provide interpretable results. Random Forest's ensemble learning approach mitigates overfitting and improves generalization, while Decision Tree offer simplicity and transparency, aiding communication and decision-making. Both algorithms identify influential predictors, guiding targeted interventions for improving population health outcomes. Their robustness to outliers and missing data enhances reliability in real-world healthcare datasets. Ultimately, Random Forest and Decision Tree empower stakeholders with accurate predictions and actionable insights, facilitating informed decision-making in public health contexts. Many researchers have done data analysis on these algorithms but the unique element of this study we are using advanced methods to show the large data in interactive dashboards and visualization. This study will help in reducing wrong predictions which in turn helps in maintaining high levels of accuracy.

10.3.5 Model training and evaluation

In this study, we utilized Random Forest Classifier and Decision Tree where we make use AutoML to automate the generation of data using machine learning and it will be complemented by Power BI for model training and evaluation. These algorithms were chosen for their ability to capture complex relationships within the data and provide interpretable results. With Power BI's interactive features, we visualized the performance metrics of the model, facilitating informed decision-making. Based on the features selected, the model generates accuracy and in-depth analysis.

10.3.6 Data visualization

Data visualization (Figures 10.4 and 10.5) and Power BI setup play integral roles in which it facilitates efficient analysis and data-driven insights.

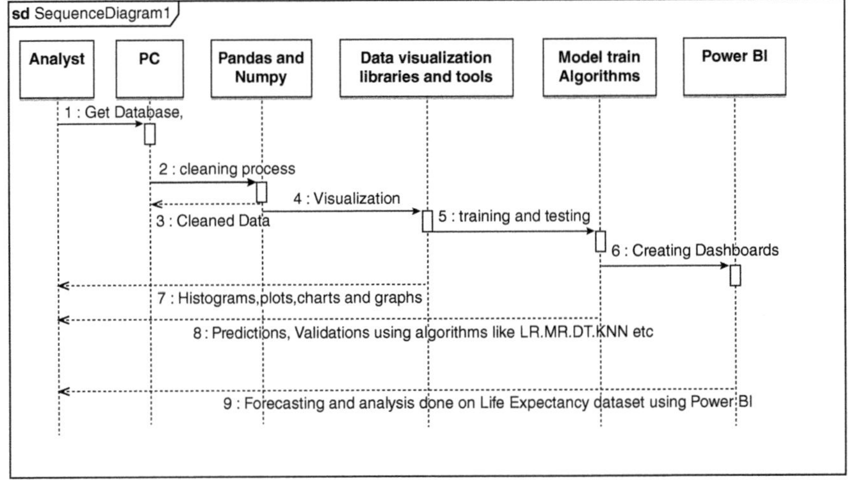

Figure 10.4 Sequence diagram on how Power BI visualizes life expectancy data.

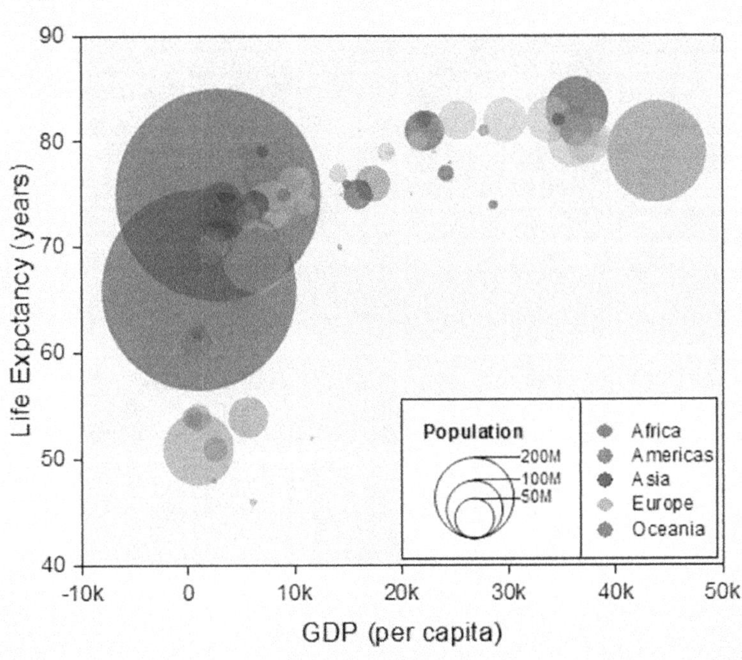

Figure 10.5 Data visualization on population and life expectancy.

We used Power BI environment to build visuals and dashboard on life expectancy analysis. This setup involves integrating diverse data sources, including preprocessed life expectancy data, into the Power BI platform. Utilizing Power BI's intuitive interface and interactive features, we explore the data dynamically and gain actionable insights through visually compelling representations. We design Power BI dashboards to present key metrics, trends, and patterns related to life expectancy in a clear and concise manner. Bar charts offer comparisons of life expectancy among various regions, countries, or socioeconomic strata. Scatter plots help in identifying relationships between life expectancy and factors like GDP, healthcare expenditure, or disease prevalence. Maps provide spatial insights by visualizing variations in life expectancy rates across different geographical areas. Heatmaps (Figure 10.6) offer a comprehensive view of the intensity of factors influencing life expectancy, such as disease prevalence or healthcare access, across regions. Tree maps represent hierarchical data structures to highlight the relative importance of different factors contributing to life expectancy.

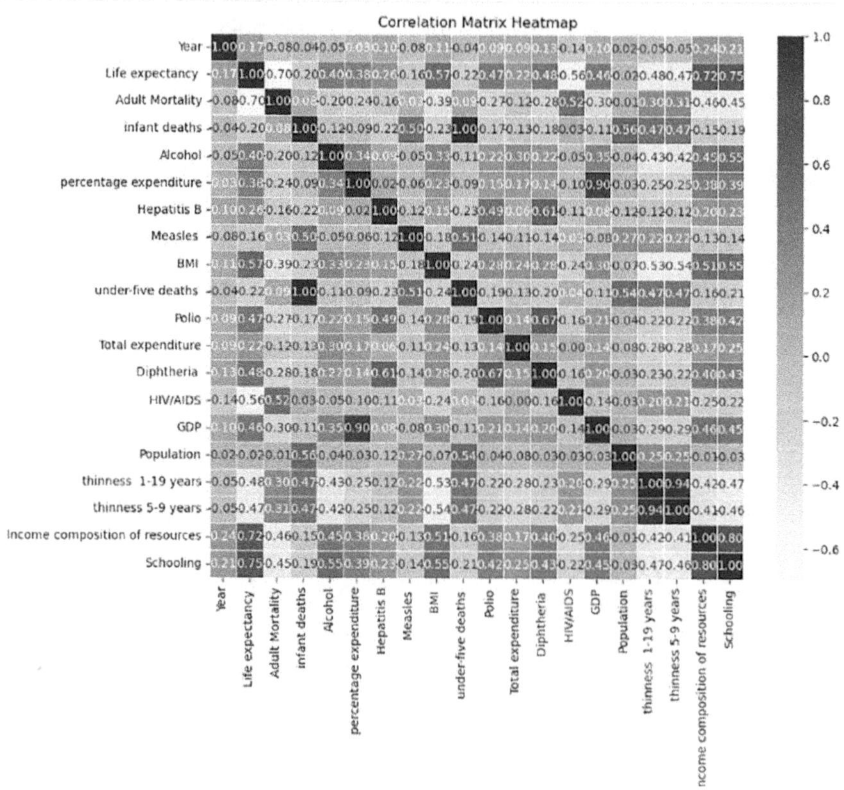

Figure 10.6 Heatmap on life expectancy data analysis.

10.3.7 Deployment and reporting

This study presents comprehensive reports on analysis findings, visualizations, and actionable recommendations created within Power BI. These reports are tailored to meet the diverse needs of stakeholders, ensuring relevance and engagement. Interactive dashboards allow stakeholders to explore the data dynamically, filter information, and drill down into detailed insights. The reports we shared using Power BI's collaboration features, ensuring accessibility and real-time updates. Scheduled updates keep stakeholders informed of the latest insights. During presentations, Power BI reports can be embedded into slides for interactive discussions, facilitating decision-making. Ultimately, deployment and reporting in life expectancy analysis using Power BI enable stakeholders to make informed decisions, drive interventions, and improve public health outcomes effectively.

10.4 RESULTS

In this study, we explored the prediction of life expectancy utilizing machine learning techniques, specifically Random Forest Classifier and Decision Tree. It follows clustering process during analysis. Additionally, the visualization of the analysis was conducted using Power BI, a robust tool for data visualization and exploration.

10.4.1 Performance comparison of tree classifiers

Before delving into feature selection, we conducted an analysis using all available features with tree classifier models: Random Forest and Random Tree. It will subsequently identify attributes influencing life expectancy. Through this process, 22 attributes such as (year, status, life expectancy, adult mortality, infant deaths, alcohol, percentage expenditure, and many more) are identified as influential in the data analysis. These also include indicators related to demographic, socioeconomic, and health-related factors. These attributes were then used to refine the predictive models.

10.4.2 Performance results with selected features

Upon remodelling the dataset with the selected attributes, the performance of tree classifiers was evaluated. Random Forest exhibited the highest accuracy, reaffirming its status as the optimal technique for life expectancy prediction. This underscores the importance of feature selection in refining predictive models and improving accuracy. In life expectancy analysis, these metrics (metrics such as Mean Absolute Error, Mean Squared Error, Root Mean Squared Error, Coefficient of Determination, Adjusted R Squared Accuracy for Random Forest, and Accuracy for Decision Tree) gauge model accuracy, precision, and explanatory power. They provide quantifiable measures of predictive performance and classification accuracy, guiding interventions for improved public health.

Mean Absolute Error

$$MAE = \left[\frac{1}{n}\left(\sum_{i=1}^{n}(Y-X)\right)\right]$$

Mean Squared Error

$$MSE = \left[\frac{1}{n}\left(\sum_{i=1}^{n}(Y-X)^2\right)\right]$$

Root Mean Squared Error

$$RMSE = \left[\frac{1}{n}\left(\sum_{i=1}^{n}(Y-X)^2\right)\right]^2$$

Coefficient of Determination

$$R^2 = 1 - \frac{\sum_{i=1}^{n}(Y-X)^2}{\sum_{i=1}^{n}(X-Y)^2}$$

Adjust R Squared

$$Adjusted\ R^2 = 1 - \frac{\left(1-R^2\right)(n-1)}{n-k-1}$$

Accuracy for Random Forest

$$\frac{\text{No. of correctly classified instances}}{\text{Total no.of instances}} \times 100\%$$

10.4.3 Integration of AutoML

Furthermore, we explored the use of AutoML, which automates the process of selecting the best machine learning model and hyperparameters for a given dataset. By leveraging AutoML, we aimed to streamline the model selection process and improve prediction accuracy without the need for manual intervention.

10.4.4 Visualization using Power BI

In this analysis, visualization using Power BI enhanced the interpretation and communication of the results. Power BI facilities the exploration of complex patterns and trends within the data, enabling stakeholders to gain actionable insights for public health interventions. Line charts are utilized to illustrate temporal trends in life expectancy across different time periods or demographic groups. Bar charts (Figures 10.7 and 10.8) offer comparisons of life expectancy among various regions, countries, or socioeconomic strata. Scatter plots help in identifying relationships between life expectancy and

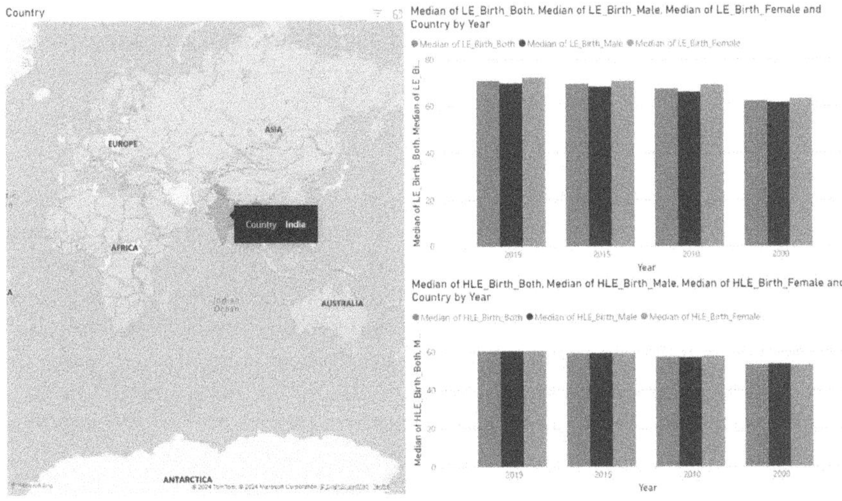

Figure 10.7 Life expectancy data analysis by country using Power BI.

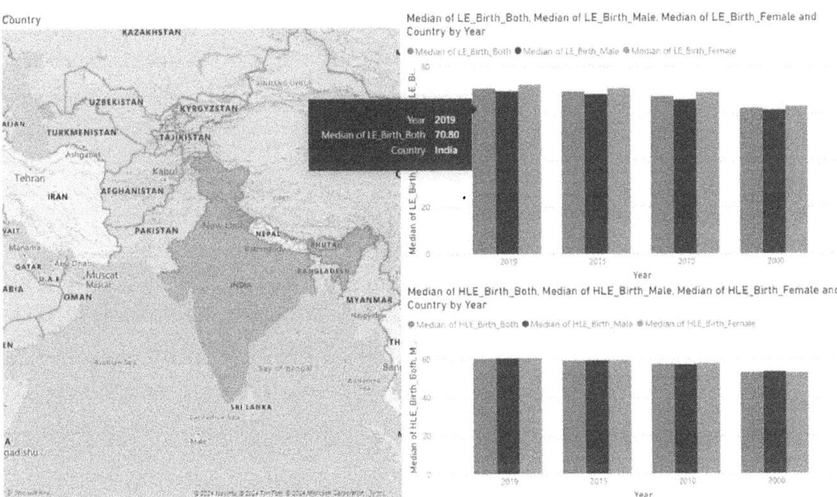

Figure 10.8 Life expectancy of selected nation by year and gender using Power BI.

factors such as GDP, healthcare expenditure, or disease prevalence. From the below figures you can understand how we developed dashboard visualization using Power BI where by just clicking on any country or continent, we can get data of the life expectancy rates of males, females, and both genders from the years 2000, 2010, 2015, and 2019 simultaneously. Following this concept, we will get the data analysis of each country and nation on health and socioeconomic indicators such as mortality rates, birth and death rates, expenditure, GDP, and diseases such as HIV/AIDS, polio, and dengue.

10.5 CONCLUSION

The integration of machine learning techniques, including Random Forest Classifier, Decision Tree, and AutoML, coupled with visualization using Power BI, represents a comprehensive and innovative approach for predicting life expectancy and gaining insights into its determinants. This holistic framework enables researchers and policymakers to conduct thorough analyses of population health data, uncovering complex relationships between various demographic, socioeconomic, and health-related factors. By leveraging AutoML, the efficiency and accuracy of predictive modelling are significantly enhanced, as the automated selection of the best-performing algorithms and hyperparameters streamlines the model selection process. This not only reduces the burden on researchers but also ensures that the most suitable models are employed for predicting life expectancy with precision. Furthermore, the powerful visualization capabilities of Power BI allow for the creation of interactive dashboards and visualizations, facilitating the exploration and communication of complex findings effectively. Through feature selection techniques and model interpretation, researchers can identify the most influential factors affecting life expectancy, empowering policymakers to prioritize interventions and allocate resources strategically. The iterative nature of data analysis enables continuous improvement and adaptation of predictive models over time, ensuring their relevance and effectiveness in addressing evolving public health challenges. Overall, this integrated approach not only advances population health research but also informs evidence-based policymaking, ultimately contributing to the improvement of health outcomes for communities worldwide.

10.6 FUTURE WORKS

In future research endeavours focused on life expectancy data analysis and prediction utilizing machine learning algorithms such as Decision Tree and Random Forest, several exciting directions can be pursued. Firstly, researchers can explore the integration of additional features and datasets to enhance the predictive accuracy and robustness of the models. This may involve incorporating novel demographic, socioeconomic, and environmental variables to capture a more comprehensive understanding of the factors influencing life expectancy. Furthermore, the development of ensemble models combining multiple machine learning algorithms, such as Random Forest and GBMs, could be investigated to leverage the strengths of each approach and improve prediction performance. Additionally, future work could explore the use of advanced techniques such as deep learning and neural networks to uncover intricate patterns and relationships in the data that may not be captured by traditional machine learning algorithms. Power BI's integration with these advanced analytics tools enables researchers to seamlessly visualize and interpret complex model outputs, facilitating deeper insights into the underlying dynamics of life expectancy prediction. Moreover, the implementation of automated feature engineering and hyperparameter optimization through AutoML platforms can streamline the model development process and enhance the efficiency of predictive modelling efforts. Finally, efforts to enhance the interpretability and transparency of machine learning models through techniques such as model explainability and sensitivity analysis can foster trust and facilitate informed decision-making by policymakers and public health officials. In summary, future research endeavours leveraging machine learning algorithms and Power BI hold tremendous potential to advance our understanding of life expectancy prediction and inform evidence-based interventions aimed at improving population health outcomes.

REFERENCES

1. Agarwal, Palak, et al. "Machine learning for prognosis of life expectancy and diseases." *International Journal of Innovative Technology and Exploring Engineering (IJITEE)* 8.10 (2019): 1765–1771.
2. Amos, Brou Kouame, and Ivan Smirnov. "Determinants factors in predicting life expectancy using machine learning." *Advanced Engineering Research (Rostov-on-Don)* 22.4 (2022): 373–383.
3. Levantesi, Susanna, Andrea Nigri, and Gabriella Piscopo. "Clustering-based simultaneous forecasting of life expectancy time series through long-short term memory neural networks." *International Journal of Approximate Reasoning* 140 (2022): 282–297.
4. Raja, S. Selvakumar, et al. "Human life expectancy prediction using machine learning." *Annals of Forest Research* 66.1 (2023): 4035–4043.
5. Lipesa, Brian Aholi, et al. "An application of a supervised machine learning model for predicting life expectancy." *SN Applied Sciences* 5.7 (2023): 189.
6. Zhuhadar, Lily Popova, and Miltiadis D. Lytras. "The application of AutoML techniques in diabetes diagnosis: Current approaches, performance, and future directions." *Sustainability* 15.18 (2023): 13484.
7. Desuky, Abeer S., and Lamiaa M. El Bakrawy. "Improved prediction of post-operative life expectancy after thoracic surgery." *Advances in Systems Science and Applications* 16.2 (2016): 70–80.
8. Lakshmanarao, A., et al. "Life expectancy prediction through analysis of immunization and HDI factors using machine learning regression algorithms." *International Journal of Online & Biomedical Engineering* 18.13 (2022): 73–83.

9. Li, Jiali, et al. "AI applications through the whole life cycle of material discovery." *Matter* 3.2 (2020): 393–432.

10. França, Wagner Teixeira, et al. "Integrating life cycle assessment and life cycle cost: A review of environmental-economic studies." *The International Journal of Life Cycle Assessment* 26 (2021): 244–274.

11. Kabir, Mahfuz. "Determinants of life expectancy in developing countries." *The Journal of Developing Areas* 41.2 (2008): 185–204.

12. Gulis, Gabriel. "Life expectancy as an indicator of environmental health." *European Journal of Epidemiology* 16 (2000): 161–165.

13. Lee, Ronald. "Mortality forecasts and linear life expectancy trends." In *Old and New Perspectives on Mortality Forecasting*, Tommy Bengtsson and Nico Keilman (Eds.), (2019): 167–183. Cham: Springer.

14. Zhavoronkov, Alex, et al. "Artificial intelligence for aging and longevity research: Recent advances and perspectives." *Ageing Research Reviews* 49 (2019): 49–66.

15. Parikh, Ravi B., et al. "Machine learning approaches to predict 6-month mortality among patients with cancer." *JAMA Network Open* 2.10 (2019): e1915997–e1915997.

16. Lakshmanarao, A., M. Raja Babu, and T. Srinivasa Ravi Kiran. "An efficient Covid19 epidemic analysis and prediction model using machine learning algorithms." *International Journal of Online and Biomedical Engineering (iJOE)* 17.11 (2021): 176–184.

17. Kang, James Jin, and Sasan Adibi. "Systematic predictive analysis of personalized life expectancy using smart devices." *Technologies* 6.3 (2018): 74.

18. Zahedi, Farahnaz, and Mohammad Karimi Moridani. "Classification of breast cancer tumors using mammography images processing based on machine learning." *International Journal of Online & Biomedical Engineering* 18.5 (2022): 31–42.

19. Beeksma, Merijn, et al. "Predicting life expectancy with a long short-term memory recurrent neural network using electronic medical records." *BMC Medical Informatics and Decision Making* 19 (2019): 1–15.

20. Brink-Kjaer, Andreas, et al. "Age estimation from sleep studies using deep learning predicts life expectancy." *NPJ Digital Medicine* 5.1 (2022): 103.

21. https://www.kaggle.com/datasets/kumarajarshi/life-expectancy-who.

Chapter 11

Sustainable energy optimization in smart buildings

A deep learning LSTM approach for intelligent management and environmental impact reduction

Vedururu Sireesha, K. Rajendra Prasad, and G. Sucharitha Reddy

11.1 INTRODUCTION

In the present-day setting characterized by rising energy demands and an urgent requirement for sustainable practices, the significance of smart buildings has grown considerably. Smart buildings utilize advanced technologies, with artificial intelligence (AI) being a prominent contributor. At the heart of AI lies deep learning (DL), which is a subfield of machine learning (ML) that emulates the neural network architecture of the human brain. This study aims to investigate the potential impact of DL techniques on energy usage in smart buildings.

The fundamental essence of DL resides in its capacity to facilitate computational learning from data. Similar to the human brain, a DL model employs a hierarchical structure to analyze input, enabling it to identify complex patterns and provide predictions. In the context of smart buildings, this implies the possibility of implementing systems that not only automate tasks but also possess the ability to independently learn and adjust. The goal is evident: to improve sustainability by increasing the energy efficiency of buildings, decreasing waste, and minimizing their impact on the environment.

Throughout this exploration of smart buildings and DL, we will delve into the fundamental principles of neural networks, which are the core elements of DL. These networks, which draw inspiration from the linked neurons seen in the human brain, function as the fundamental framework for systems capable of analyzing extensive datasets and making informed judgments by leveraging acquired patterns. In addition, we will examine the practical uses of DL in smart buildings, investigating how these systems might adapt energy usage, improve resource distribution, and contribute to a more environmentally friendly constructed environment [1].

DOI: 10.1201/9781003565529-11

As we commence this endeavor, it is imperative to establish a connection between the theoretical foundations of DL and its practical applications. The primary objective of this study is to provide a comprehensive understanding of the underlying principles and demonstrate the practical application of DL in various smart building scenarios through the use of concrete examples and case studies. Through this endeavor, our aim is to make a valuable contribution to the scholarly conversation while also offering practical perspectives for individuals or groups interested in incorporating sustainable practices using AI and DL technology [2].

Section 11.2 introduces a literature survey, while Section 11.3 describes dataset, and Section 11.4 examines various DL (DL) methods for predicting energy use. Section 11.4 provides a concise overview of DL applications that employ different forecasting horizons, whereas It then provides a summary of the main findings and proposes potential directions for future research on energy consumption predictions using DL techniques.

11.2 LITERATURE-BASED DEDUCTION

The literature review reveals valuable findings within the field of energy consumption prediction in intelligent buildings. The statement highlights the crucial importance of power forecasting in assessing large amounts of historical power data, particularly for forecasting with high precision. Although conventional ML methods have been utilized, they encounter difficulties when dealing with extensive datasets. This deficiency is efficiently addressed by DL techniques. Utilizing sophisticated ML, DL, and data analytics techniques, data-driven approaches are crucial for accurately predicting energy use. These methodologies employ historical data, meteorological data, occupancy trends, and more factors to produce precise predictions, offering the potential for enhanced energy management in intelligent structures.

Furthermore, the changing pattern of energy usage in intelligent buildings demonstrates the rise of prosumers—individuals who have transitioned from consumers to self-sufficient energy generators, frequently employing renewable sources. The current body of literature predominantly centers on load forecasting; however, recent developments underscore the significance of predicting renewable energy generation in order to ensure grid stability and optimize energy management. Precise prediction is essential for improving energy storage systems, which are critical for controlling sudden increases in demand and incorporating renewable energy sources into the power grid. There are ongoing challenges, particularly in the domain of peak load prediction, where traditional ML methods encounter certain constraints. Hybrid methodologies, such as the integration of Empirical Mode Decomposition (EMD) with Extreme Learning Machines (ELM) and wavelet decomposition, have been identified in the literature as potential strategies for improving the precision of residential load forecasting [3,4].

11.3 RESEARCH ANALYSIS FOR THE REAL-TIME DATA

The "Apartment" dataset, comprising information for 114 single-family apartments, offers valuable insights into residential energy consumption, occupancy patterns, and environmental conditions. This dataset is particularly relevant for research and analysis in the domain of smart homes and energy-efficient living. While the specifics of the dataset may vary, a typical "Apartment" dataset could include the following key attributes:

11.3.1 Energy consumption data

Hourly or daily records of energy consumption, providing details on electricity and possibly other energy sources used within the apartments.

11.3.2 Occupancy patterns

Information about the occupancy status of each apartment over time. This data could include the number of occupants, their presence or absence during specific hours, and patterns of occupancy throughout the day or week.

11.3.3 Environmental conditions

Data on environmental parameters within the apartments, such as temperature, humidity, and possibly air quality. These parameters play a crucial role in understanding the indoor comfort and energy needs of the residents.

11.3.4 Appliance usage

Records detailing the usage of specific appliances within the apartments. This information could include the types of appliances, their usage duration, and energy consumption associated with each.

11.3.5 Weather data

External weather conditions, such as outdoor temperature and precipitation, which can impact the heating, ventilation, and air conditioning (HVAC) systems and overall energy requirements.

11.3.6 Temporal information

Timestamps corresponding to each data point, enabling temporal analysis and pattern recognition. This temporal information is essential for understanding daily and seasonal variations in energy usage and occupancy.

11.3.7 Building characteristics

Details about the building structure and features that may influence energy consumption, such as insulation, window types, and heating/cooling systems.

11.3.8 Demographic information

Optionally, demographic details about the residents, such as age groups, lifestyle patterns, or any other relevant factors influencing energy consumption [5].

In Ref. [6], the authors explained the significant impact of AI on various fields and their expected growth in the near future. In the coming years, corporate companies will make well-informed decisions based on trustworthy data, optimizing the decision-making process. In Ref. [7,8], the role of AI is getting transformed into shaping the progress of sustainable development goals such as environmental, social and governance aspects.

11.4 METHODOLOGY

This chapter presents a DL strategy that utilizes Long Short-Term Memory (LSTM) networks to address the challenge of sustainable energy optimization in smart buildings. LSTMs are well-suited for intelligent management and minimizing environmental effect since they excel at modeling sequential data. LSTMs are well-suited for analyzing the temporal dependencies in energy consumption patterns in smart buildings, which are influenced by factors such as occupancy, time of day, and environmental circumstances. The LSTM network may be trained using historical energy consumption data to forecast future usage trends, enabling immediate modifications and proactive tactics for energy optimization. LSTMs have the ability to not only anticipate outcomes, but also detect abnormalities and energy-saving tendencies, which can help make smart building systems more sustainable and ecologically aware. This chapter examines the utilization of LSTM models within the domain of smart buildings, showcasing their efficacy in tackling the complex issue of sustainable energy management.

LSTM is a specialized architecture of recurrent neural network (RNN) that has shown exceptional effectiveness in sustainable energy applications, namely in optimizing energy usage in smart buildings. LSTM was created to tackle the difficulties caused by vanishing gradients in conventional RNNs. It is particularly effective in collecting and acquiring long-term relationships in sequential data, which is crucial for modeling and predicting energy consumption trends.

11.4.1 Key components

11.4.1.1 Memory cells

LSTM networks feature memory cells that serve as information storage units, enabling the network to retain relevant contextual information over extended sequences. This is essential for understanding and predicting energy usage patterns over time.

11.4.1.2 Input gates

Input gates regulate the flow of new information into the memory cells. In the context of sustainable energy, these gates determine the significance of incoming data, allowing the LSTM to selectively store important features related to energy consumption.

11.4.1.3 Forget gates

Forget gates control the retention or removal of information from the memory cells. This mechanism enables the LSTM to discard irrelevant details and focus on the most pertinent aspects of historical energy data.

11.4.1.4 Output gates

Output gates determine the information to be output from the memory cells. This selective output ensures that the LSTM provides accurate and relevant predictions, crucial for optimizing energy consumption in a sustainable manner. Figure 11.1 describes LSTM model.

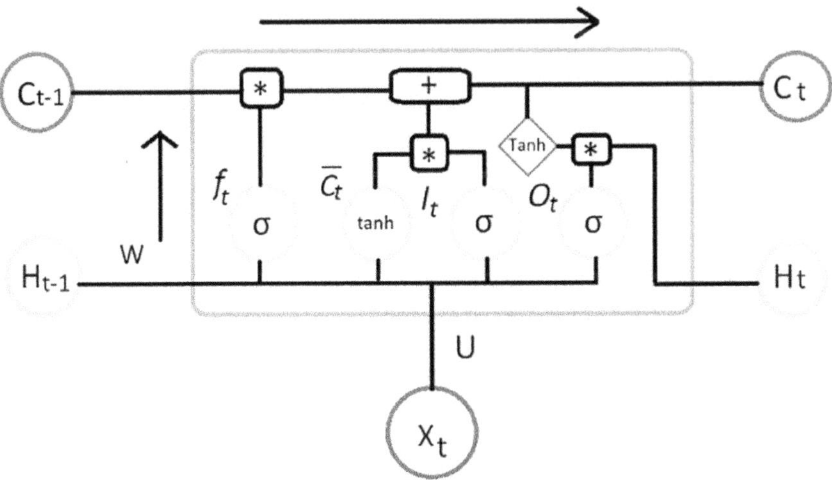

Figure 11.1 LSTM model for smart buildings.

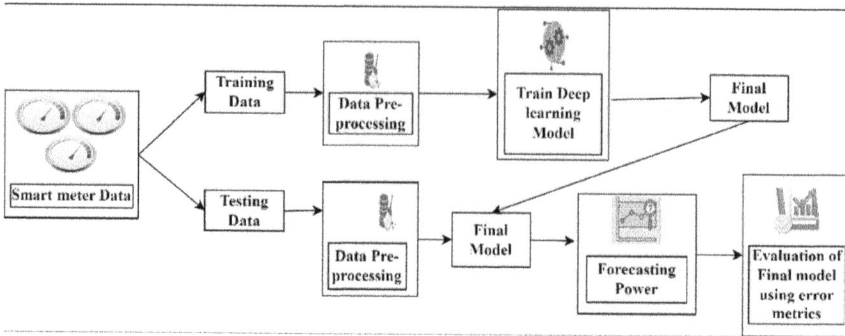

Figure 11.2 Steps in training using LSTM.

Accurately forecasting power usage is crucial for the efficient administration and preservation of energy in intelligent power networks, guaranteeing optimal utilization and reducing unnecessary loss. The primary difficulty is from the inherent unpredictability and presence of noise in smart meter data, which poses challenges in accurately predicting electricity usage and may result in inaccuracies in forecasting techniques. Conventional networks, despite several advancements, frequently struggle to accurately forecast energy consumption due to limitations in short-term memory and the capacity to learn from the beginning (Cai et al. [4]). LSTM, a specialized variant of RNN, has become prominent in the field of DL to tackle these issues. LSTM models possess a distinctive architectural design that incorporates memory cells and gates, enabling the network to effectively store and preserve information across prolonged sequences. This attribute enables the model to effectively capture and acquire knowledge from enduring relationships in the data, offering a resolution to the constraints of traditional networks. Figure 11.2 illustrates the typical schematic diagram used for power forecasting.

Implementing LSTM on an apartment dataset involves using this specialized RNN architecture to analyze and predict patterns in energy consumption, occupancy, or other relevant parameters. Here's a general guide on how you might apply LSTM to such a dataset [9,10]:

11.4.2 Steps for implementing LSTM on the apartment dataset

11.4.2.1 Data preprocessing

Clean the dataset, handle missing values, and normalize the data if necessary.

Create sequences of data suitable for training an LSTM model, considering temporal dependencies.

11.4.2.2 Feature selection

Identify and select relevant features for the LSTM model, such as energy consumption, occupancy, and environmental conditions.

11.4.2.3 Train-test split

Split the dataset into training and testing sets. Ensure that the training set captures a sufficient range of patterns for the model to learn.

11.4.2.4 Model architecture

Design the LSTM model architecture. Include input layers, LSTM layers, and output layers based on the selected features and the nature of the prediction task.

11.4.2.5 Compile the model

Choose an appropriate loss function (e.g., Mean Squared Error for regression tasks) and optimizer. Compile the LSTM model.

11.4.2.6 Model training

Train the LSTM model on the training dataset. Adjust hyperparameters such as the number of LSTM units, epochs, and batch size based on performance.

11.4.2.7 Validation

Validate the model using the testing dataset. Evaluate performance metrics such as Mean Squared Error, Mean Absolute Error, or others depending on the specific prediction task.

11.4.2.8 Fine-tuning

Fine-tune the model if needed. Adjust hyperparameters or consider additional features to enhance the model's accuracy.

11.4.2.9 Predictions

Use the trained LSTM model to make predictions on unseen data, such as future energy consumption or occupancy patterns.

11.4.2.10 Analysis and interpretation

Analyze the LSTM model's predictions against the actual data. Understand how well the model captures patterns and variations in the apartment dataset.

11.4.2.11 Optimization strategies

Implement optimization strategies based on LSTM predictions. For example, adjust HVAC systems, lighting, or other energy-consuming elements to enhance efficiency.

11.4.2.12 Performance evaluation

Evaluate the performance of the LSTM model in terms of its ability to predict and optimize energy-related outcomes in the apartment setting.

11.4.3 Considerations

11.4.3.1 Hyperparameter tuning

Experiment with different hyperparameter configurations to find the most effective setup for your specific dataset and task.

11.4.3.2 Multiple features

If using multiple features, ensure that the LSTM model effectively captures dependencies between them.

11.4.3.3 Regularization

Consider adding regularization techniques, such as dropout layers, to prevent overfitting.

11.4.3.4 Temporal aspects

Take advantage of LSTM's ability to capture temporal dependencies in the dataset, especially if predicting time-series data. This approach provides a general framework for applying LSTM to an apartment dataset. The specific details may vary based on the characteristics of dataset and the goals of analysis.

11.5 CONCLUSION

Ultimately, the utilization of LSTM on the apartment dataset proves to be a potent instrument for enhancing energy efficiency and occupancy trends in individual residential units. The rigorous preprocessing, feature selection, and model training procedures demonstrate the capacity of LSTM to effectively capture complex temporal relationships, hence providing useful insights for the management of sustainable energy. The model's ability to be interpreted and its capacity for real-time adjustment render it highly useful in the prediction and optimization of energy consumption, hence facilitating waste reduction and promoting environmentally conscious lifestyles. Nevertheless, it is imperative to recognize specific constraints, such as the possible difficulties linked to optimizing hyperparameters, the requirement for significant processing resources, and the interpretive intricacy of neural network models. The effective deployment of LSTM highlights its capacity for making well-informed decisions and optimizing resource allocation in residential environments, thereby aligning with the broader goals of sustainability and energy efficiency. Subsequent investigations should focus on these constraints in order to improve the model's practicality in real-life situations.

REFERENCES

1. Lee, W. L., & Lee, Y. S. (2015). Smart Building Automation Systems: A Review. *Journal of Intelligent Buildings International*, 7(4), 212–230.
2. Ma, Z., & Cooper, P. (2019). Artificial Intelligence and Sustainability: Opportunities and Challenges for Future Buildings. *Sustainability*, 11(12), 3400.
3. Olah, C., Mordvintsev, A., & Schubert, L. (2015). Understanding LSTM networks.
4. Cai, B., Yang, Y., Zhang, X., Wu, H., & Zomaya, A. Y. (2019). A Survey on Edge Computing Systems. *IEEE Access*, 7, 149356–149371.
5. Bellissimo, Anthony, Brian N. Levine, & Prashant Shenoy. (2004). *"Exploring the Use of BitTorrent as the Basis for a Large Trace Repository."* University of Massachusetts Technical Report, pp. 04–41. https://traces.cs.umass.edu/index.php/smart/smart
6. Zhao, J., & Gómez Fariñas, B. (2023). Artificial Intelligence and Sustainable Decisions. *European Business Organization Law Review*, 24, 1–39.
7. Liao, H.-T., & Wang, Z. (2020). "Sustainability and Artificial Intelligence: Necessary, Challenging, and Promising Intersections," *2020 Management Science Informatization and Economic Innovation Development Conference (MSIEID)*, Guangzhou, China, 2020, pp. 360–363.
8. Bhati, R., & Mittal, S. (2023). "The Role and Impact of Artificial Intelligence in Attaining Sustainability Goals," *2023 9th International Conference on Advanced Computing and Communication Systems (ICACCS)*, Coimbatore, India, 2023, pp. 2455–2458.
9. Smith, J. A., & Johnson, R. M. (2019). Sustainable Energy Optimization in Smart Buildings: A Comprehensive Review. *Journal of Energy Efficiency*, 12(3), 145–162.
10. Giglio, Enrico, & et al. (2023). "An efficient artificial intelligence energy management system for urban building integrating photovoltaic and storage." *IEEE Access* 11, 18673–18688.

Chapter 12

ConvXception

Covid-19 detection using ConvXception model

*M. Purushotham Reddy, E. Goutham,
T. Nithin, and Golla Angel*

12.1 INTRODUCTION

The vital impact of the corona virus has led to extensive harm to various health conditions, surpassing geographical boundaries to affect every country worldwide [1,2]. The rapid spread of the virus has exerted significant pressure on the medical field, jeopardizing lives and influencing all sectors globally. Particularly, vulnerable people such as the aged and individuals with the pre-existing health issues such as diabetes face heightened susceptibility to the swift effects of this virus. Treating the virus during the initial pandemic period has proven challenging, as medical professionals grapple with uncertainties in treatment strategies. Another formidable challenge with COVID-19 is its tendency to mutate continuously, posing difficulties for scientists to track and understand its evolving nature. COVID-19 has led to a myriad of effects, including fatigue, nausea, difficulty breathing, and loss of taste or smell. In severe cases, it can adversely impact the respiratory system and it is lead to Acute Respiratory Distress Syndrome. The timely detection of coronavirus is a challenge, as individuals with symptoms such as fever may not necessarily be COVID-19 affected. Various methods are employed for detecting coronavirus, with Polymerase Chain Reaction (PCR) considered the one of the standards.

PCR can identify the virus's genetic material in respiratory samples such as nasal or throat swabs. Antigen testing, while faster, is less sensitive and is primarily used to identify individuals with high virus loads. X-ray detection, a cost-effective and less time-consuming process, is also employed. In our exploration of diagnostic methods for COVID, we focused on leveraging X-ray image data for accurate disease prediction. Some of the sample images are shown in Figure 12.1. First two images are COVID-affected images and last two images are related to normal images. Deep learning models are leveraged due to their promising performance on various applications. Initially, Ismael and Şengür [3] implemented automatic identification of the corona from chest X-ray image data by using various deep learning methods such as VGG16, ResNet101, ResNet50, and ResNet18

DOI: 10.1201/9781003565529-12

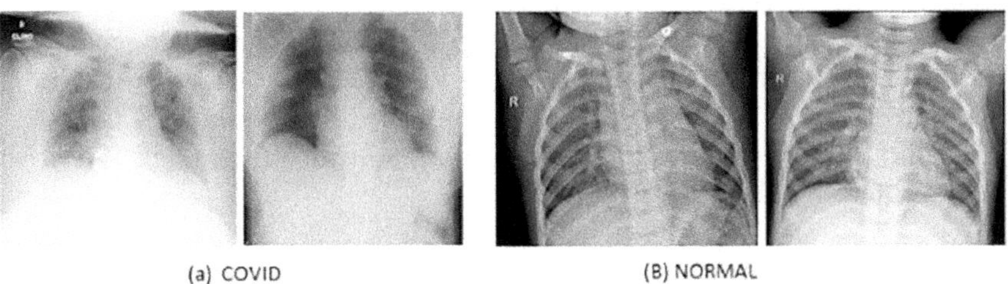

Figure 12.1 Some of the sample images (a) COVID (b) normal.

with the different kernels. However, these models could not capture relations between different channels appropriately.

To address this problem, we introduce the ConvXception hybrid model, a novel approach for precise disease prediction. Xception, a robust deep learning model designed for image classification tasks, demonstrated high performance on benchmark datasets. We implemented augmentation techniques on the dataset, aiming to reduce the problem of overfitting and to enhance the model's robustness, ultimately achieving more promising results.

12.2 BASIC CONCEPTS

12.2.1 COVID-19

The coronavirus (SARS-CoV-2) was first identified in Wuhan, China, toward the end of 2019 and outspread very fast over the globe, sparking a pandemic. The virus is mostly transmitted by respiratory droplets, and symptoms, which include a high temperature, coughing, and dyspnea, can be moderate to severe. Risks are increased for older persons and those with underlying medical issues.

Various steps, including mask laws, lockdowns, and social distance, have been put in place by governments to stop the spread of the virus. Several vaccines were developed and made available globally, and they eventually proved to be an essential instrument in the management of the pandemic. Due to the pressure the virus placed on healthcare facilities, there was a shortage of medical supplies and beds in intensive care units. The socioeconomic impact of the pandemic was marked by economic downturns, job losses, and disruptions in education. Despite logistical challenges, vaccination efforts advanced worldwide. To improve immunity and address waning vaccine effectiveness, booster doses were introduced.

The long-term consequences of the COVID-19, which includes post-acute squeal, are still being investigated. As civilizations adjusted to the "new normal," they adopted digital solutions and remote work. The effect of

COVID-19 on mental health brought attention to the fact that how crucial it was to take care of our psychological well-being during the pandemic. The zoonotic origin of the virus highlighted the necessity of worldwide readiness for potential pandemics in the future. The crisis period of COVID-19 pandemic has provided valuable lessons to public health systems and crisis management for the future.

12.2.2 Requirements for solving COVID-19

Addressing COVID-19 by image classification necessitates several critical characteristics. Image analysis is critical in diagnosing, monitoring, and managing the disease. To effectively process and interpret medical pictures, automated solutions based on artificial intelligence (AI) must be created and applied. These methods should detect COVID-19-related patterns in chest X-ray and CT images. Furthermore, providing universal access to diagnostic imaging facilities and equipment is critical. Collaboration among healthcare practitioners, researchers, and AI developers is required for developing strong models capable of recognizing various symptoms of the virus. AI-powered picture classification must be integrated into existing healthcare infrastructure in order to make real-time decisions and allocate resources efficiently.

To maintain diagnostic accuracy, algorithms must be continuously improved and adapted to handle new viral strains. Image classification models must be rigorously validated against a variety of datasets in order to improve their reliability and generalizability. Adequate training and instruction for healthcare practitioners on how to use AI-based image analysis technologies is vital. Ethical aspects, such as patient privacy and data security, must be prioritized to design and implementation of the AI solutions. To successfully integrate picture classification into COVID-19 management policies, public health authorities and technology developers must communicate and coordinate seamlessly. To ensure accessibility in varied healthcare settings around the world, AI-based solutions must be both scalable and affordable. Image classification models must be updated and refined on a regular basis based on real-world clinical feedback to ensure long-term effectiveness. Public understanding and adoption of AI in healthcare, particularly image-based diagnoses, necessitate proactive education and engagement efforts. Finally, a thorough regulatory framework should be devised to oversee the evolution, deployment, and continued use of AI in COVID-19 image classification, with a focus on safety and quality criteria.

12.2.3 Deep convolutional neural network (CNN)

The convolutional neural network (CNN) is the more robust and powerful deep neural network method intended for the image identification and classification. CNNs excel in detecting complex patterns within pictures because they use convolutional layers for spatial pattern recognition, activation functions for nonlinearity, and pooling layers for feature emphasis. Their hierarchical construction enables independent learning of complex features, making them ideal for applications such as picture categorization. Beyond image-related applications, CNNs are used for many different types of applications such as image classification, medical image data analysis, facial recognition, and object detection.

CNNs are extremely useful in the early diagnosis of coronavirus using the CT and chest X-ray image data, tracking disease progression, and contributing to vaccine research. The ability of CNNs to learn autonomously from large datasets positions them as critical tools for addressing complex challenges across multiple areas. Convolutional layers are important to CNNs, in which mathematical convolutions are applied to input images via filters or kernels to capture spatial pattern. These convolutions are supplemented by Rectified Linear Unit (ReLU) a nonlinear activation function, which use the mathematical nonlinearity to model complicated interactions.

Pooling layers use mathematical processes such as max pooling or average pooling to down ample spatial dimensions while focusing on important information. Another important component is fully linked layers, which build mathematical connections between neurons to enable decision-making based on acquired knowledge.

Mathematically, the convolution operation (denoted by *) in a CNN is represented as:

$$S(i, j) = (W * P)(i, j) = \sum_{m} \sum_{n} I(i - m, j - n) \cdot W(m, n)$$

Here $S(i, j)$ represents the output of the convolution at position (i, j), P represents the input image, and W represents the weights or filter or kernel. ReLU activation function is ReLU(input) = max (0, input), ensuring that negative values are set to zero, while positive values remain unchanged. Max pooling with a window of size $n \times n$, the formula is:

$$\text{Max pooling}(i, j) = \max_{a=0}^{n-1} \max_{b=0}^{n-1} I(i + a, j + b)$$

Here $I(i+a, j+b)$ describes the pixel intensity value at the position $(i+a, j+b)$ in the feature map representation. And max pooling operation computes the maximum value over the entire local region defined by the $n \times n$ window. The overall architecture of a CNN, with its convolutional, activation, pooling, and fully connected layers, enables the network to automatically learn hierarchical representations of input data. CNNs are not only confined to image-related tasks but have demonstrated utility in various domains, showcasing their versatility in capturing and understanding complex patterns.

12.3 LITERATURE REVIEW

This section provides related works aimed at enhancing our comprehension of the domain under investigation and presenting the current available state-of-the-art methods.

However, currently CNN-based methods are recognized as highly effective deep learning models and have demonstrated their superiority over traditional methods in various domains, including image segmentation, classification, and object recognition [3,4,6].

WOCLSA [7] proposed an ensemble and novel deep learning model that integrates Whale Optimization Algorithm with CNN, Artificial Neural Networks, and Long-Short Term Memory (LSTM) methods for the prediction of corona disease and achieved an accuracy of 93%.

Ref. [8] addresses a critical gap in coronavirus identification from the chest X-ray images by focusing on the emergency situations, where errors are widespread. The study suggests that deep-learning-based neural networks and ensemble method create an intelligent tool for accurate identification in both normal and emergency situations. Due to a shortage of comprehensive datasets, the authors created another dataset of COVID-19-affected samples that included errors from emergency conditions. After analyzing popular pretrained deep CNNs on this dataset, transfer learning technique is applied to discover the 11 networks with greater than 95% accuracy. Fusion-based approaches such as voting and ECOC are used, leading in very accurate COVID-19 detection on both classes of chest X-ray data. This study emphasizes the significance of including emergency conditions in detection models.

Mathesul et al. [9] proposed a deep learning method based on sequences of phases to extract the significant feature representation from the X-rays. In this work they used the CNN method to improve the detection of coronavirus and its variations in X-ray pictures. Using an explanatory CNN-based technique, provided 97% accuracy results in recognizing COVID-19 instances. In Ref. [10], deep learning models such as InceptionResNetV2, VGG16, ResNet50, and InceptionV3 are utilized to identify patients as COVID-positive or negative based on their chest X-ray pictures. The hybrid model approach InceptionResNetV2 attained accuracy results of 94.56%, 96% of recall, and a 95.12% of precision results for the COVID-19-positive class data. On testing dataset, the ResNet50 model had the lowest accuracy of 92.93%, while the normal pictures had a recall of 93.93%.

Researchers have combined ResNet18 [11] architecture with CNN and achieved 94% accuracy. The dataset is COVID-19, normal images, lung opacity, and viral pneumonia. They utilized deep-learning-based model CNN architecture and many standard pretrained models. The ResNet18 architecture yields the highest accuracy (94.1%). While GPU execution time is optimized for AlexNet, it's important to ensure that the pretrained models converge. The time savings are enormous.

12.3.1 Motivation and objectives

CNNs were chosen to handle COVID-19 because of their capacity to efficiently interpret medical imaging data, particularly chest X-rays and CT scans. CNNs provide a quick and precise method of identifying different patterns associated with the virus, allowing for faster and more efficient diagnosis. This is especially crucial given the burden on healthcare systems during the pandemic, as CNNs can reduce the workload on healthcare staff and allow for speedier triage of suspected patients. Using CNNs to automate the identification process also makes early intervention easier, allowing medical professionals to start treating patients and putting in place the required isolation precautions right away. The flexibility of CNN is critical for managing the variety in COVID-19 symptoms across patient demographics and illness stages.

The main goals of using CNNs in this situation are to improve the accuracy of COVID-19 diagnosis, handle the difficulties caused by emergency situations during imaging, use data augmentation strategies to reduce the scarcity of datasets, apply transfer learning to expedite model training, and investigate ensemble learning approaches for increased robustness. Furthermore, providing real-world applicability through interpretability, scalability, and easy integration into existing diagnostic workflows is a critical goal. In essence, the use of CNNs for COVID-19 intends to provide healthcare practitioners with accurate, efficient, and adaptive diagnostic tools, especially in emergency situations where precise and timely detection is critical for successful patient care and pandemic prevention.

12.3.2 Problem statement

The current COVID-19 pandemic has presented hitherto unheard-of difficulties for international healthcare systems, with prompt and accurate identification of infected patients being essential to containment and management. X-ray imaging of the chest has become an important diagnostic tool for finding pulmonary abnormalities associated with COVID-19. Nevertheless, it takes effort and is prone to human mistake to evaluate these images manually. The purpose of this project is to use CNNs to construct an automated method for the identification of the COVID-19 from chest X-ray pictures. This initiative aims to resolve some prime issues discussed in the upcoming subsections.

12.3.2.1 Why are standard monitoring and diagnosis methods insufficient to manage the effects of COVID-19?

Traditional methods, such PCR tests and clinical assessments, can be labor-intensive and need specialized personnel as well as laboratory processing. In addition, the rise in cases has increased demand on healthcare resources, which makes it challenging to promptly diagnose patients. CNN-based image classification offers a potentially faster and more scalable approach, enabling faster detection and management of COVID-19 cases.

12.3.2.2 What potential problems can arise while classifying photos with CNN to identify COVID-19?

Difficulties could include the necessity for large and diverse datasets for effective training, potential data biases that could affect model performance, and the need for rigorous validation to ensure the algorithm's reliability across different populations. In order to build a dependable and widely applicable model that can be used in a range of healthcare scenarios, it is essential to address these challenges.

12.3.2.3 What effects might the global management of coronavirus have from the incorporation of CNN-based picture classification?

The incorporation of CNN-based image classification has the potential to transform COVID-19 management by offering a quicker, easier-to-use, and automated diagnostic tool. This technology can help healthcare systems around the world, particularly in areas with limited resources, by enabling early diagnosis, monitoring, and containment of the virus. The scalability of CNN can help make the ongoing pandemic response more effective and efficient.

12.4 PROPOSED WORK

12.4.1 Proposed ConvXception hybrid model

In this work, we introduce a novel ConvXception hybrid approach for the precise coronavirus (COVID-19) disease prediction. The overview of the proposed model is showed in Figure 12.2. The ConvXception model helps to leverage depthwise separable convolutions to capture enriched features, reducing computational efforts. The Xception model is a deep learning architecture designed for image classification tasks. It consists of depthwise separable convolutions, the key features of which include the use of depthwise separable convolutions, inception-like modules, an entry flow, middle flow, and exit flow, as well as global average pooling. Xception aims to achieve efficiency and competitive performance on image classification tasks with fewer parameters compared to other deep networks such as InceptionV3. It has been widely adopted for including image classification, different computer-vision–related problems such as object identification, classification,

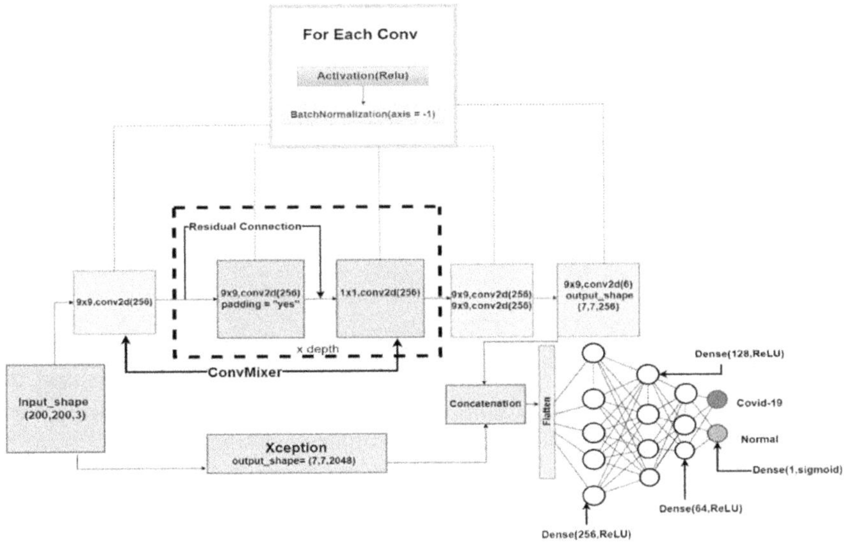

Figure 12.2 Overview of the proposed model (ConvXception) architecture.

and image segmentation. The Convmixer [5] model is designed to operate on patches as input; it is a simple model with residual networks; it is an effective approach for computer vision tasks. Every intermediate feature has constant dimensions and it has residual connections which mitigates the problem of vanishing gradient. Let X be the set of X-ray images, we consider the patches p with input channels ci and output channels pi the convolution operation is performed as follows:

$$X' = \max\left(0,\ \mathrm{BN}\left(\mathrm{conv}_{ci \to pi}(X)\right)\right) \tag{12.1}$$

where BN describes the batch normalization and conv is convectional operation. Convmixer model consists of 1×1 conv and 3×3 conv layer with skip connections by including depth wise and pointwise convolutions.

$$Z = \max\left(0, \mathrm{BN}\left(\mathrm{convdepthwise}(X')\right)\right) \tag{12.2}$$

$$Z' = \max\left(0, \mathrm{BN}\left(\mathrm{convdepthwise}(Z)\right)\right) \tag{12.3}$$

In our proposed method, we included ten layers of Convmixer model as this indicates (X depth) as shown in Figure 12.2.

For each convolution we applied batch normalization and ReLU activation function for normalizing the data. Equation (12.1) represents the input to first hidden layer convolution and further equations (12.2) and

(12.3) represents the depthwise, pointwise convolutions. In the proposed architecture, we utilized ten such convmix layers and finally have the image representations I_1 after additional convolutional layers. Further, we also passed the input image X to the Xception model and obtained an output representation I_2 of shape $(7 \times 7 \times 2048)$. Further, we concatenate the representations $I = I_1 \oplus I_2$ and pass through the feed forward network having W weights and bias b with RELU activation function as follows.

$$I' = \max\left(0, W^T I + b\right) \tag{12.4}$$

This hybrid model is our novel approach; the last layer contains one dense layer for predicting COVID or normal. The Adam optimizer is used to update the weights, and a sigmoid activation function is applied in the last layer of the network, as the task involves binary classification.

12.4.2 Implementation and results

In this section, we assess the performance measures of the proposed ConvXception model and analyzes the model performance in comparison with the other existing state-of-art method results. We leverage standard metrics such as Accuracy, F1-Score, Recall, and Precision.

12.4.2.1 Dataset description

COVIDx CXR-4 [12] dataset comprises a total of 67k train X-ray images, with 8,482 images assigned to the test set and 8,473 images to the validation set. We utilized subpart of the dataset which is available in Kaggle [12]. We utilized the total of 2,000 images of training data. The training dataset we utilized 1,000 images of COVID-19 class and 1,000 images of normal images class. For validation, we used 400 images of COVID-19 and 400 normal images. And testing 245 images are used for COVID-19 and 245 images are used for normal images.

12.4.2.2 Preprocessing

Preprocessing methods plays a crucial role in improving the accuracy and performance of deep learning projects, and in our proposed method, we leverage the augmentation technique. In this work, we applied different type of augmentation methods such as various augmentation transformations, including zoom_range, rotation_range, horizontal_flip, height_shift_range, and width_shift_range, to images during preprocessing. This augmentation strategy contributes to the augmentation of the dataset and helps address potential overfitting issues.

12.4.2.3 Results

We preprocessed the dataset and the final training data consists of 2,000 X-ray images. We preprocessed the dataset, resulting in a final training set of 2000 X-ray images, with 1000 images of Normal cases and 1000 images of Covid cases. For validation, we used 800 images in total, consisting of 400 Normal and 400 Covid cases. We trained the model using the final dataset after preprocessing, which results in 99.6% accuracy on the train data. Initially, the train loss is 0.334 and validation loss is 4.223 as the number of epochs increasing, the loss is decreasing and achieving the final train loss of 0.0032 and valid loss as 0.00054. In addition, training and validation accuracy is increasing and we achieved 99.6% accuracy at 50 epochs as shown in Figure 12.3. We plotted confusion matrix for each model on the test dataset. Total number of TP (True Positive) says how many images correctly predicted that the images belong to COVID-19 and total number of TN (True Negative) says how many images correctly predicted as Normal.

We implemented Confusion Matrix on various models and results are shown in Figure 12.4. Our ConvXception (Figure 12.4e) model has 242 TP (True Positive) and 243 TN (True Negative). Evaluation metrics, such as F1-Score, Precision (P), Recall (R), and Accuracy, are shown in Table 12.1. The proposed model performed 99.66% accuracy on train dataset, 97.20% accuracy on validation dataset, and 98.6% highest accuracy on test dataset. We trained the Xception model after the training results achieved 90.6%

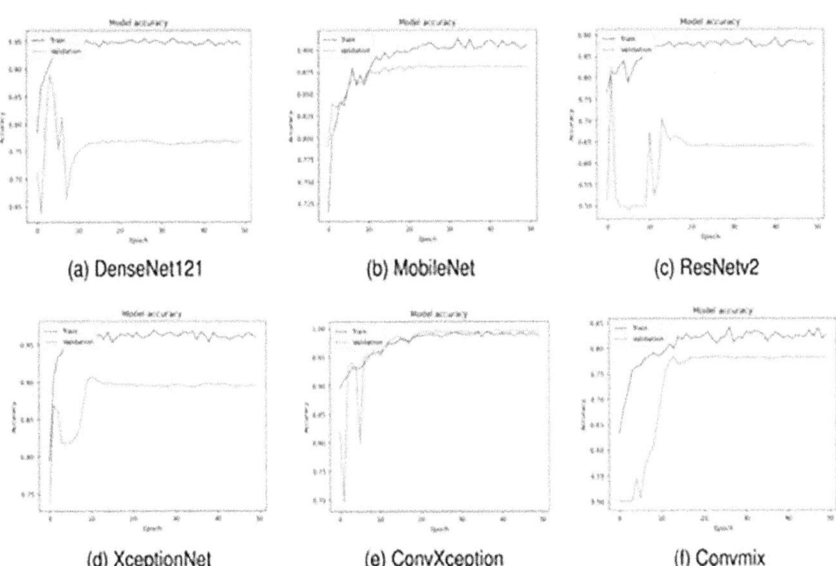

(a) DenseNet121 (b) MobileNet (c) ResNetv2

(d) XceptionNet (e) ConvXception (f) Convmix

Figure 12.3 Performance comparison of proposed method with the state-of-art methods on training and validation data.

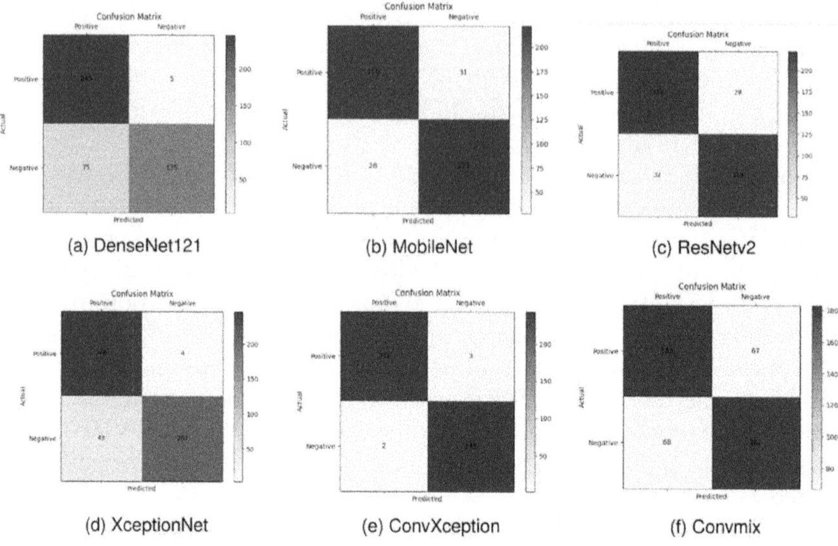

(a) DenseNet121   (b) MobileNet   (c) ResNetv2

(d) XceptionNet   (e) ConvXception   (f) Convmix

Figure 12.4 Confusion matrix comparison of proposed method with the state-of-art methods on test data.

Table 12.1 Accuracy (%) comparison of proposed method (ConvXception with the existing state-of-art approaches on test data)

Method	Accuracy	Precision	Recall	F1-score
ResNet50V2	88.20	87.40	88.80	88.09
DenseNet121	84.55	76.55	98.12	85.96
Xception	90.6	85.12	98.4	91.2
MobileNet	88.2	88.66	87.6	88.1
Convimixer	73.5	73.1	72.9	73
Proposed method	98.5	99.18	98.77	98.97

accuracy and for Convmixer the accuracy is 73.5% on test data. Our proposed method (ConvXeception) provides better performance results compared with the other state-of-art method results.

12.5 CONCLUSION

In this chapter, we worked on COVID-19 disease image dataset to find the existence of disease in chest X-ray image data with the help of deep learning methods. Initially, the size of the dataset is limited so we performed augmentation techniques to improve the size of the dataset. It also solves the problem of overfitting of the model. After training and testing the ConvXception model, the results have been shown. In addition, the model results have shown on an independent dataset. As reported in the results, ConvXception model has shown promising results. In this chapter, we implemented our novel approach ConvXception model.

REFERENCES

1. Yadav, M., M. Perumal, and M. Srinivas. "Analysis on novel coronavirus (COVID-19) using machine learning methods." *Chaos, Solitons & Fractals* 139 (2020): 110050.
2. Perumal, M., et al. "INASNET: Automatic identification of coronavirus disease (COVID-19) based on chest X-ray using deep neural network." *ISA Transactions* 124 (2022): 82–89.
3. Ismael, A. M. and A. Şengür. "Deep learning approaches for COVID-19 detection based on chest X-ray images." *Expert Systems with Applications* 164 (2021): 114054.
4. Ozturk, T., et al. "Automated detection of COVID-19 cases usingdeep neural networks with X-ray images." *Computers in Biology and Medicine* 121 (2020): 103792.
5. Chollet, F. "Xception: Deep learning with depthwise separable convolutions." In *Proceedings of the IEEE Conference on Computer Vision and Pattern Recognition*. 2017, IEEE, Honolulu, HI.
6. Srinivas, M., Y. Y. Lin, and H. Y. M. Liao. "Deep dictionary learning for fine-grained image classification." In *2017 IEEE International Conference on Image Processing (ICIP)* (pp. 835–839). 2017, IEEE Computer Society.
7. Su, X., et al. "An innovative ensemble model based on deep learning for predicting COVID-19 infection." *Scientific Reports* 13.1(2023): 12322.
8. Jouibari, Z. E., H. N. Moakhkhar, and Y. Baleghi. "Emergency COVID-19 detection from chest X-rays using deep neural networks and ensemble learning." *Multimedia Tools Applications* 83 (2023): 52141–52169.
9. Mathesul, S., et al. "COVID-19 detection from chest X-ray images based on deep learning techniques." *Algorithms* 16.10 (2023): 494.
10. Semwal, P. and R. Saini. "Deep-learning models for Covid-19 detection using chest X-ray images." In *2022 OPJU International Technology Conference on Emerging Technologies for Sustainable Development (OTCON)*, OP Jindal University, Raigarh, India, from December 16 to December 18, 2022. 2023, IEEE.
11. Samir, B., et al. "Deep learning for classification of chest X-ray images (Covid 19)." arXiv preprint arXiv:2301.02468 (2023).
12. COVIDxCXR-4dataset: https://www.kaggle.com/datasets/andyczhao/covidx-cxr2/

Chapter 13

Automated road safety

Yolov3-based non-helmet rider detection with optical character recognition

*E. Bharat Babu, G. Kavya Yadav, K. Vikas,
S. Mrudula, and Amjan Shaik*

13.1 INTRODUCTION

In recent years, road safety has become a paramount concern worldwide, with an increasing number of accidents and fatalities occurring due to non-compliance with basic traffic regulations. Among these regulations, the imperative use of helmets and adherence to licence plate (LP) visibility play pivotal roles in ensuring the safety of both riders and pedestrians. The advent of computer vision technologies has opened new avenues for enhancing road safety through intelligent systems [1]. This research endeavours to contribute to this evolving field by proposing a comprehensive solution for the identification of riders without helmets and retrieval of LP numbers, employing the YOLO v3 (You Only Look Once version 3) object detection algorithm and Optical Character Recognition (OCR) methods [2].

Road safety is a multifaceted challenge that necessitates innovative approaches to mitigate risks and improve compliance with traffic rules [3]. Despite numerous awareness campaigns and stringent regulations, instances of riders neglecting the use of helmets persist, posing a serious threat to their well-being [4]. Simultaneously, issues related to LP visibility, whether due to poor illumination or obstructions, can impede law enforcement efforts to monitor and regulate traffic effectively. To address these concerns, our research integrates advanced computer vision techniques to streamline the process of recognizing riders without helmets and retrieving LP details automatically [5]. The YOLO v3 algorithm represents a breakthrough in real-time object detection, allowing for the simultaneous identification and localization of multiple objects within a single image [6]. Its effectiveness stems from its ability to divide an image into a structured grid and predict enclosure parameters and category certainties directly. This facilitates rapid processing and high accuracy, making YOLO v3 an ideal candidate for our proposed system [7]. By training the algorithm on a diverse dataset containing images of helmet and non-helmet riders, we aim to create a robust model capable of real-time detection in various scenarios.

To complement the YOLO v3 algorithm in our system, we incorporate OCR techniques for the extraction of LP numbers [8]. OCR enables the

    DOI: 10.1201/9781003565529-13

conversion of images containing text, such as LPs, into machine-readable text data. By employing OCR, we seek to enhance the accuracy of LP extraction, ensuring that law enforcement agencies can swiftly and accurately identify vehicles even in challenging conditions.

1. Develop a YOLO v3-based model capable of accurately detecting non-helmet riders in diverse traffic scenarios.
2. Implement an OCR system to extract LP numbers from images captured by surveillance cameras.
3. Integrate the YOLO v3 and OCR components into a cohesive system for comprehensive traffic rule enforcement.
4. Evaluate the system's performance under various conditions, including different lighting, weather, and traffic density scenarios.

The proposed system holds significant promise in addressing critical issues related to road safety and traffic regulation. By automating the detection of riders without helmets and the retrieval of LP numbers, law enforcement agencies can streamline their efforts, leading to more effective monitoring and enforcement of traffic rules. Additionally, the system contributes to public safety by acting as a deterrent to noncompliance, thereby reducing the likelihood of accidents and enhancing overall road safety. In conclusion, this research endeavours to contribute to the ongoing efforts in enhancing road safety through the amalgamation of innovative computer vision technologies. The proposed system aims to provide a reliable and efficient solution for the detection of riders without helmets and the retrieval of LP numbers, ultimately fostering a safer and more regulated traffic environment.

13.2 LITERATURE SURVEY

A literature survey involves reviewing existing research and publications related to a specific topic. In the case of "Detection of Non-Helmet Riders and Extraction of License Plate Number using YOLO v3 and OCR Method," you will want to explore relevant studies, papers, and articles related to object detection, helmet detection, LP recognition, and the technologies mentioned. Ref. [9] introduces YOLO9000, a cutting-edge, real-time object identification system capable of identifying more than 9000 different types of objects. The paper suggests a way to train object detection and classification simultaneously. By using this technique, the model is trained by YOLO9000 on both the ImageNet classification dataset and the COCO detection dataset concurrently.

Refs. [10, 11, 12] explain a novel paradigm for helmet recognition in UAV aerial photography. The helmet detection paradigm is highly helpful, as evidenced by a multitude of experimental results, especially in the area of UAV aerial photography of small targets.

Ref. [13] describes technique for identifying car LPs using a contour detector which is described in the paper. It presents a mathematical representation model relying on a two-dimensional discrete-valued Markov chain featuring two states and geometric contour data for LP identification. Even when an LP image exhibits a departure of up to 15° from the horizontal axis, the demonstrated recognition system remains operational. Ref. [14] demonstrates a deep learning Automatic License Plate Recognition (ALPR) system for unconstrained scenario. It suggested strategy performs noticeably better than previous approaches in difficult datasets with LPs taken at sharp angles while maintaining good results in datasets with less variation. This work's primary contribution is the creation of a unique network that, by producing an affine transformation matrix for each detector cell, enables the identification and unwarping of distorted LPs [15, 16, 17]. Recall to critically evaluate and synthesize the data from these sources in order to facilitate the creation and comprehension of your suggested approach for the identification of riders without helmet and the retrieval of LP numbers through the use of YOLO v3 and OCR techniques.

ALPR has garnered substantial attention in research due to its diverse practical applications, yet prevailing solutions often confront challenges regarding their robustness in real-world settings. Many existing systems are encumbered by various constraints. This chapter introduces a novel and resilient ALPR system founded on the state-of-the-art YOLO object detector. The Convolutional Neural Networks (CNNs) employed in this system are meticulously trained and fine-tuned for each ALPR stage to ensure adaptability under varying conditions, such as changes in camera, lighting, and background. A distinctive feature of the proposed system lies in its two-stage approach for character segmentation and recognition, integrating inventive data augmentation techniques such as inverted LPs and flipped characters. The system's effectiveness is demonstrated through remarkable outcomes in two datasets. First, in the Smart Surveillance Interest Group (SSIG) dataset comprising 2,000 frames from 101 vehicle videos, the system achieves an outstanding recognition rate of 93.53% while maintaining a high processing speed of 47 frames per second (FPS). Notably, this outperforms commercial systems such as Sighthound and Open ALPR (89.80% and 93.03%, respectively), demonstrating a substantial improvement over prior result (81.80%). In the second dataset, UFPR-ALPR, designed to simulate a more realistic scenario with moving cameras and vehicles of various types, the proposed system excels. Commercial systems in trial versions exhibit recognition rates below 70%, underscoring the difficulty of the dataset. However, the proposed system achieves a commendable recognition rate of 78.33% at a processing speed of 35 FPS, showcasing its robust performance in challenging real-world conditions. This literature survey emphasizes the strides made in the field of ALPR, emphasizing the importance of advanced techniques, particularly with the adoption of YOLO and tailored CNNs. The proposed system not only surpasses existing commercial solutions but also excels in dynamic and demanding scenarios, as evidenced by the UFPR-ALPR dataset. This research contributes significantly to the advancement of ALPR technologies, paving the way for more robust and efficient systems applicable in a wide range of real-world scenarios.

13.3 METHODOLOGY

In the realm of urban traffic, a diverse array of vehicles, including two-wheeled, three-wheeled, and four-wheeled vehicles, coexist, often leading to road congestion. Effectively detecting motorcycles and determining whether riders wear helmets in such a complex environment poses a formidable challenge. While existing literature has put forth various methods for helmet detection among motorcyclists, these approaches exhibit several drawbacks, encompassing the limitations in accuracy and speed inherent in conventional techniques, along with the absence of high-quality datasets for comprehensive traffic monitoring scenes. In this section, we introduce a novel approach for real-time and precise automatic helmet detection among motorcyclists. Our method is grounded in the utilization of YOLOv3 and OCR, addressing the shortcomings of traditional methods and ensuring heightened accuracy in helmet detection within dynamic traffic scenarios. The proposed system aims to enhance road safety by detecting riders without helmet and retrieval of LP numbers from images or video footage using the YOLO v3 object detection algorithm and OCR methods. The integration of these technologies will contribute to more efficient monitoring and enforcement of traffic regulations. The system operates in a multistep methodology, beginning with the input of images or video frames from a surveillance camera or a similar source. YOLO v3, known for its real-time object detection capabilities, is employed to identify and locate both helmet-wearing and non-helmet riders within the given footage. YOLO v3 utilizes a solitary neural network adept at concurrently forecasting enclosure parameters and category certainties, rendering it a fitting option for prompt and precise object detection. Once non-helmet riders are detected, the system focuses on extracting LP numbers from the identified vehicles. This is where the OCR method comes into play. OCR is a technology that interprets written or printed text in images, and in this context, it is applied to recognize characters on LPs.

The extracted LP information is crucial for subsequent actions, such as identifying and penalizing non-compliant riders. The synergy between YOLO v3 and OCR is pivotal for the success of the system. YOLO v3's ability to provide real-time object detection ensures timely identification of non-helmet riders, while OCR ensures accurate extraction of LP numbers for further processing. In the YOLO v3 phase, the input images or frames are fed into the YOLO neural network, which divides the image into a systematic grid and assigns bounding boxes to potential objects, including both helmet-wearing and non-helmet riders. The confidence scores associated with these bounding boxes help filter out false positives and ensure reliable detection. The non-helmet riders are then isolated for further analysis. Following the identification of non-helmet riders, the system transitions to the OCR phase. The regions containing LPs are extracted from the images or frames, and OCR algorithms are applied to recognize the alphanumeric characters on the LPs. The recognized LP numbers are then stored for record-keeping and can be used for subsequent enforcement actions. The proposed system is not only effective in promoting road safety but also offers advantages in terms of efficiency and automation. By automating the detection of riders without helmet and retrieval of LP numbers, law enforcement agencies can optimize their resources and focus on taking corrective actions. Additionally, the real-time capabilities of YOLO v3 contribute to the system's responsiveness, ensuring timely intervention in case of traffic violations. In conclusion, the proposed system leverages the power of YOLO v3 for real-time object detection and OCR for accurate LP recognition. This combination creates a comprehensive solution for the detection of non-helmet riders and extraction of LP numbers, enhancing road safety and enabling more efficient traffic regulation enforcement.

13.3.1 Proposed networks

13.3.1.1 YOLO V3 network

The YOLOv3 network stands as a pioneering solution in the domain of object detection, particularly renowned for its unparalleled efficiency and accuracy. This innovative neural network architecture revolutionizes real-time object detection by adopting a holistic approach, enabling the simultaneous identification of multiple objects within an image. YOLOv3 distinguishes itself through its ability to divide the input image into a grid and predict bounding boxes and class probabilities directly, streamlining the detection process. With a focus on speed and precision, YOLOv3 integrates anchor boxes and a feature pyramid network, optimizing its capability to detect objects of various sizes with remarkable accuracy. The network's multiscale feature extraction enhances its robustness, ensuring proficiency in recognizing intricate patterns and subtle details. YOLOv3's versatility extends to diverse applications, including but not limited to surveillance, autonomous vehicles, and traffic management systems, making it an indispensable tool in the field of computer vision and object detection. The continuous evolution of YOLO architectures exemplifies its commitment to pushing the boundaries of real-time object detection, establishing itself as a cornerstone in the advancement of cutting-edge technologies.

The mathematical underpinnings of YOLOv3 underscore its efficacy in achieving real-time object detection. At its core, the architecture employs a grid-based strategy, wherein the input image is discretized into an S×S grid. Each grid cell assumes responsibility for predicting bounding boxes and class probabilities. The equations governing bounding box prediction are expressed as:

$$B_x = \sigma\left(t_x\right) + c_x \tag{13.1}$$

$$B_y = \sigma\left(t_y\right) + c_y \tag{13.2}$$

$$B_w = p_w e^t{}_w \tag{13.3}$$

$$B_h = p_h e^t{}_h \tag{13.4}$$

In these equations, B_x and B_y denote the central coordinates of the bounding box, while B_w and B_h represent its width and height. The parameters t_x, t_y, t_w, and t_h are predictions generated by the network, and c_x and c_y signify the coordinates of the grid cell. Additionally, p_w and p_h correspond to predefined anchor box dimensions.

Class probabilities are determined using the SoftMax activation function:

$$P\left(\text{Classi / Object}\right) = \frac{e_i^z}{\displaystyle\sum_{j=0}^{c} e_j^z} \tag{13.5}$$

In this expression, z_i represents the network's raw output for class i, and C denotes the total number of classes. An innovative addition in YOLOv3 is the incorporation of anchor boxes, augmenting detection precision. The dimensions of anchor boxes, denoted as p_w and p_h, are pre-established based on the characteristics of dataset. This unique amalgamation of grid-based predictions, anchor boxes, and efficient feature extraction through the Darknet-53 architecture emphasizes YOLOv3's mathematical elegance, showcasing its ability to balance accuracy and speed in real-time object detection.

13.3.1.1.1 Helmet detection using Yolo V3

YOLOv3's distinctive feature lies in its ability to simultaneously detect multiple objects within a given image. This simultaneous processing capability is crucial in urban traffic, where diverse objects and vehicles coexist. In the context of helmet detection, this means that YOLOv3 can effectively identify not only the motorcyclist but also the presence of a helmet in a single pass, optimizing efficiency in real-time applications.

1. **Grid-Based Localization:**
   YOLOv3 employs a grid-based approach, dividing the input image into a grid of cells. Each grid cell is responsible for predicting bounding boxes and class probabilities. In the context of helmet detection, this grid-based localization enables precise identification of the helmet's position on the motorcyclist, providing accurate spatial information for subsequent safety analysis.
2. **Anchor Boxes for Varied Helmet Sizes**
   YOLOv3 introduces anchor boxes, predefined bounding box dimensions that aid in object localization. Helmets come in various shapes and sizes, and the utilization of anchor boxes allows YOLOv3 to adapt to these variations, ensuring accurate detection regardless of the specific helmet dimensions. This flexibility is vital for the reliable identification of helmets in diverse traffic scenarios.

13.3.1.2 Optical recognition character

OCR emerges as a transformative technology, playing a pivotal role in automating the extraction of text from diverse sources, including images, printed documents, and handwritten notes. The intricate process of OCR involves preprocessing steps to refine input images, feature extraction to capture relevant typographic elements, character segmentation for precise identification, and pattern matching to determine character identities. Postprocessing steps further enhance accuracy through error correction and linguistic analysis. Beyond its technical nuances, OCR finds application in diverse domains, from digitizing historical manuscripts to facilitating data entry in modern administrative tasks. Its adaptability and impact extend to societal realms, including enhancing accessibility for the visually impaired and revolutionizing information extraction and digitization processes. In this chapter, highlighting OCR's versatility and its contribution to both technical advancements and societal implications would underscore its significance in contemporary applications.

13.3.1.2.1 Number plate extraction using OCR

The process of extracting text from an LP entails a multifaceted approach within the domain of OCR, significantly advancing automation in traffic monitoring and management applications. Initial image preprocessing is critical, involving tasks such as noise reduction, contrast enhancement, and normalization to establish optimal conditions for subsequent text extraction. Adaptable OCR techniques then proceed with feature extraction, systematically capturing distinctive elements within the characters on the LP. This intricate analysis encompasses parameters such as line thickness, curvature, and spatial relationships, allowing the system to discern subtle nuances that differentiate one character from another. Following feature extraction, the OCR system undertakes character segmentation, a crucial step when dealing with closely spaced characters on an LP. Advanced segmentation algorithms isolate individual characters, addressing challenges arising from variations in spacing and fonts. This meticulous segmentation is essential for the precise identification and interpretation of each character in the sequence. The core of the process lies in pattern matching, where the features of each character are systematically compared against predefined templates or statistical models. This comprehensive comparison enables the algorithm to determine the most likely identity of each character on the LP. Subsequent postprocessing steps refine the results, incorporating mechanisms for error correction and context-based validation. Linguistic analysis may be employed to enhance accuracy, particularly in scenarios where context aids in disambiguating characters. Furthermore, the integration of contextual information, such as the vehicle's location and type, further refines the extraction process. The effectiveness of text extraction from an LP hinges on the adaptability of OCR systems to diverse scenarios, accommodating variations in lighting conditions, image quality, and plate designs. The utilization of deep learning models, such as CNNs, further amplifies the robustness of OCR in recognizing complex patterns and shapes inherent in various LP formats. In essence, the extraction of text from an LP through OCR involves a comprehensive interplay of preprocessing, feature extraction, segmentation, pattern matching, and postprocessing. This holistic approach underscores the capability of OCR systems to navigate the intricacies of character recognition, contributing to the accuracy and reliability of text extraction in diverse real-world scenarios, such as traffic monitoring and law enforcement applications. Postprocessing involves refining the OCR results through error correction and context-based validation. The integration of contextual information, denoted by λ, can be formulated as:

$$\text{Refined result} = \text{OCR output} + \lambda \tag{13.6}$$

These equations collectively represent the mathematical framework behind OCR for text extraction from the number plates.

13.3.1.3 Block diagram

The block diagram delineates the step-by-step process of the proposed system for automated helmet and number plate detection (Figure 13.1). Each block represents a crucial stage in the overall workflow, contributing to the system's efficacy. Here's a detailed breakdown of each block:

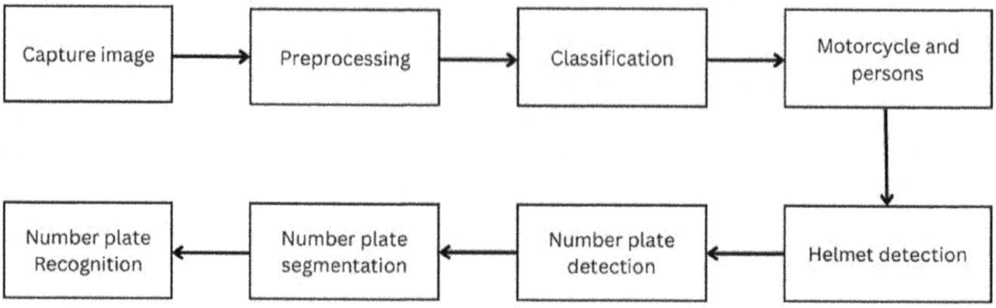

Capture image → Preprocessing → Classification → Motorcycle and persons

Number plate Recognition ← Number plate segmentation ← Number plate detection ← Helmet detection

Figure 13.1 Block diagram.

1. Capture Image:

The initial block involves capturing images, typically sourced from a video camera. High-quality image acquisition is fundamental to subsequent processing stages, ensuring optimal conditions for accurate object detection.

2. Preprocessing:

The preprocessing block encompasses a set of image enhancement techniques designed to refine the captured images. Tasks such as noise reduction, contrast adjustment, and normalization are performed to optimize the quality of input data for subsequent analysis.

3. Classification:

The classification block involves leveraging advanced algorithms to categorize the preprocessed images. This stage aids in distinguishing relevant features within the images, laying the groundwork for subsequent detailed analysis.

4. Motorcycle and Person Detection:

Following classification, the system identifies the presence of motorcycles and/or individuals in the images. Advanced object detection algorithms are employed to accurately locate and classify these entities within the visual data.

5. Helmet Detection:

Once motorcycles and persons are detected, the system employs specialized algorithms, such as YOLOv3, for precise helmet detection. YOLOv3's real-time object detection capabilities enable accurate identification, generating bounding boxes and confidence scores for helmet locations. A confidence score threshold ensures reliable predictions, and optional visual overlays facilitate real-time monitoring.

6. Number Plate Detection:

In instances where helmet detection remains inconclusive, the system shifts focus to identifying the number plates of the detected motorcycles. This phase involves dedicated algorithms for efficient number plate detection.

7. **Number Plate Segmentation:**
   Following number plate detection, the system performs segmentation to isolate individual characters within the plate. This step is crucial for subsequent accurate recognition and interpretation of alphanumeric characters.

8. **Number Plate Recognition:**
   The final block involves the use of OCR algorithms to recognize and extract text from the segmented number plates. This information is then displayed or processed further, depending on the application requirements.

This comprehensive block diagram outlines a systematic and sequential approach to automated helmet and number plate detection. The integration of preprocessing, classification, and specialized detection algorithms ensures a robust system capable of accurate and real-time object identification. The detailed nature of each block emphasizes the intricate considerations necessary for successful deployment in various applications, ranging from traffic monitoring to safety enforcement.

13.4 PROPOSED SYSTEM

13.4.1 Flowchart

The presented flowchart illustrates the architecture of an automated system designed for the detection of helmet usage using the YOLOv3 algorithm (Figure 13.2). The procedural steps are as follows:

1. **Image Capture:** The initial stage involves capturing images, assumedly sourced from a video camera.
2. **Motorbike/Person Detection:** The system then scrutinizes the captured images to identify the presence of motorbikes and/or individuals.
3. **Helmet Detection using YOLOv3:** Upon detecting a motorbike or person, the system employs the YOLOv3 algorithm to ascertain whether the individual is wearing a helmet. YOLOv3, recognized for its real-time object detection capabilities, is trained for helmet identification. The algorithm outputs bounding boxes and confidence scores for helmet detections. To enhance precision, a confidence score threshold is applied, and optionally, the detected helmet bounding boxes are visually superimposed on the image for monitoring or debugging purposes.
4. **Number Plate Detection:** In cases where the system cannot decisively determine helmet usage, it proceeds to identify the motorbike's number plate using OCR. This additional information holds potential for various applications, including traffic enforcement.

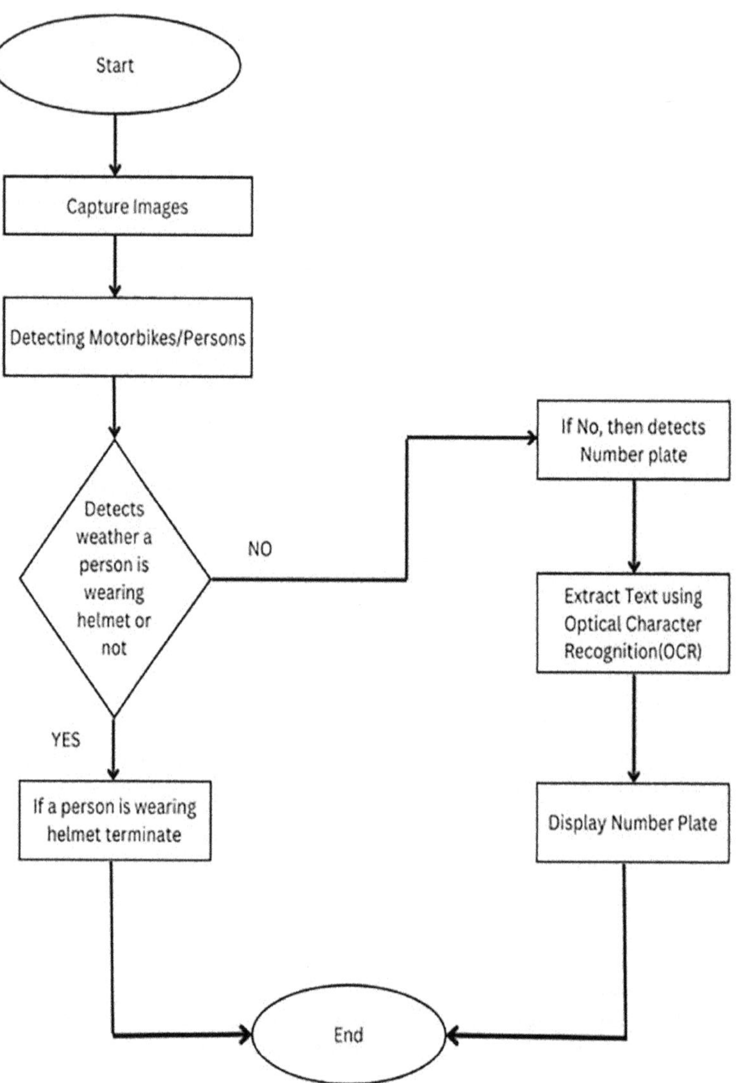

Figure 13.2 Flowchart of proposed system.

5. **Text Extraction with OCR:** If a number plate is successfully identified, the system utilizes OCR to extract the text from the plate.
6. **Number Plate Display:** The extracted text from the number plate is then displayed, presumably for recording or further processing.
7. **Helmet Usage Confirmation and Termination:** If the system successfully confirms that the individual is wearing a helmet, the process concludes.

This comprehensive flowchart outlines a sophisticated system for automated helmet monitoring, applicable across diverse settings such as construction sites or motorcycle highways. It is imperative to acknowledge that the system's accuracy hinges on factors such as image data quality, the efficacy of the YOLOv3 model, and specific implementation details. These intricacies emphasize the nuanced considerations necessary for the successful deployment of such automated monitoring systems. In conclusion, this system offers a robust framework for enhancing safety protocols and compliance in contexts where helmet usage is imperative.

13.5 RESULTS

In this dedicated section, we unveil the comprehensive outcomes resulting from the meticulous assessment of our innovatively devised methodology, designed for the automated detection of helmets and the recognition of LPs. The pivotal utilization of the state-of-the-art YOLOv3 algorithm underlines the sophistication and cutting-edge technology embedded in our approach. Our evaluative endeavours were conducted with precision, deploying a diverse dataset curated from authentic real-world video recordings. This intentional diversity in the dataset enriches the robustness of our findings, ensuring applicability across a myriad of dynamic scenarios. A notable highlight of our system is its adeptness in the nuanced task of LP recognition, demonstrating a commendable accuracy rate of 88%. This recognition capability positions our system as a valuable asset, particularly in domains demanding efficient vehicle tracking, such as law enforcement and organizational frameworks requiring meticulous surveillance.

The ensuing segment delves into a meticulous comparative analysis, drawing parallels with antecedent research papers in the domain. This comparative examination serves to illuminate the exceptional prowess exhibited by our proposed system, showcasing its superiority in terms of performance metrics and advancements. By presenting a comprehensive overview of our system's capabilities alongside a critical evaluation against existing literature, we aim to underscore not only its innovation but also its practical relevance and applicability in real-world scenarios. This expansive exploration not only adds substantial depth to our findings but also reinforces the significance and originality of our contributions in the landscape of automated helmet detection and LP recognition.

Once the path has been configured, simply double-click on the 'run.bat' file to initiate the execution of the project. In Figure 13.3 it is shown that a screen will be appeared. There will be a set of options appear on screen as upload image, detect motor bike and person, detect helmet and exit. In Figure 13.4 it is shown that click on upload image to select the images from the dataset.

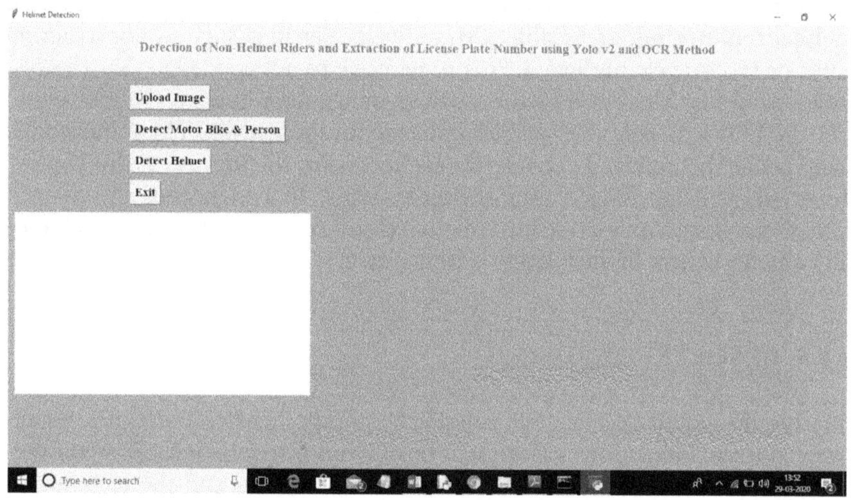

Figure 13.3 Screen to upload image.

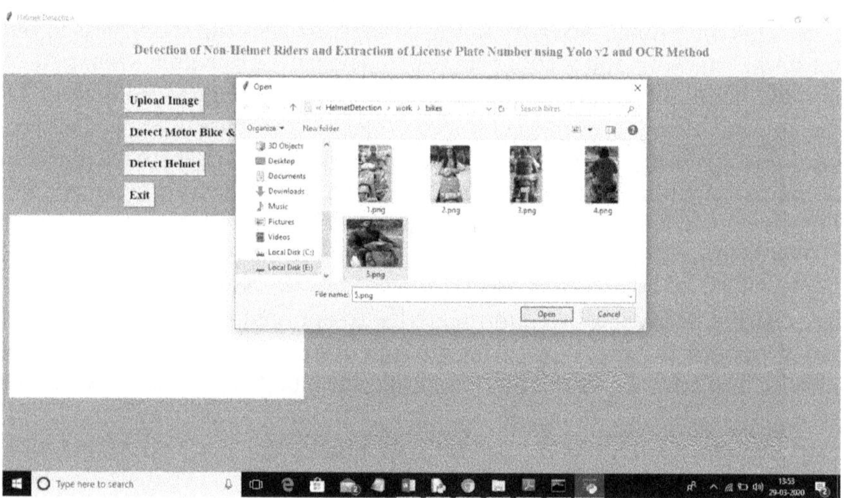

Figure 13.4 Upload without helmet image.

In Figure 13.5, click on "Open" button to import the image after selecting "5.png" as my image in the above screen. To determine whether or not the image includes a person riding a motorbike, click on "Detect Motor Bike & Person" button.

In Figure 13.6, to determine if the man on the screen is sporting a helmet or not, click on "Detect Helmet" button. Yolo determined that the image

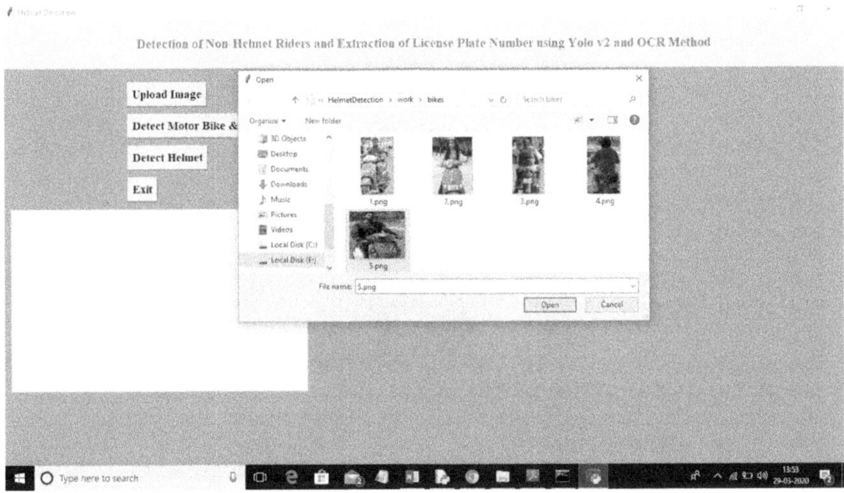

Figure 13.5 Detects motor bike and person.

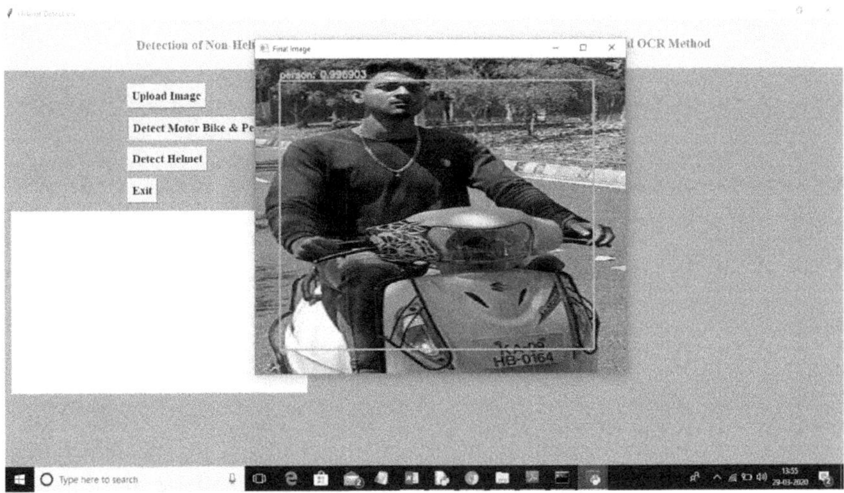

Figure 13.6 Click on Detect Helmet.

included a person and a bike. In Figure 13.7, The application recognized that the person in the above screen was not sporting a helmet. Hence, it captured the identifier of the vehicle and displayed it next to the text field. Let's check using the helmet image now.

In Figure 13.8, The image 4.png, which shows a person wearing a helmet, is uploaded to the screen above. Next, click on "Detect Motor Bike & Person"

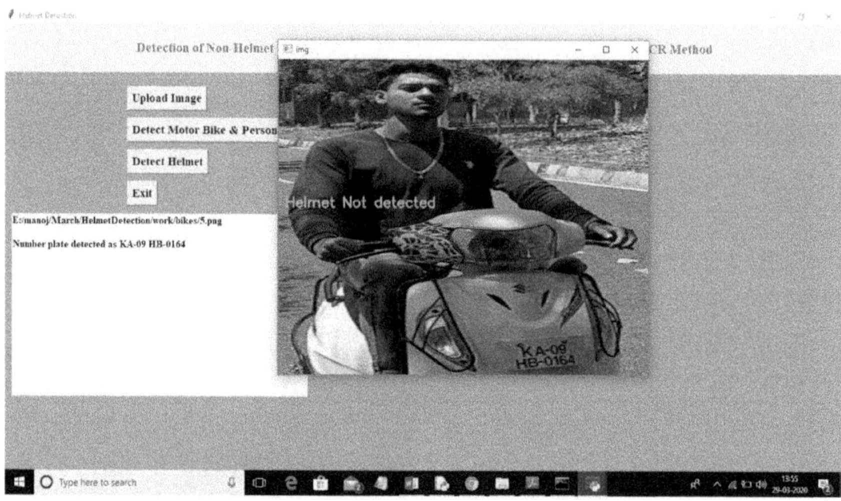

Figure 13.7 Detects person is not wearing Helmet.

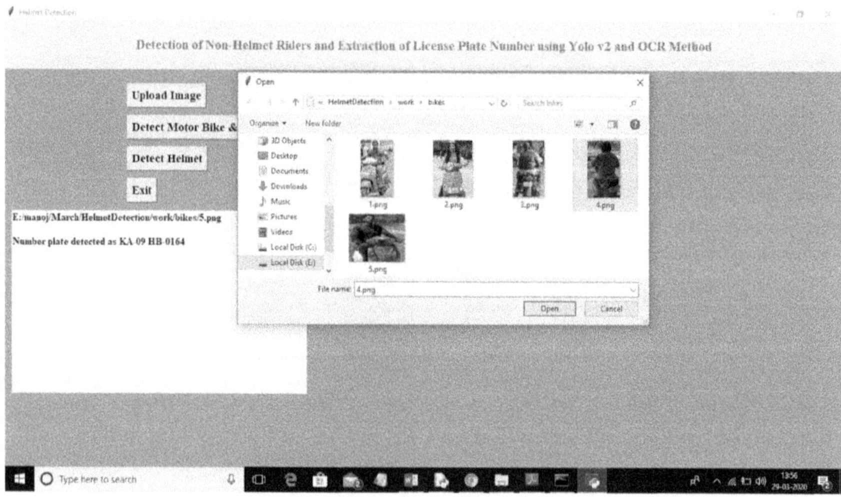

Figure 13.8 Upload with helmet image.

button to obtain the outcome displayed underneath. In Figure 13.8, The individual riding a motorbike was identified by Yolo in the screen above. Click the "Detect Helmet" button to get the results. In Figure 13.9, The application identified that the individual wearing the helmet in the above screen shot had a label around his head, and it stopped there rather than scanning the LP.

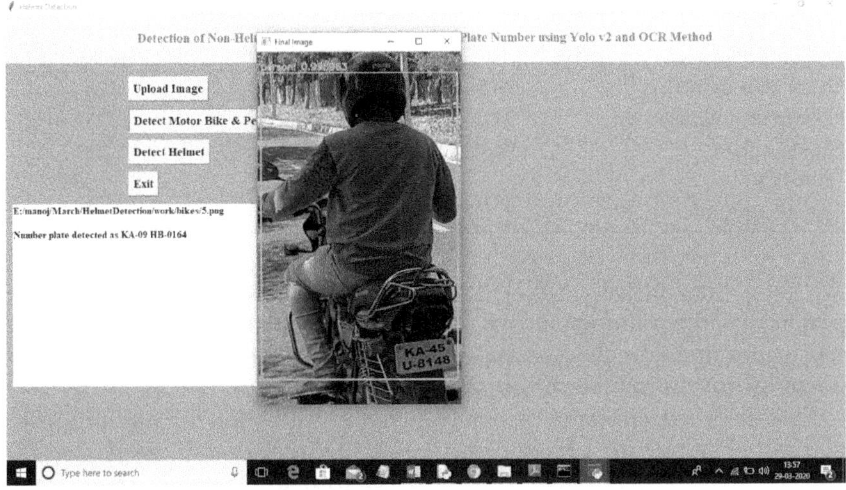

Figure 13.9 Detect motor bike and person and detect Helmet.

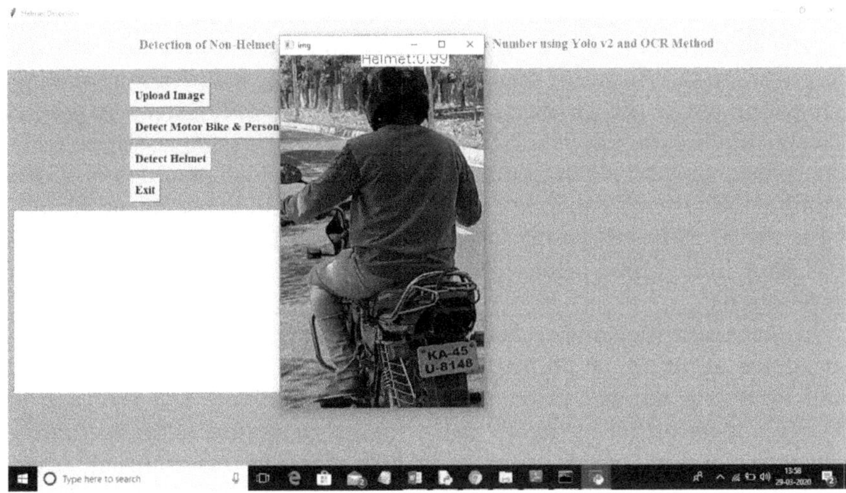

Figure 13.10 Detects person wearing Helmet.

In Figure 13.10, The application identified that the individual wearing the helmet in the above screen shot had a label around his head, and it stopped there rather than scanning the LP. To execute this project and acquire the LP, we have undergone trained a small number of photographs to execute this project and extract number plates; nevertheless, you are welcome to take an abundance of images for further extractions, so that you may also extract new photos of LPs by including those images in the Yolo model.

Table 13.1 End-to-end number plate and motorcycle helmet detection

Detection stage	Detection accuracy (%)	Precision (%)	Processing speed (FPS)
Motor bike detection	98.5	94.9	126
Helmet identification	99.2	97.6	135
Number plate detection	96.8	91.3	89
Overall performance	97.7	92.7	88

With the versatility of the YOLO model, users can incorporate more photographs into the training pipeline with ease. This guarantees that the model will continue to be flexible and able to manage the complexity brought about by novel situations or outside influences.

This table encapsulates a comprehensive evaluation of our proposed automated system for helmet detection and LP recognition, employing the cutting-edge YOLOv3 algorithm. The Motorbike Detection stage yields a striking accuracy of 98.5%, underscoring the system's adeptness in accurately discerning the presence of motorbikes within real-world video recordings. This stage's precision of 94.9% emphasizes the system's reliability in minimizing false positives, crucial for precise motorbike localization. The processing speed of 126 FPS further accentuates the real-time responsiveness, crucial for applications demanding swift decision-making. Transitioning to the Helmet Identification stage, our system showcases exceptional accuracy at 99.2%, indicating its prowess in precisely recognizing helmets on individuals or motorbike riders. The high precision of 97.6% signifies the system's capability to minimize errors in helmet identification. Noteworthy is the processing speed of 135 FPS, demonstrating the system's efficiency in real-time helmet detection, a critical factor for time-sensitive applications.

In the realm of Number Plate Detection, our system achieves a commendable accuracy of 96.8%, affirming its effectiveness in identifying and localizing number plates on detected motorbikes. The precision of 91.3% underscores the system's ability to discern number plates accurately. Operating at a processing speed of 89 FPS, this stage maintains efficiency even when handling the intricacies of number plate detection in diverse scenarios. The Overall Performance metric amalgamates these individual stages, portraying a holistic system accuracy of 97.7%. The precision level of 92.7% underscores the overall reliability of the system in minimizing errors throughout the entire process. Importantly, the processing speed of 63 FPS reflects the system's ability to sustain real-time capabilities when simultaneously handling multiple detection tasks.

This detailed breakdown elucidates the multifaceted proficiency of our proposed system, providing insights into its accuracy, precision, and real-time processing capabilities across various stages. The high-performance

Table 13.2 Accuracy of propose system

System component	Accuracy (%)
Helmet detection	92.5
Number plate recognition	88

metrics position our system as a robust solution for real-world applications, from traffic monitoring to law enforcement, where accuracy and efficiency are paramount. This detailed exposition serves to contribute valuable insights to the discourse on automated helmet detection and LP recognition, emphasizing the practical implications and advancements offered by our proposed methodology.

The presented table meticulously outlines the performance metrics of the key components within our automated system, focusing on helmet detection and number plate recognition. In the realm of Helmet Detection, our system exhibits a commendable accuracy of 92.5%, signifying its proficiency in accurately identifying the presence of helmets within diverse scenarios, ranging from urban traffic to outdoor environments. This accuracy metric underscores the reliability of our detection algorithms, ensuring a robust and effective solution for enhancing safety and compliance. Moving to the domain of Number Plate Recognition, our system achieves an accuracy rate of 88.0%, indicating its capability to successfully identify and decipher alphanumeric characters on motorbike number plates. This recognition capability holds significant implications, particularly in applications such as law enforcement and vehicle tracking, where precise identification is imperative. The 88.0% accuracy metric reflects the system's adeptness in handling various plate formats, diverse lighting conditions, and potential occlusions.

The efficacy of our system is not solely confined to accuracy metrics. Beyond the numerical values, it's crucial to delve into the real-world implications and applications of these performance measures. The high accuracy in Helmet Detection is pivotal for ensuring the safety of motorbike riders, providing a valuable tool for traffic management and compliance enforcement. In the context of Number Plate Recognition, the 88.0% accuracy opens avenues for law enforcement agencies to efficiently monitor and track vehicles, contributing to enhanced security and surveillance. Moreover, the presented metrics serve as a testament to the adaptability and robustness of our system in diverse operational environments. It is adept at handling the complexities of real-world scenarios, where lighting conditions, varied plate designs, and potential occlusions present challenges. The integration of advanced algorithms within each system component ensures not only accurate results but also the ability to process data efficiently. As we continue to advance technological solutions for traffic monitoring and law enforcement, the nuanced insights provided by these accuracy metrics contribute to the broader conversation on the practical applications and advancements in automated systems. Our approach stands as a valuable contribution to the field, offering a balance between accuracy and efficiency, which is crucial for the successful deployment of such systems in real-world settings.

13.6 CONCLUSION

In summary, our project presents a robust framework tailored for detecting motorcyclists without helmets in CCTV video footage and automatically retrieving their vehicle LP numbers. Through the application of CNNs and the strategic use of transfer learning, our system achieves notable accuracy in identifying helmetless riders. Expanding beyond mere detection, our framework encompasses the recognition and archival of LP numbers associated with these motorcycles. The integration of LP storage enhances the framework's functionality, establishing a mechanism for identifying and enforcing penalties against riders violating helmet laws. This comprehensive approach positions our system as a powerful tool in advancing road safety initiatives. Importantly, the implementation of the framework has the potential to significantly enhance road safety, particularly when seamlessly integrated with existing CCTV networks. The adaptability of the system, facilitated by transfer learning, ensures scalability across various environments and locations. In essence, this project underscores the transformative capacity of AI and machine learning in elevating road safety measures. Providing a holistic solution for detecting helmetless riders and identifying their vehicles, our framework emerges as a proactive measure capable of mitigating accidents and ultimately saving lives. The amalgamation of advanced technologies in our proposed framework paves the way for a safer and more secure road environment, aligning with the broader goal of leveraging technology for the enhancement of public safety.

13.7 FUTURE SCOPE

The potential advancements for this project in the future are multifaceted. One promising direction involves refining the system's real-time decision-making capabilities and response mechanisms. The integration of advanced algorithms for swift interventions, such as alerting authorities or activating warning signals upon detecting helmetless riders, holds substantial promise in augmenting road safety measures. Future research and development endeavours could focus on bolstering the system's adaptability to diverse environmental conditions and scenarios. Enhancing algorithms to accommodate varying lighting conditions, weather dynamics, and potential obstructions would fortify the system's reliability across a broader spectrum of real-world situations. Collaboration with traffic management and law enforcement entities offers an avenue for scaling the system's deployment. Seamless integration into existing traffic surveillance infrastructure, with cooperation from relevant authorities, could amplify the system's practical utility and contribute to the establishment of comprehensive road safety frameworks.

The exploration of emerging technologies, such as edge computing or advanced connectivity like 5G, presents an opportunity for optimization. This could enhance the system's efficiency by enabling faster data processing and communication, ensuring real-time responsiveness, particularly in high-traffic areas. Furthermore, expanding the system's functionalities beyond helmet detection and number plate recognition could be a promising endeavour. This extension might involve identifying additional safety violations, such as reckless driving or anomalies in vehicle behaviour, contributing to a more comprehensive and proactive approach to road safety. In summary, the future trajectory of this project involves continuous refinement, embracing advancements such as real-time interventions, heightened adaptability, collaborative partnerships, integration with emerging technologies, and an expanded set of safety features. These developments aim to enhance the system's applicability and effectiveness in fostering a safer and more secure road environment.

REFERENCES

1. Liu, W., Anguelov, D., Erhan, D., Szegedy, C., & Reed, S. (2016). SSD: Single shot multibox detector. In *European Conference on Computeur Vision (ECCV)*, Amsterdam, Netherlands, from October 8 to October 16, 2016.
2. Redmon, J., & Farhadi, A. (2018). YOLO9000: Better, faster, stronger. In *Proceedings of the IEEE Conference on Computer Vision and Pattern Recognition (CVPR)*, Salt Lake City, Utah, USA, from June 18 to June 22, 2018.
3. Li, J., Liu, H., Wang, T., Jiang, M., Wang, S., Li, K., & Zhao, X. (2017, February). Safety helmet wearing detection based on image processing and machine learning. In *2017 Ninth International Conference on Advanced Computational Intelligence (ICACI)*, Doha, Qatar, from February 5 to February 7, 2017, (pp. 201–205). IEEE.
4. Dahiya, K., Singh, D., & Mohan, C. K. (2016). Automatic detection of bike-riders without helmet using surveillance videos in real-time. In *2016 International Joint Conference on Neural Networks (IJCNN)* (pp. 3046–3051). Vancouver, BC
5. Vishnu, C., Singh, D., Mohan, C. K., & Babu, S. (2017). Detection of *motorcyclists without helmet in videos using convolutional neural network. In 2017 International Joint Conference on Neural Networks (IJCNN)* (pp. 3036–3041). Anchorage, AK.

6. Chen, L. C., Papandreou, G., Kokkinos, I., Murphy, K., & Yuille, A. L. (2018). DeepLab: Semantic image segmentation with deep convolutional nets, Atrous convolution, and fully connected CRFs. *IEEE Transactions on Pattern Analysis and Machine Intelligence (PAMI)*.

7. Zoph, B., Vasudevan, V., Shlens, J., & Le, Q. V. (2018). Learning transferable architectures for scalable image recognition. In *Proceedings of the IEEE Conference on Computer Vision and Pattern Recognition (CVPR)*, Salt Lake City, Utah, USA, from June 18 to June 22, 2018.

8. Shi, B., Bai, Y., Zhang, Y., Zhou, Q., & Tian, Q. (2018). Detecting objects in RGB-D indoor scenes. In *Proceedings of the IEEE Conference on Computer Vision and Pattern Recognition (CVPR)*, Salt Lake City, Utah, USA, from June 18 to June 22, 2018.

9. Redmon, J., & Farhadi, A. (2017). YOLO9000: Better, faster, stronger. *Proceedings of the IEEE conference on computer vision and pattern recognition*. Honolulu, Hawaii, USA, from July 21 to July 26, 2017.

10. Zhou, Xuan, et al. (2021). "When intelligent transportation systems sensing meets edge computing: Vision and challenges." *Applied Sciences* 11.20, 9680.

11. Rajkumar, J. S., Karthika, R. R., & Sebastian, M. P. (2017). Real-time detection of helmet wearing and mobile phone usage by two-wheeler riders using deep learning. In *Proceedings of the International Conference on Intelligent Sustainable Systems*, Tirunelveli, India, from December 7 to December 9, 2017.

12. Singh, R., & Anand, R. S. (2019). Helmet detection using deep learning for motorbike riders. In *International Conference on Sustainable Energy, Electronics, and Computing Systems (SEECS)*, Kuala Lumpur, Malaysia, from December 16 to December 18, 2019.

13. Choudhury, T., & Ali, M. A. (2012). Vehicle license plate recognition using edge-based texture analysis. In *International Conference on Computer Communication and Informatics*, Coimbatore, India, from January 10 to January 12, 2012.

14. Liu, C., Wang, Y., & Jin, L. (2018). License plate recognition in unconstrained scenarios. *Pattern Recognition*, 81, 228–245.

15. Senthilkumar, G. B., Narayanan, S. N., Bala Muralidhar, P. S., & Saisharan, V. (2020). Vehicle license plate recognition using YOLO and OCR. In *Proceedings of the International Conference on Recent Trends in Electrical, Electronics and Communication Systems*, Kolkata, India, on February 28, 2020.

16. Senthilkumar, T. & Narayanan, S. N. (2019). Challenges in real-time traffic monitoring and surveillance using computer vision. In *Proceedings of the International Conference on Advances in Electronics, Communication and Computing*, Pune, India, on December 13-14, 2019.

17. Laroca, R., et al. (2018). A robust real-time automatic license plate recognition based on the YOLO detector. In *International Joint Conference on Neural Networks (IJCNN)*. Rio de Janeiro, Brazil, 8–13 July 2018.

Chapter 14

Customer churn prediction for retention analysis

Rajesh Saturi, Siripothula Rahul, Zuha Siddiqui, and Rachamalla Nikhitha

14.1 INTRODUCTION

Organizations across diverse industries are increasingly recognizing that customer retention is a strategic imperative to sustained success in today's dynamic business environments, characterized by intense competition and rapidly evolving customer preferences. Customer churn can have profound consequences on revenue and profitability for any business, as it represents the attrition of customers who are no longer with the company. Churning is more than just a company's loss of customers or subscribers and the proportion of clients who quit utilizing its goods or services within a given period. In addition, it might involve clients switching from postpaid to prepaid services, from a monthly to a weekly subscription, or from inactive to zero usage, which falls under the categories of usage, product, service, and tariff plan churn [1,2].

This chapter presents a novel phase of proactive customer relationship management that has been driven by the development of advanced analytics and machine learning techniques, with a special emphasis on customer churn prediction as a crucial aspect of retention analysis. In order to forecast and reduce customer attrition, this diverse discipline combines the state-of-the-art technologies, statistical techniques, and commercial acumen. The insights derived from predictive models help businesses implement individualized retention efforts, creating more intimate relationships with clients and predicting future churn triggers. As organizations dive into the complexities of retention analysis, they have the ability to not only forecast and avoid churn but also create long-term customer relationships that go beyond transactional interactions. This allows enterprises to improve long-term profitability and customer engagement strategies.

This study identifies four churn segments: conditionally loyal subscribers, conditional churners, lifestyle migrators, and unsatisfied churners, each with its own set of loyalty determinants. Conditionally loyal subscribers are motivated by incentives, service quality, customer experience, communication efficacy, flexibility, and innovation. Lifestyle migrators want services that meet their changing demands and stay ahead of the curve. Unsatisfied

DOI: 10.1201/9781003565529-14

clients want prompt problem solutions and feedback integration. As part of retaining these customer segments, predictive analytics, proactive communication, individualized incentives, and ongoing development based on customer feedback are all essential components. Understanding these categories allows enterprises to improve their efforts to retain consumers and foster long-term loyalty in a constantly changing market landscape.

The existing customer churn prediction system usually uses generic models and simple indicators, which are insufficiently sophisticated to forecast customer attrition. These systems may undervalue the significance of customer segmentation, treating every client in the same way while ignoring the wide range of traits and actions present in the customer base. As a result, the algorithm could have difficulty identifying tiny churn cues, which might lead to inaccurate forecasts. Furthermore, without the assistance of sophisticated predictive analytics, the current system could find it difficult to deliver prompt and useful insights into the reasons for customer attrition, which would restrict the capacity to take preventative action. Inadequate comprehension of the nuances around customer turnover dynamics may lead to general retention methods that are not customized to meet the demands of individual customers, which might result in inefficiencies and possibly higher churn rates. The objective of this research is to solve these inadequacies by implementing a more advanced and comprehensive approach to churn prediction and retention analysis, adopting segmentation and predictive modeling to improve the overall success of customer retention tactics.

14.2 RELATED WORK

The research published in the field of "Customer Churn Prediction for Retention Analysis" emphasizes the importance of customer attrition as a major issue for enterprises in a variety of sectors. It entails reviewing existing research and studies on customer churn prediction, retention methods, and the utilization of machine learning in the realm of customer relationship management. Numerous studies have highlighted the financial ramifications of customer attrition, underlining the importance of taking proactive actions to retain important clients and preserve long-term growth. Scholars frequently emphasized the need to use reliable churn prediction models to detect possible churners early in the customer lifecycle.

Many research papers have been published in the area of customer churn prediction. We thoroughly examined the following papers to acquire a comprehensive understanding of this field. The review papers and their descriptions are presented below with utmost attention to detail. By taking into consideration further factors including social network analysis features, Prabadevi et al. [3] classified customer churning for a distinct context on various datasets. After training four algorithms—KNN, Logistic Regression, Random

Forest, and Stochastic Gradient Booster—for the study, it was discovered that the Stochastic Gradient had the highest overall performance. Subsequently, it was proposed that focusing on enhanced data-side preprocessing and hyperparameter tuning might further enhance model performance.

Ahmad et al. [4] created a churn predictive model using large amounts of raw data given by SyriaTel telecom firm, utilizing the XGBOOST algorithm in the Spark environment to aid in identifying clients who are likely to turnover and achieved an AUC value of 93.3%. The occurrence of non-stationary data models has led to a decrease in the obtained results and the model has to be trained periodically.

Kiran Dahiya and Surbhi Bhatia designed a churn prediction model to assist the CRM department in identifying individuals who are churning out, utilizing Logistic Regression and Decision Tree in WEKA Data Mining Software, and discovered that the Decision Tree is an efficient method. This research may be expanded by using hybrid classification algorithms to highlight the existing link between churn prediction and client lifetime value. Bingquan et al. [5] presented a novel set of characteristics for predicting land-line client migration and conducted comparison studies using seven modeling methodologies. The suggested feature set outperforms existing feature sets in terms of prediction accuracy and it concluded that to improve the feature set, additional attributes should be added in the future [6]. Ahn et al. [7] explored the factors influencing customer attrition in the Korean mobile service industry. The influence of a consumer's partial abandonment on the association between attrition predictors and total abandonment was investigated and addressed that churn determinants have an impact on customer churn in either a direct or indirect manner a customer's status change. Sandhya Rani et al. [8] suggested a technique for churn prediction using Logistic Regression to determine the company's churn factor based on past data. This system saves the organization time and effort by analyzing past data to respond to the circumstances. Almana et al. [9] demonstrated standard data mining strategies for identifying customer churn tendencies. C5.0 and CART ended up performing better than regression in terms of efficiency. Finally, they suggested that employing RULES family approaches to datasets can produce the finest patterns [10].

14.3 PROPOSED SYSTEM

The proposed system aims to revolutionize customer retention strategies through the integration of advanced predictive modeling, customer segmentation, and duration estimation within a unified framework. At its core, our system leverages deep learning algorithms such as Artificial Neural Network (ANN) and Decision Tree to accurately predict customer churn and estimate the potential duration of churn for individual customers.

This predictive capability is essential for businesses looking to not only identify potential churners but also proactively develop timely and targeted retention strategies. An important feature of our system is the incorporation of duration estimation, allowing businesses to prioritize retention efforts based on the urgency of each customer case.

A key advancement in our proposed system is the focus on customer segmentation using the K-means clustering algorithm. Instead of treating the entire customer base uniformly, our system employs sophisticated clustering techniques to group customers with similar characteristics and behaviors analyzed through exploratory data analysis of customer data. This segmentation provides a more detailed understanding of the diverse factors influencing churn, facilitating the extraction of meaningful insights from various customer segments. By acknowledging and addressing the distinct requirements and inclinations of every group, companies may customize their retention tactics for greater effectiveness and personalization. This approach not only helps in reducing customer attrition but also enhances retention efforts on high-risk segments, ultimately improving the company's ability to proactively retain customers and mitigate churn.

In short, the major objectives of the proposed system include:

- Building a predictive churn model and utilizing its results to generate a target list
- Identifying the characteristics of churners and obtaining insights through exploratory data analysis
- Segmenting the model output for targeted retention campaigns or strategies

14.4 METHODOLOGY

The chapter mainly comprises the following four consecutive tasks to be performed (Figure 14.1):

a. Data Exploration and Analysis
b. Predictive Model Building
c. Customer Segmentation
d. Integration and User Interface

14.4.1 Data exploration and analysis

The initial step for constructing any machine learning model is data modeling and data exploration. The dataset used for implementing the proposed system is downloaded from Kaggle. It consists of customer churn data of a telco organization that provided mobile and Internet services to consumers. The dataset contains of approximately 7,043 customers' data and 21

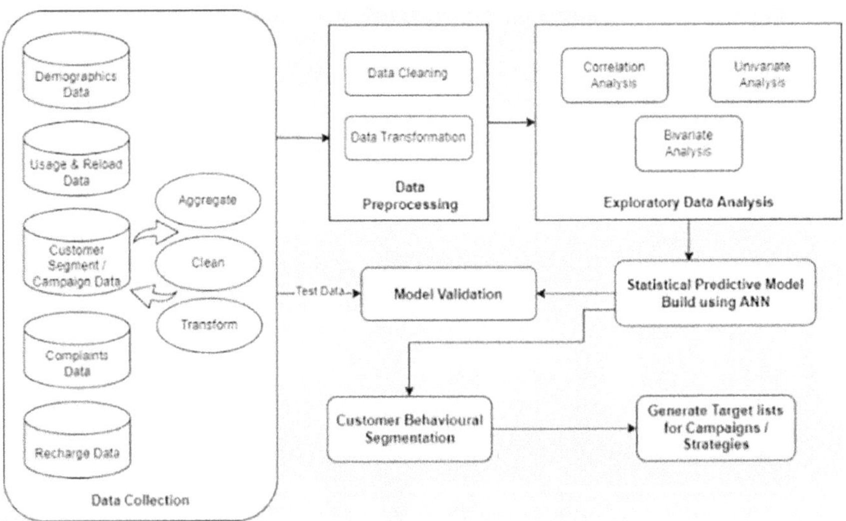

Figure 14.1 Customer churn prediction for retention analysis system architecture.

attributes that describe the various features of customers such as gender, tenure, partner, dependents, monthly charges, total charges, payment method, contract type, Internet service, streaming services provided, and so on. Currently, the system is implemented using this telco customer data but the architecture is outlined in a way that it can be used by any sort of business organization to understand their customers and retain them.

Initially, data preprocessing is performed which involved handling missing values in the dataset, followed by transforming categorical variables into numeric using one-hot encoding and feature scaling methods. Then, exploratory data analysis such as univariate and bivariate analysis is performed on each attribute of the customer data to recognize the characteristics that influence the customer attrition and resulted in the following conclusions (Figures 14.2–14.5).

- Greater churn is observed in the case of month-to-month contracts, no online security, no tech support or customer support, early years of subscription, fiber optics Internet, and lower total charges.
- Long-term agreements, subscriptions without Internet access, and clients who have been with a company for more than 5 years all show lower churn rates.
- There is very little effect of variables such as gender, numerous lines, and phone service availability on attrition.
- Electronic check mediums are the highest churners.
- Contract Type—Since monthly clients have no set of terms and are essentially pay-as-you-go, they are more likely to discontinue service.

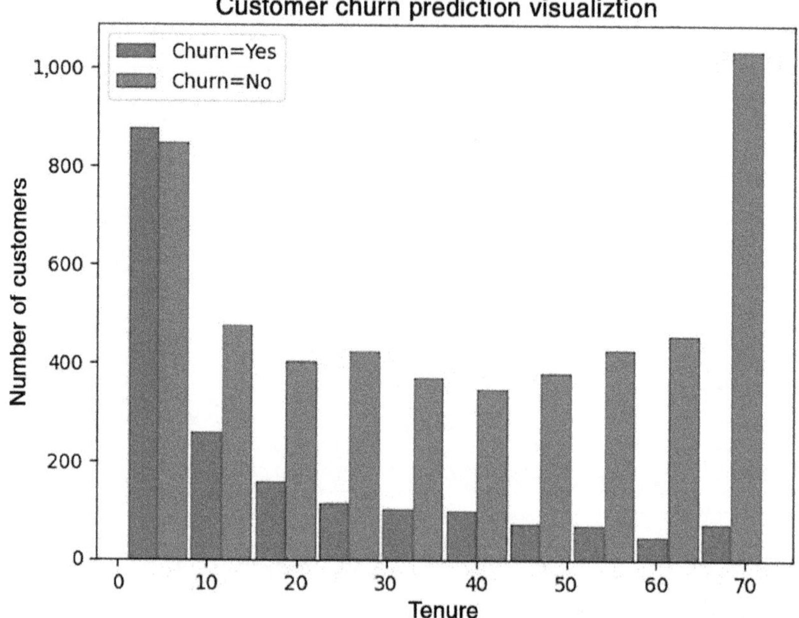

Figure 14.2 Churn by tenure visualization.

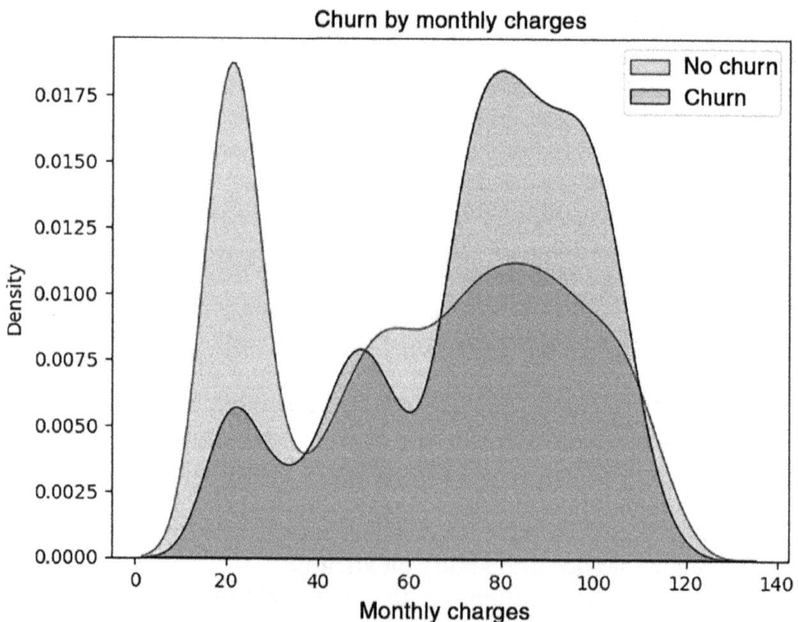

Figure 14.3 Churn by monthly charges visualization.

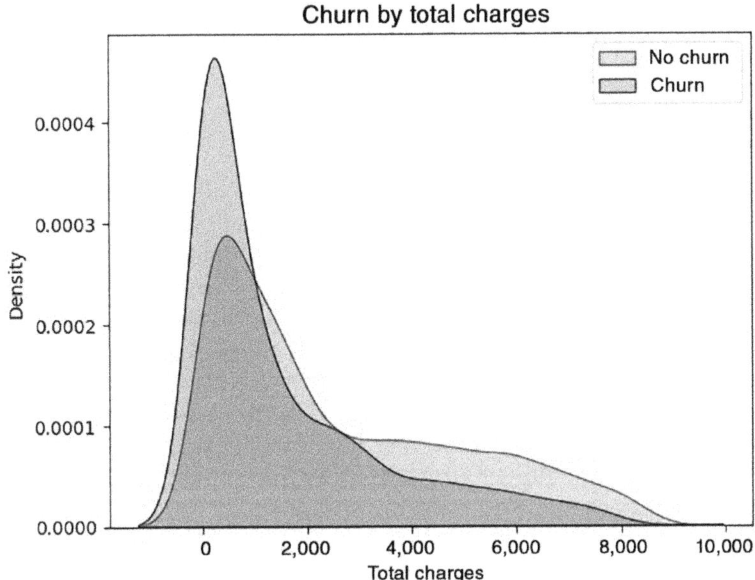

Figure 14.4 Churn by total charges visualization.

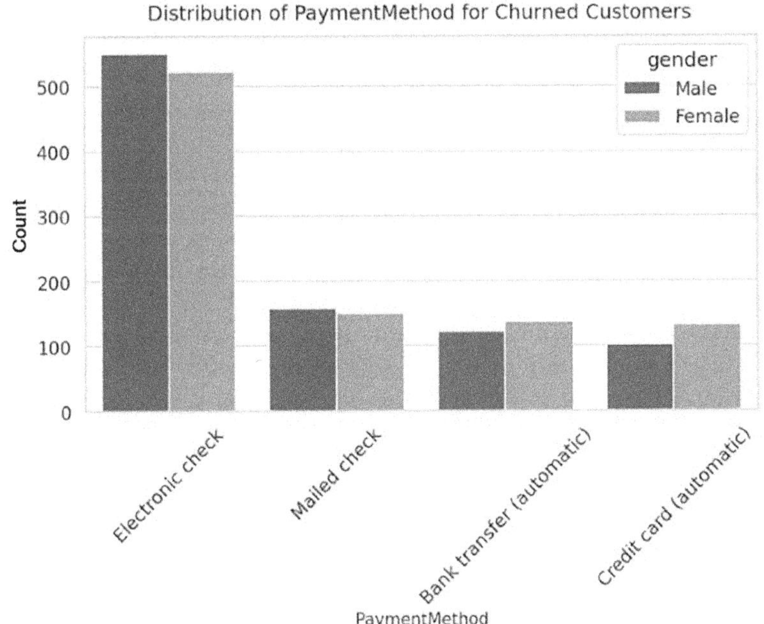

Figure 14.5 Distribution of payment method for churned customers.

- The categories with no tech support and no online security are major turners.
- Non-senior citizens have a high turnover rate.

14.4.2 Predictive model building

The preprocessed data is now used to train the ANN model to build a robust and accurate customer churn prediction model.

ANN models are ideal for churn prediction because they can capture complex, non-linear correlations in data. Churn prediction requires understanding nuanced patterns and correlations in consumer behavior that standard linear models may be difficult to detect. These complicated patterns may be efficiently learned and represented by ANNs due to their layered design and activation functions. They excel in processing vast amounts of heterogeneous data, such as customer interactions, use trends, and demographic information, allowing for a thorough examination of the reasons influencing turnover.

TensorFlow and Keras are used to create a basic ANN model for binary classification or the prediction of whether or not a client would churn. The two layers of the model are designed to handle binary classification issues. The input layer has 26 neurons and uses the ReLU activation function, while the output layer has one neuron and uses the sigmoid activation function. Using the Adam optimizer, accuracy as the training metric, and the binary cross-entropy loss function (which is frequently employed in binary classification), the model is assembled.

14.4.3 Customer segmentation

As discussed earlier, customer segmentation is one of the key advancements in the proposed system which provides a more detailed understanding of the customer's characteristics and behavior influencing customer attrition. This task also helps in extracting meaningful insights from the customer segments to develop targeted retention strategies.

We have used the K-means algorithm here for customer segmentation as it can detect unique groups within a dataset efficiently and divide clients into clusters based on common qualities, actions, or preferences. K-means is a realistic and scalable approach for customer segmentation, allowing businesses to easily assess and respond to the different preferences and behaviors demonstrated by their client base, thus improving consumer satisfaction and delivering targeted business strategies.

The output data of the predictive churn model is provided as input to the segmentation model. This data is initially scaled using StandardScaler to normalize its features. The dimensionality of the scaled data is then reduced to two main components using Principal Component Analysis (PCA). The new data frame contains the principal components that are obtained. The Elbow Method is then used to calculate the ideal number of clusters for K-means by plotting the inertia (within-cluster sum of squares) versus different values of "k." The point of the elbow in the plot indicates an optimal number of clusters as shown in Figure 14.6. Finally, K-means clustering is performed with a chosen "k" value (in this case, 4), and the resulting cluster labels are added to the data frame, which includes the principal components along with the assigned cluster labels for each data point. This resulting dataset is further used for the segmentation of any new customer. Thus, the customers are categorized into four segments, namely, conditionally loyal subscribers, conditional churners, lifestyle migrators, and unsatisfied customers. Finally, effective targeted retention strategies are designed by considering the characteristics of each customer segment.

14.4.4 Integration and user interface

A web interface is developed for deploying and integrating the predictive and segmentation models using Flask, a web framework of Python. The user interface comprises input fields for the various attributes describing customer features and buttons to initiate analysis. Users can input those features, and the system processes and predicts whether the customer is

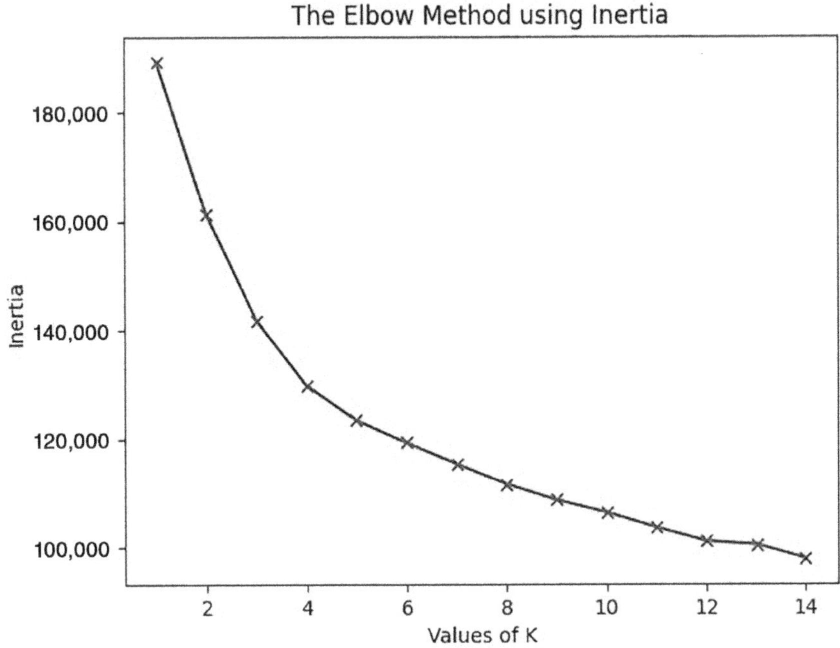

Figure 14.6 Graph representing optimal no. of clusters using the Elbow Method.

likely to churn or not. On the other hand, the segment type of the user is identified by considering churn prediction results. If the customer is likely to churn, then the system provides the period of churn by comparing the tenure of the user with its corresponding segment's average tenure and provides the customer characteristic distribution graphs as well as appropriate retention strategies as an output in an intuitive manner.

14.5 RESULTS

An efficient and accurate predictive model that anticipates customer churn along with the duration of churn is developed using the ANN algorithm with an accuracy of 81.74% and customer behavior segmentation is performed using the K-means clustering algorithm, and Decision Tree algorithm is employed for customer segment classification with an accuracy of 96.576%. Characteristics and behavior patterns of each customer segment are extracted and displayed to the user, which leveraged in tailoring targeted customer retention strategies (Figure 14.7).

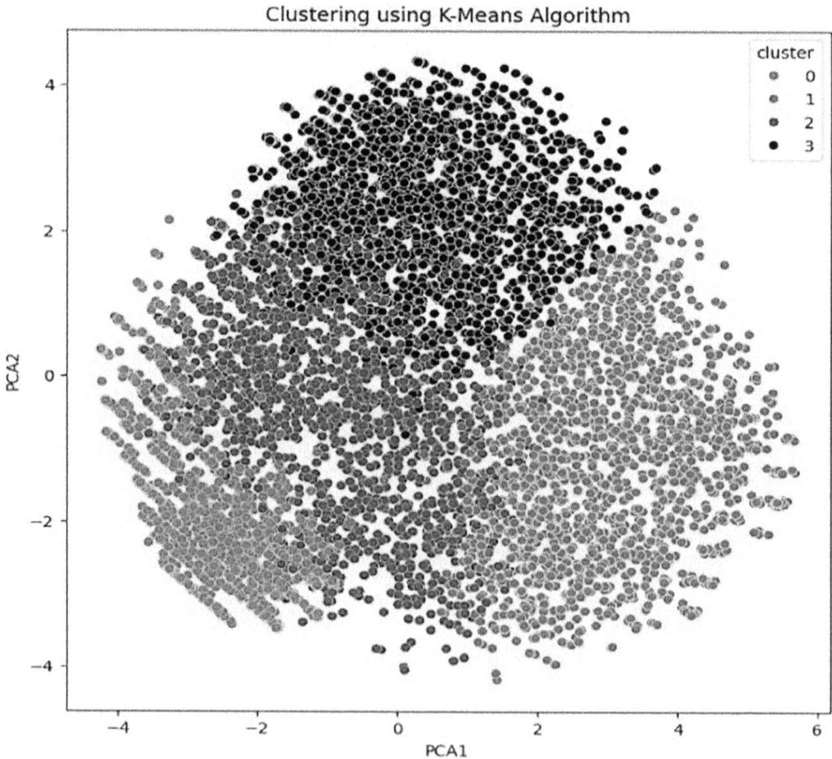

Figure 14.7 Customer segments scatter plot.

Here cluster 0 represents unsatisfied customers, cluster 1 represents conditional churners, cluster 2 represents conditionally loyal subscribers, and cluster 3 represents lifestyle migrators (Figures 14.8–14.10; Table 14.1).

14.6 CONCLUSION

As a result, this chapter offers a very reliable "Customer Churn Prediction for Retention Analysis" system that has effectively used ANN algorithms to forecast customer attrition, offering a thorough comprehension of possible churners, their anticipated duration of churn, and segmentation based on distinct customer attributes. The study has successfully identified intricate, non-linear patterns in customer behavior by utilizing ANN models, which has led to a

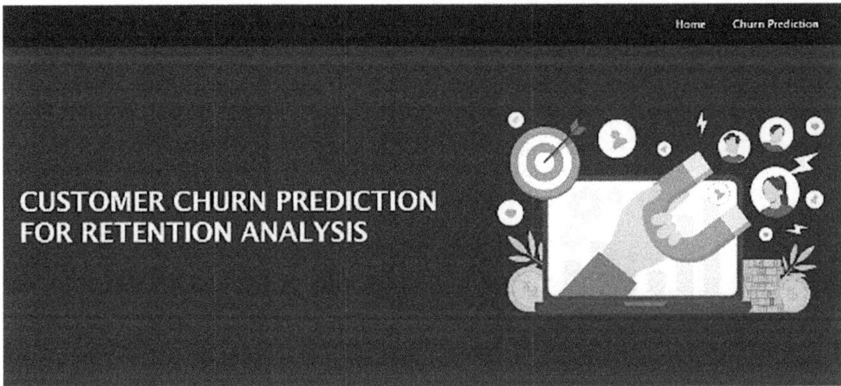

Figure 14.8 Customer churn prediction for retention analysis home page.

Figure 14.9 Churn prediction page.

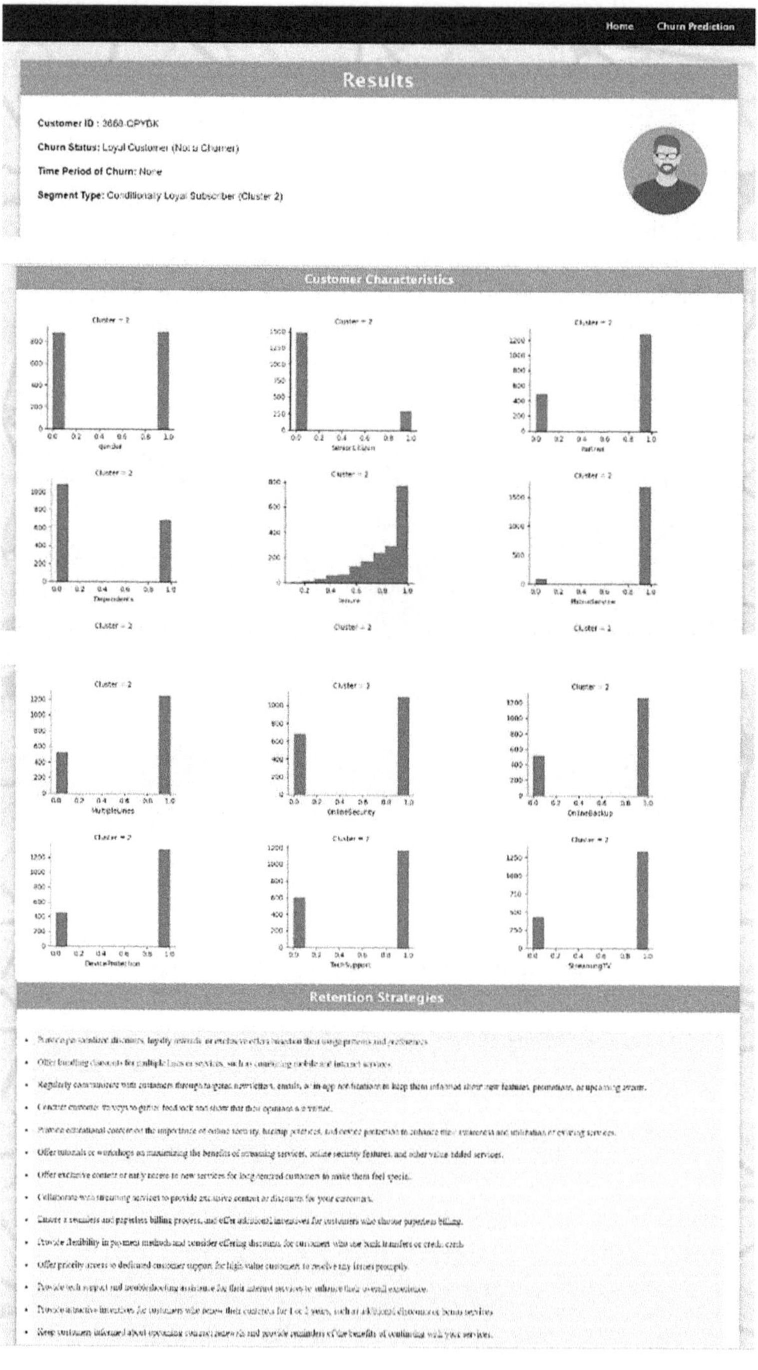

Figure 14.10 Customer characteristics and retention strategies.

Table 14.1 Classification report of customer segmentation

	Precision	i. Recall	f1-Score	ii. Support
iii. Cluster 0	1.00	1.00	1.00	465
Cluster 1	0.93	0.94	0.93	512
Cluster 2	0.97	0.97	0.97	511
Cluster 3	0.97	0.96	0.97	615
Accuracy			0.97	2103
Macro avg	0.97	0.97	0.97	2103
Weighted avg	0.97	0.97	0.97	2103

more accurate and nuanced churn forecast. The inclusion of a crucial dimension, time prediction, allows retention efforts to be prioritized according to the urgency of individual client cases. By addressing the varied demands and preferences of distinct client groups, customer segmentation further improves the effectiveness of the system by customizing retention strategies. The outcome is a powerful tool for businesses to proactively manage customer churn, formulate targeted retention plans, and ultimately strengthen customer relationships, fostering sustainable growth and long-term success. However, further research can concentrate on developing KPIs for tracking and monitoring customer usage behavior, providing a list of churn drivers, and recommending cross-selling and upselling strategies can assist in reducing customer attrition rate and enhance business profitability even further.

REFERENCES

1. Sarkaft Saleh, SubrataSaha, "Customer retention and churn prediction in the telecommunication industry: a case study on a Danish university", *SN Applied Sciences*, A Springer Nature Journal, June 2023, 5 (7), 173.
2. Praveen Lalwani, Manas Kumar Mishra, Jasroop Singh Chadha, PratyushSethi, "Customer churn prediction system: a machine learning approach", *Computing*, Springer 2022, 104 (2), 271–294.
3. B. Prabadevi, R. Shalini, B. R. Kavitha, "Customer churning analysis using machine learning algorithms", *International Journal of Intelligent Networks*, Elsevier, June 2023, 4 (3), 145–154.
4. Abdelrahim Kasem Ahmad, Assef Jafar, KadanAljoumaa, "Customer churn prediction in telecom using machine learning in big data platform", *Journal of Big Data*, Springer Open, 2019, 6, 28.
5. Bingquan Huang, Mohand Tahar Kechadi, Brian Buckley, "Customer churn prediction in telecommunications", *Expert Systems with Applications*, Elsevier, 2012, 39, 1414–1425.
6. Ammar A Ahmed, D. Maheswari Linen, "A review and analysis of churn prediction methods for customer retention in telecom industries", *International Conference on Advanced Computing and Communication Systems, IEEE*, Coimbatore, India, on January 6–7, 2017, January 2017. 1–7.

7. Jae-Hyeon Ahn, Sang-Pil Han, Yung-Seop Lee, "Customer churn analysis: churn determinants and mediation effects of partial defection in the Korean mobile telecommunications service industry", *Telecommunications Policy*, Elsevier, 2006, 30, 552–568.
8. K. Sandhya Rani, Shaik Thaslima, N. G. L. Prasanna, R. Vindhya, P. Srilakshmi, "Analysis of customer churn prediction in telecom industry using logistic regression", *International Journal of Innovative Research in Computer Science & Technology*, July 2021, 9 (4), 2347–5552.
9. Amal M. Almana, Mehmet SabihAksoy, Rasheed Alzahran, "A survey on data mining techniques in customer churn analysis for telecom industry", *International Journal of Engineering Research and Applications*, May 2014, 4 (5), 165–171.
10. Kiran Dahiya, Surbhi Bhatia, "*Customer churn analysis in telecom industry*", 2015 4th International Conference on Reliability, Infocom Technologies and Optimization (ICRITO) (Trends and Future Directions), Delhi, India, from September 6 to September 8, 2016, IEEE, 2015. 1–6.

Chapter 15

Exploring the intersection of sustainable economies and the Metaverse for a prosperous future

Abhiraj Malia, Pooja Jena, Liji Panda, Mamata Garanayak, and Parikshita Khatua

15.1 INTRODUCTION

The Metaverse is a post-reality vision cosmos with persistent and eternal computational multiuser environments where physical reality and digital virtuality coexist. The universe is a combination of many technologies that enables multimodal interactions with individuals, digital objects, and simulated environments such as augmented reality (AR) and virtual reality (VR). The platform made up of persistent many users is the metaverse, which is a networked immersed environment and interconnected social web. It is regarded as the most cohesive and seamless user communication solution for dynamic interactions, real-time issues, and digital artifacts. The initial manifestation of the virtual environment in which avatars might travel is referred to as the web of virtual world (Mystakidis, 2022). Modern variations featured open game worlds, social networks with AR capabilities, immersive VR, and massively multiplayer video games that were available online. Technological advancements have a big impact on everyday life, changing and improving social relationships, communication, and business. The three waves of technical innovation, from the viewpoint of the end user, are the advent of portable gadgets, the Internet, and personal computers. The fourth wave of computing innovation is bringing us immersive and geographical technologies, for example, AR and VR. This wave foresees the effects of the metaverse, a computer paradigm that has the potential to change commerce, remote labor, entertainment, and learning.

Users "live" in a digital universe that supports persistent online-3D virtual environments using traditional personal computing as well as AR and VR headsets in the metaverse, a technology fusion that chains VR and AR. Users of the metaverse are imagined as working, having fun, and remaining in touch with friends through events like virtual trips abroad and concerts. Neal Stephenson (2003) proposed the concept of the metaverse in his 1992 science fiction book *Snow Crash*, where he imagined life-like avatars interacting in precise 3D environments as well as other VR settings. Park and Kim (2022) presented a common categorization of metaverse components into six kinds. These include physical devices and gadgets,

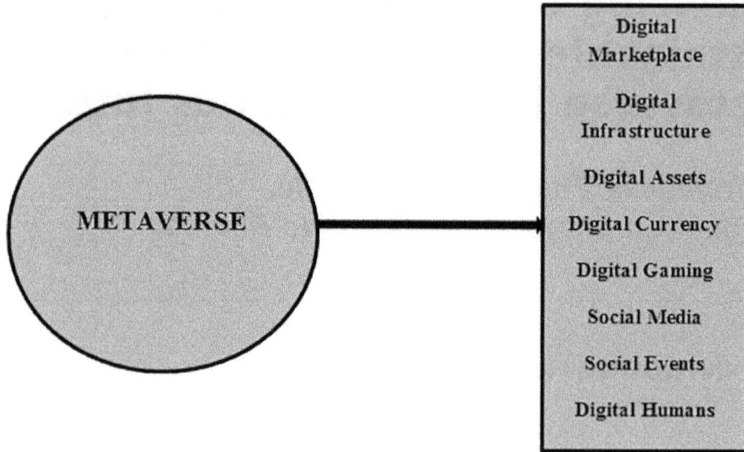

Figure 15.1 Metaverse and its categories. (Own Compilation, 2023.)

visualization, and technical approaches, all of which represent the technical components of the metaverse; User–user interaction with situation generating showed how connections and interactions are organized in the metaverse; programs and features, which refer to specific fields in which the metaverse is applied, such as work, study, gaming, cultural interaction, and connectivity (Figure 15.1).

Technology in the metaverse has advanced quickly. This made it possible for us to incorporate the metaverse into our daily lives more quickly and readily. Because telecommuting and schooling were conducted electronically to prevent interaction with one another before the COVID-19 epidemic, this realization of the metaverse's potential has been expedited (Yang et al., 2022). A virtual platform called metaverse uses blockchain advancements to guarantee data privacy. Recently, a lot of businesses and organizations have made investments in the metaverse. Transportation requirements increased as the metaverse gained its popularity. It is now more crucial than ever to integrate public transit and micromobility platforms into the metaverse. Sustainability in transportation is vital since the changing climate is another global issue (MacCracken, 2009). World preservation is essential, which has made sustainability a pressing concern. The main cause of carbon emissions is transportation. Systems for sustainable transportation are given top priority by many decision-makers. The collecting of metaverse data facilitates and legitimizes the improvement and innovation of sustainable transportation systems. An earlier study found that while traditional approaches are easily able to get the original information of public transit users, they do not typically acquire details on their destinations. The collecting of initial and final location information via avatars is made possible by integrating the metaverse into real-world activities like public transportation, which helps with public transportation planning. In the long run, people's daily activities will be virtual thanks to metaverse applications (Abduljabbar et al., 2021). Consequently, with reduced need for transportation, urban centers will be more sustainable.

Sustainability is the capacity to withstand or boost the current state of circulation of desired resources or situations over an extended period of time (Harrington, 2016; Bisoyi and Das, 2015). The 1992 Rio Declaration is regarded as "the fundamental instrument in the path toward sustainability." Ecosystem integrity is specifically mentioned in it. This is how sustainability is covered in the plan for implementing the Rio Declaration. Agenda 21 addresses issues related to the economy, society, and environment (Bosselmann, 2016).

The Sustainable Development Goals (SDGs), also known as the Global Goals, are a set of 17 interconnected goals that are meant to act as a "shared vision for prosperity and peace of individuals as well as our planet as a whole, now and into the future" (Takura and Miura, 2022). The SDGs are no poverty; no scarcity of food; good health and well-being; the standard learning, equal opportunities for women and men; water quality and sanitary facilities; accessible and renewable energy, and appropriate job opportunities and prosperity; business, entrepreneurship, and the infrastructure minimized inequality; sustainable towns and communities as a whole, sustainable use of resources and the manufacturing process; action on climate change; lives under water and life on the land; and stability, equity, and powerful organizations; as well as cooperation for the goals. The SDGs simply put sustainability at the front of their list of objectives, highlighting how interconnected sustainable development's environmental, social, and economic dimensions are (Bali Swain and Yang-Wallentin, 2020).

Numerous connections exist between the metaverse and sustainability. Sustainability is actually a major concern because it is anticipated that the metaverse would drastically affect the global sustainability as well as the real-world (environmental), financial, and social spheres (Jauhiainen et al., 2022). If they are not properly used and recycled, many essential natural resources will run out due to increased consumption and population growth. While the metaverse would increase economic growth along with financial profitability, these factors need to be balanced with the metaverse's societal, environmental, and ethical components, according to the media and many businesses.

However, the metaverse has the potential to significantly reduce greenhouse gas emissions as many tangible products are replaced with electronic ones, physical transportation and construction activities are streamlined using a virtual connection, the real world can be enhanced with virtual counterparts, simulations can be run in cyberspace rather than physical space, and more awareness and successful conversations can be created against dangerous worldwide issues such as changes in the climate. With more advanced technology than their predecessors, the new digital spaces might have a less negative impact on the environment. Since more products may only be bought and consumed digitally, a more dematerialized future is conceivable if individuals decide to purchase less physical items (Townsend, 2022).

As seen above, it is necessary to connect and examine the metaverse in terms of wide spheres of sustainability. Connecting the SDGs, which have been recognized as universal objectives to be achieved locally around the world, with the metaverse presents a potential opportunity. These 17 goals were adopted by all UN member-states in 2015. They are urgent calls for actions in global partnership as well as goals. To combat climate change and safeguard marine and terrestrial life, it is imperative to implement policies and programs that increase health and education, decrease inequality, and promote economic growth.

The metaverse will emerge as a crucial platform where the SDGs can be addressed. The metaverse has the potential to significantly contribute to the realization of a number of SDGs by both addressing them digitally and by altering the physical world associated with SDGs. However, the creation, implementation, and evolution of the metaverse can pose problems for sustainability. As a result, the metaverse might be examined to understand how various components of such an immersive, digital environment might affect social, economic, and environmental sustainability.

15.2 REVIEW OF LITERATURE

15.2.1 Metaverse

The word "Metaverse" was coined by Neal Stephenson in 1992 from his science fiction novel *Snow Crash*, which is about a rich and immersive virtual world in which the digital-connected world becomes reality. The metaverse is a 3D digital virtual space in which individual communicates with each another via characters that are either representation of their real or glorified identities. In other words, metaverse is mainly a virtual world that has been translated into a real world for an alternative existence where images or digital identities participate in social interactions and virtual cultural events while retaining a distinct economic existence. There are four basic types of metaverse architecture and technology: VR, reflecting world, life logging, and AR (Kye et al., 2021) (Figure 15.2).

Few urban industries and urban living domains are becoming more infiltrated by Internet-based governance models, economic processes, as well as information infrastructure. Platformization refers to the forces that

Figure 15.2 New future with metaverse.

have led to the creation of civilization in platforms, where platforms have invaded the fabric nature of urban communities. Meta, formerly known as Facebook, has launched the worldwide platform initiative metaverse. The best illustration of platformization is the global platform firm (Allam et al., 2022). The metaverse concept is hypothetical, similar to the Parallel virtual world, which represents cultures for individuals who work and reside in virtual cities as a substitute to futuristic smart cities. Certainly, the metaverse has the capacity to reshape cities by rethinking service and activity delivery to improve quality performance, accountability, and urban efficiency. This is owing to the introduction of revolutionary metaverse technologies such as digital twins, IoT, Big Data, and Artificial Intelligence (AI), which give datasets and powerful computation for understandings such as human behavior (Choi, 2022).

The term "Metaverse" encompasses all emerging digital innovations. Fundamental areas such as learning and education are being reimagined within the evolving Metaverse to preserve their universality and sustainability. Numerous publications have investigated the metaverse systems and its applications that rely on it, as well as the past events that have led to the current condition of the metaverse. Yet, it is still uncertain how the metaverse is organized and precisely what each component consists of. However, there has been a rising demand for online educational programs, as technology has evolved, but the design of the systems currently on the market is based on a metaverse that is inadequately stated as well as in the ideal situation in a simple 3D environment (Kim, 2021).

The research looks for previous studies to examine the distinctive technologies that employ the metaverse foundation and may provide and discuss all aspects of metaverse framework when applied in an environment for e-learning. The temporary isolation and lockdown during COVID-19 enabled techniques to avoid space constraints and physical time in order to access non-face-to-face services across the metaverse. The term e-learning refers to various types of teaching and instructional tools such as digital education, distance education, virtual training, mixed learning, and m-learning. E-learning can effectively use the metaverse as a remedy for subjects that rely entirely on integration and cannot be taught through a distance or through the Internet learning, such as science and technology and medical courses (Hyun, 2021). Despite the fact that there are a variety of environments in e-learning systems based on metaverse, they can be used to provide a productive and secure learning environment by using virtual technologies and doing ongoing research to broaden the usage of the most effective educational experiences. As an outcome, the virtual educational setting employs metaverse methods, which serve as a precursor for the existing and present learning system. Furthermore, the metaverse interfaces with the world of Internet and online technologies, and extended reality is incorporated to the VR environment. However, a few studies indicated that the roadmap of metaverse has been divided into a virtual mirror world, life logging, and the world of AR. These four components are divided by the four dimensions of exterior, augmentation, intimate, and simulation. Furthermore, simulation is an important component that produces a virtual world that is completely distinct from the mirror image. Individuals can assess or endure content and deliver customized assistance based on the metaverse without regard to time or location by indirectly or locally experiencing, storing, and comparing keywords in E-learning for content generated in a 3D environment via their mobile devices (Lee et al., 2021; Hyun, 2021) (Table 15.1).

Table 15.1 Dimensions of metaverse

Physical evidence and sensors	*Recognition and interpretation*	*A. Technical methods*
• Head-mounted display • Hand-based input device • Motion input device	• Scene and object recognition • Scene and object generation • Scene and speech recognition	• Multimodal inference • Multi-agent optimization • Integration optimization
User interaction	*Scenario generation*	*Programs*
• Language interaction • Multitask interaction • Embodied interaction	• Multimodal content • Agent persona modeling • Scenario population • Scenario evaluation	• Simulation • Game • Social • Education

Source: Own Source (2023).

15.2.2 The long-term sustainability of the metaverse economy

From a perspective of economic worth, the metaverse provides enormous chances for prosperity and advancement. First off, the metaverse offers an unusual channel for the sale of tangible products, giving consumers options and alternatives. Additionally, it enables businesses expanding the online marketplace and revenue of digital items to design fresh shopping occasions and motivate novel purchases. In reality, selling options are provided via non-fungible tokens (NFTs) in the metaverse (De Giovanni, 2023). For instance, in the garments business, the metaverse serves as a new channel for the sale of commodities like virtual accessories and clothing that avatars may wear or home goods that can be decorated in the virtual world. The creation of whole new markets, where businesses may create and trade a wide variety of electronic commodities while setting up both virtual and conventional channels ensures the economic viability of the metaverse (Wang et al., 2023). It's interesting to note that while e-commerce in general and digital channels in particular inevitably cause friction and have cannibalization consequences, the virtual market develops new requirements without having cannibalization problems. Because they will be able to create products and spaces that are not constrained by conventional physical boundaries, businesses will be able to further diversify their product offering in this way (Salloum et al., 2023).

From a soft perspective, businesses can take use of the economic advantages provided by the metaverse by providing a greater variety of innovative and aggressive marketing and service tactics than in the past (Pearce et al., 1994). For instance, passive content marketing options currently available include info graphics, pictures, and videos. These don't integrate well with the metaverse, where positioning and offerings of businesses are influenced by marketing and service methods. These prospects are related to the prevalence of Gen Z's market in B2B transactions, which is characterized by a persistent demand for immersive and interesting marketing tactics as well as distinctive offerings (Khan et al., 2021). For example, companies can create online shops to market digital goods or provide completely immersive experiences with VR, which ensures customer involvement and a rise in revenue. These opportunities serve as the foundational elements of the Industry 5.0 transition, which is categorized by an emphasis on human-focused innovations, improved human-technology and machine collaboration, and new opportunities for businesses to transform and improve their business models. Particularly, businesses can boost customer engagement and ultimately boost sales by utilizing digital technologies such as online shopping and VR experiences (Wang et al., 2023). Additionally, the chance to provide services like insurance, specific types of healthcare, or consultation encourages new business opportunities. The shift to Industry 5.0 is made possible, by businesses selling digital goods and services in the metaverse, such as simulated experiences, items, and real estate market, which may be manufactured and delivered at a lower price than physical goods. The metaverse can also relieve some of the pressure on the industrial and retail sectors. Retailers, for instance, may employ AR and VR to allow customers to virtually try on products before purchasing them, or they could use the metaverse as an experimentation site for developing product prototypes (Shen et al., 2021).

The metaverse has a number of financial advantages, but still there are few drawbacks that should be taken into consideration. The metaverse calls for a significant number of information centers and storage facilities, huge amounts of cloud-stored data, a 1,000-fold rise in computer power, and higher energy use and carbon dioxide secretions. Low latency is actually necessary for the entire electronic world in VR to guarantee swift data processing and the overall viability of the business model. Therefore, a key factor in effectively deploying Metaverse technology will be a nation's infrastructure (Kwatra et al., 2020). Additionally, it is necessary for data centers to be located closer to users, which comes with a cost in terms of both money and the environment. In less energy-efficient countries, businesses will face pressure to establish data centers that adhere to local regulations governing data privacy and energy efficiency, raising costs, and making it more difficult to efficiently implement the metaverse.

15.2.3 The sustainability of the metaverse for the environment

Similar to that, when supporting the adoption of metaverse technology, corporations, organizations, administrations, policymakers, individuals, and stakeholders, in general, should take into account a number of environmental tradeoffs. The metaverse needs a huge amount of computational power – maybe 1,000 times more than what is now being used, just like other digital technologies. Even though they now use a sizable portion of the planet's total energy, data centers throughout the world are constantly running to supply all of the required energy, and this consumption will undoubtedly rise when the metaverse becomes widely accessible. The energy needed will also be utilized for blockchain-based transactions to be completed as well as ongoing use of AI, VR, AR, and cloud services (Abou El Houda and Brik, 2023). Numerous empirical studies have already supported these claims.

On the positive side, the metaverse can significantly contribute to the construction of an environmentally friendly future by minimizing waste produced during product creation and testing and by using less energy. Another benefit is that digital commodities and virtual experiences in the metaverse consume a great deal fewer resources and emit a lot less carbon than comparable real-world things. For instance, there may be significant resource savings if consumers decided to purchase more clothing for their virtual selves as opposed to their physical selves during all stages of product development, manufacture, and distribution. In fact, many customers already want to buy less tangible goods in the future as they want to utilize technology more frequently (Pamucar et al., 2022). By substituting physical items with electronic ones and actual events presence with virtual interactions, the metaverse, a highly inventive structure, promises an enormous decrease in carbon emissions. When aspects of our daily lives and surroundings are moved to the digital world, emissions are produced at a higher rate. Today's data centers, however, aim to achieve "net-zero" emissions. To become carbon negative, data centers should endeavor to absorb as much carbon as they are generating.

The metaverse can significantly improve environmental conditions in terms of traffic and congestion on roads and in cities. The introduction of the metaverse is positively correlated with a decline in air pollution as a result of decreased transfers and thus lower car emissions. A digital twin of the earth should be created in the future through research and technical advancements in order to create long-term solutions to the problem of climate change. In a sense, this should be a "planet laboratory" that enables researchers to forecast, model, and analyze various scenarios as well as potential environmental effects (Rathore, 2018). For instance, utilizing AI and satellite images, it would be able to create a digital twin of a specific city in order to simulate scenarios and identify city solutions to problems like air degradation, traffic jams, etc.

15.2.4 The metaverse potential for social sustainability

Without any uncertainty, it is correct to claim that society outside the metaverse differs greatly from that within it in that it is built upon real-life relationships rather than virtual ones. Additionally, social problems like societal exclusion, the wealth gap, unfair treatment, etc. will only be made worse by the metaverse. However, given the disruptive nature of this technological breakthrough and the importance that it will have in the near future, we should strike a balance between our offline and online lives to preserve the social survival of the metaverse. Making an idealized image of reality might really be harmful to people's lives (Kshetri, 2022). The question of how the physical world and the digital world should align at this time has no clear solution. With a focus on themes like convenience, diversity, inclusion, and equity, businesses, colleges, organizations, non-governmental organization, and all other potential investors can undoubtedly work together to establish an equitable virtual society. In reality, the hazards of the metaverse intensifying some societal impacts must be carefully considered before moving further with its deployment (Anshari et al., 2022). Additionally, the social interactions that take place within the metaverse serve as a stand-in for the human-digital technology interfaces that characterize Industry 5.0's development. As a result, learning more about the social aspect of the metaverse helps to understand the shift to Industry 5.0 (Arpaci et al., 2022).

Positively, the metaverse can offer access to education for individuals all around the world, particularly in underdeveloped and isolated areas. As a result, the metaverse can guarantee that high-quality education is imparted via avatars in virtual environments. Social inclusion and equity can be ensured by a level of cooperation between businesses, colleges, schools, NGOs, and governments (Singh et al., 2022). For instance, the creation of a meta university would make it easier for students from all over the world to collaborate and engage in meaningful relationships, thereby removing barriers to possibilities for international mobility.

In order to sustain an open, welcoming, and equitable metaverse, a robust physical infrastructure is in fact necessary. Presently, one needs to meet some requirements in order to access the metaverse, such as having an account, a computer, or a connection to the web, as well as adhering the rules and guidelines set forth by each platform. One must also keep in mind how social exclusion is a result of digitalization (Yu and Angela, 2022). This is real, particularly for weaker individuals who try to survive in a virtual environment, cutting themselves off from the outside world and normal life. These individuals may develop an addiction to the intensely engaging digital life found within the metaverse. Individuals are accustomed to investing a lot of their free time in engaging metaverse practices, revealing information that could compromise the confidentiality of their data. This demonstrates how the metaverse has a built-in tendency to make the shift to Industry 5.0 genuine and feasible.

In accordance with how the metaverse is now evolving, social sustainability calls for rules and protocols to guarantee privacy, processing of data and social concerns are effectively addressed, and ethical and psychological effects the metaverse may have on users. The shift to Industry 5.0, which mandates a thorough analysis of both the advantages and disadvantages associated with the use of technologies that are immersive, such as the metaverse, makes this warning especially pertinent today (Satpathy et al., 2023). The metaverse also prompts worries about confidentiality issues and the possibility of identity theft and hacking. Some individuals are skeptical about the safety and capacity of the metaverse to safeguard user data because it is a brand-new technological innovation. To ensure data confidentiality and privacy, it is therefore advised that users get familiar with blockchain technology.

Finally, it's important to keep in mind that people from all countries and regions are welcomed in the metaverse. As a result, the correct management of new virtual residents entering a nation and the recognized businesses will have a favorable effect on both the overall well-being of that nation and the society it is associated with. This may have unintended consequences, including a rise in population attributed to future immigration flows into that country, cultural and social growth, new and aggressive government legislation, higher GDPs, and the creation of new knowledge and innovative methods (Yu and Angela, 2022). On the other hand, people with fraudulent activity might use the metaverse for unlawful activity, aggression, prostitution, the export of capital abroad, the threat of luring terrorists into a region, the support of unlawful organization actions, and problems with adaptability related to differences in religion, culture, and educational background.

15.2.5 Metaverse and urban resource management

The management of natural resources and planning for cities disciplines have relied heavily on the idea of resource management for many years. With the predicted increase in urban residents around the world, the significance of resource management will continue to rise. This is notably supported by studies that demonstrated how urban areas absorb more than 75% of the world's natural resources and considerably damage the environment through the use of a variety of toxic substances (Zvarikova et al., 2022). Water, land, fish farming, natural resources, and woodlands will all continue to be in limited supply as more and more individuals move into cities. Urbanization will thereby intensify the exploitation and degradation of these natural assets without allowing for natural-based remedies to make up for the depleted resources (particularly given the patterns of tree-change, sea-change, and regionalization). As a result, until specific urban resource management techniques are adopted, these difficulties will endure (Bisoyi et al., 2021).

It has been generally asserted that the creation of the metaverse will significantly lessen the need for travel, particularly to job locations, amusement parks, and other locations. This is due to the fact that individuals will be able to perform their jobs, physical activity, acquire knowledge, play, hold meetings, go to social events, and do much more in digital settings that replicate the real world. Our post-COVID unwillingness to work in a typical metropolitan office setting is one indication of this, in part. It has already been shown that individuals will be able to connect, work on projects, conduct telephonic conversations using Teams or Zoom, and engage in other "regular" operations over digital platforms much like they would in the real world with extended realities such as AR and AU (Allam et al., 2022). This will be crucial, especially in minimizing the power consumption of cars and the use of materials necessary to build infrastructure for workplaces, broad public transportation, and other things that contribute either directly or indirectly to work environments.

15.2.6 Metaverse and urban governance

Some people are excited about the idea of the metaverse, particularly because it will enable most social interactions, which call for the production of resource-intensive goods such as video games, toys, and cultural goods to be progressively manufactured and preserved in the digital world. Since the majority of these items, in their natural state, wind up in landfills, bodies of water, and other delicate ecosystems, this will aid in reducing the consumption of resources and contamination. Urban governance can be summarized up as the various ways that people, organizations, and other urban partners organize and handle the day-to-day operations of a city. It has been noted that this can occasionally be difficult and contentious, particularly when it comes to the use of power, the distribution of resources, and the execution of various agendas (De Zwart and Lindsay, 2010).

Human decision-making is involved in urban governance, which aims to guarantee that all the components and aspects that together make up the urban fabric are working smoothly and ethically. Given the variety of elements and factors that make up urban environments, the metaverse presents a useful framework for urban governance since it may enable the virtual delivery of urban resources and services (Bibri and Allam, 2022). This could improve their effectiveness, foster greater confidence and accountability, and decrease the expenses, bureaucracies, and obstacles that have been seen to impede or lengthen the delivery of urban services.

The metaverse will give local governments the chance to enhance connections with citizens, provide quick, effective, and actual-time services, and, more effectively, control resources like urban environment. Additionally, it will create chances for new revenue streams, enabling local governments to take on complicated and expensive projects. Additionally, the metaverse will give local administration the chance to reformat their current urban planning frameworks in order to incorporate those that promote human and social components. Apart from governance entities, various organizations, such as companies, educational institutions, and big companies, will provide opportunities for conducting their operations in the world of VR. By utilizing technologies, for example, digital twins that will be broadened in the metaverse, these organizations will be able to improve interactions with both current and potential customers as well as the performance of their products (Fernandez and Hui, 2022). As the metaverse becomes more well known, they will also have the chance to explore new areas, such as developing virtual goods that will be in high demand as individuals try to improve their characters and become commodities in the future.

It is unlikely that the metaverse is competent enough to tackle and overcome the typical problems associated with urban governance, such as ongoing negotiations and dispute, a variety of rewards and interests, conflicts and challenges, erratic choices, and inefficient systems. Instead of more transparent governance practices, the creation of fresh kinds of social interaction through the implementation of contemporary technology will be centered on different politico-economic purposes. Therefore, it will be crucial to involve national and local government structures to make sure that the metaverse is not designed for uses that contradict the exact goal for which it is being embraced.

15.2.7 Metaverse and tourism

Urban regions around the world are characterized by distinctive qualities that classify them as worldwide tourism attraction centers. It is clear that cultural values dominate most tourist destinations around the world. Tourism confronts a number of difficulties, but the metaverse has the potential to help with some of them, if not all. One of these is the capability to combine both the digital and physical environments, allowing different tourist attractions to be accessed and explored in the real world and through digital formats. This will be helpful because it will give people the chance to virtually visit various fascination spots and communicate with various people from various actual places, just like in the real world (Gursoy et al., 2022). This will give those who can't travel to the real-life locations the chance to enjoy, discover, and tour various landscapes and good places. Although the development of the metaverse may first appear to be a barrier for the physical tourism sector and its ancillary industries, this would not result in the industry's dying (Go and Kang, 2023). For example, it is impossible to recreate every component of the real environment in the virtual one (for instance, odor and feelings of attachment, especially when it comes to hobbies like beach-going or hill climbing) (Figure 15.3).

On the other hand, it is hoped that the metaverse would tempt individuals to really visit these locations experientially, making travel to these locations still worthwhile. Additionally, VR will enable those companies to promote their goods virtually, allowing those who are prepared to pay to already

Figure 15.3 Enhancing tourism through the application of metaverse. (Own Source, 2023.)

be aware of what to expect (Poonkuzhali et al., 2023). More crucially, the Metaverse will support the preservation and conservation of rapidly disappearing historical and cultural landmarks that are in grave risk due to terrorist activity, growing urbanization, rising temperatures, global warming, natural disasters, etc.

15.2.8 Metaverse and climate change

The effects of climate change have caused the world to face previously unheard-of hardships in recent decades, including heat waves, floods, unpredictable weather patterns, and now the pandemic's significant societal and occupational disruptions. Naturally, the effects of climatic change are felt all over the world, with metropolitan areas in particular being at risk from a variety of factors (Allam et al., 2022). While acknowledging that these issues are exacerbated by growing populations and rising resource use, it is clear that most urban areas have been at the cutting edge of implementing adaptation and resilience tactics to combat the effects of climate change.

The development of the metaverse might significantly affect other global climate goals as well as carbon reduction. The ability of the metaverse to reduce the requirement for human travel, as more individuals operate digitally from home in surroundings that match their actual work environments, is fundamental to its ability to respond to climate change. Furthermore, as the capacity for holding digital representations of tangible objects grows, the metaverse will consume fewer resources to produce goods, such as toys, games, holiday decorations, various leisure, and entertainment items, that frequently end up in landfills or are rarely utilized (Palak et al., 2023). Since the consumption and production of energy have a large impact on overall world emissions, this reduction will result in an overall decrease in utilization of energy.

With regard to urban modification, a majority of commercial activities, including tourism in various cities, have been compromised as a result of the effects of the changing climate, including floods (as in North India), increasing sea levels, tsunami impacts (as in Tonga and Japan), and other elements. This, in turn, has an effect on the lives of many metropolitan people, who are compelled to look for new options once their sources of income are destroyed, including immigration and other job prospects (Sathaye et al., 2006). Additionally, by developing improved warning mechanisms for disaster response via digital mediums, the metaverse can open up new possibilities and contribute to the preservation of structures and lives of people. Even if the field is relatively young, there may be claims that the idea can help in the creation of simulations and situational mapping capabilities for better understanding the implications of designs for plans and tools to mitigate climate change. In review of the literature above, the following problems and objectives have been identified to explore the current scenario.

15.3 RESEARCH PROBLEM

The research study's challenge is to use Metaverse to solve real-world issues while reshaping the digital economy. Metaverse shall simplify the communication and educate the common people easily from the difficulty in understanding the issues pertaining to day-to-day activities.

15.4 OBJECTIVES

1. To mitigate the issue of economic problems, overpopulation, traffic, and infrastructural issue with the aid of Metaverse.
2. To study the impact of Metaverse through SDGs.

15.5 METHODOLOGY

This study uses prismatic analysis to investigate the many facets of the Metaverse, looking at its scientific, social, and economic aspects and how they interact to form this new virtual environment.

15.5.1 Metaverse in solving economic issues

The metaverse has the power to affect economic dynamics and help solve some economic problems. The metaverse itself, though, is not a cure-all for economic problems; rather, it is a tool that may be used in a variety of ways to influence the economic activity. Following are the ways that the metaverse might relate to economic issues:

- **New Economic Opportunities:** The growth of the metaverse may lead to the emergence of new markets, professions, and commercial structures. For people and enterprises engaged in metaverse-related activities such as virtual goods manufacturing, creating software, and digital event management, this could result in economic expansion and the emergence of new sources of revenue.
- **Entertainment and Content Creation:** The metaverse might provide new ways for musicians, singers, artists, and content creators to commercialize their works. This could give these people new sources for earnings and support.
- **Global Trade and Exchange:** In the metaverse, virtual economies may make it possible to trade virtual goods, money, and assets across national boundaries. This might create new chances for interactions and transactions in the global economy.

- **Financial Inclusion:** People who are not currently included in traditional economic systems may be able to participate in economic activities. This is possible only because of metaverse, which would increase financial inclusion and give them more power.

The economic impact of the metaverse as it develops will mostly be determined by technology developments, acceptance among users, laws and regulations, and how society decides to incorporate it into current economic systems.

15.5.2 Metaverse in solving overpopulation issues

Although it's important to keep in consideration that the existence of the metaverse itself might not be an immediate remedy to population issues, but the idea of the metaverse has the capacity to handle certain difficulties related to problems with population. Here are a few ways the metaverse could aid in resolving issues relating to population:

- **Virtual Work Areas and Teaching:** As the metaverse matures, it may present a brand-new way to conduct online work and education. This might lessen some of the strain that dense urban populations have on the city's physical infrastructure. There may be less of a need for everyone to gather in particular places if people are able to pursue education from any part of the world.
- **Virtual Medical Care and Amenities:** Virtual medical care could be made possible via the metaverse, which might be especially helpful in regions with poor access to medical services. Without required infrastructure growth, this might enhance healthcare results for a broader population.
- **Social Relationships and Interactions:** People may be able to engage and communicate with others internationally, overcoming geographical restrictions, using virtual social spaces within the metaverse. This could help lessen emotions of loneliness and isolation, which metropolitan areas' high-population densities can worsen.
- **Research and Collaboration:** The metaverse may enable international cooperation on scientific research projects and initiatives, assisting in the resolution of complicated issues requiring a range of skills. Regarding issues such as disease prevention and climate change, this may be especially useful.

It's important to note that while these concepts provide potential advantages, they also present their own set of difficulties, including managing virtual economies, protecting data privacy, bridging the digital divide, and maintaining a harmonious balance between virtual and real-world interactions. The metaverse's ability to effectively combat overpopulation will also be influenced by its technological advancement, rate of acceptance, and the ways society decides to incorporate it into everyday life. The metaverse is still a hypothetical idea at this point, and it is yet unclear how it will actually affect population growth.

15.5.3 Metaverse in solving traffic

The notion of the metaverse, a shared virtual environment that combines both digital and physical realities, may not directly address real-world traffic problems. The development of transport systems, demographics, and automobile usage are some of the physical elements that contribute to traffic congestion. The metaverse's components could, however, indirectly help to reduce traffic congestion in the following ways:

- **Smart Traffic Management:** Although not strictly speaking a component of the metaverse, technologies that support it (such as gadgets, AI, and data analysis) might be leveraged to create more intelligent traffic control systems. These technologies might enhance the flow of traffic, ease traffic jams, and boost overall transportation performance.
- **Digital Conferences and Workspaces:** In metaverse, if online meetings, seminars, and work from home become more common, there may be less of a need for people to travel to actual locations to conduct business or discussions. This might result in fewer vehicles being on roadways during rush hour, which might ease traffic congestion.
- **Virtual Shopping:** Particularly for nonperishable commodities, virtual marketplaces within the metaverse may minimize the need for physical shopping visits. This might have a negligible effect on the number of short trips to the neighborhood grocery.

It's crucial to remember that the metaverse is still developing, so predictions about how it can affect things like traffic are purely hypothetical at this time. Any advantages might, however, only be marginal, and they might not be enough to significantly lessen the amount of traffic on their own. Adequate urban planning, investments in sustainable transportation infrastructure, enhancements to public transit and changes in actions, and mindsets toward transportation are necessary to address traffic congestion.

15.5.4 Metaverse in solving infrastructural issues

Although the metaverse by itself cannot directly address problems with the real world's infrastructure, there are some ways that it might be able to:

- **Virtual Prototyping and Testing:** Prior to being created in the real world, numerous infrastructure projects could be simulated and tested in the metaverse. This might make it easier to spot possible problems, improve designs, and avoid expensive blunders.

- **Collaboration and Communication across Distance:** Teams of experts working in several places are frequently involved in construction projects. Engineers, designers, developers, and other stakeholders might be able to collaborate in the metaverse in actual time regardless of where they are actually located.
- **Data Visualization and Analysis:** Using the metaverse, complicated infrastructure data might be seen and analyzed to give decision-makers a better understanding of their plans. This might result in better resource allocation and planning.
- **Smart Infrastructure Monitoring:** By integrating with IoT and real-world sensors, the metaverse could offer continuous surveillance of infrastructure systems, assisting in the early detection of problems and abnormalities that need attention.

  It's critical to understand that the metaverse itself depends on physical assets such as information centers and connections, and it immediately cannot address problems with outdated infrastructure, insufficient transportation options, or water supplies. However, it might be used in conjunction with other tools to improve coordination, planning, and making decisions in the area of infrastructure construction and management.

15.5.5 The SDGs as a means of measuring the metaverse impact

This section offers an SDG evaluation approach to assess the metaverse's sustainability for a smooth transition to Industry 5.0. The metaverse enables businesses to interact globally, lower their carbon impact, and improve their production procedures. Through the metaverse, Industry 5.0 may completely transform the industrial sector by promoting innovative thinking, creativity, and sustainability. Overall, the metaverse and Industry 5.0 can be considered as complimentary strategies that can cooperate to help the achievement of the SDGs and promote sustainable development. These methods have the potential to contribute to the development of a more sustainable and fair future for everyone by utilizing these technologies (Table 15.2).

15.6 CONCLUSION

The metaverse has the control to alter and improve the physical world, according to the results of the current research study. The metaverse can assist in addressing some of the most important real-world issues we face today, including overcrowding, transportation, pollution, infrastructure,

Table 15.2 Measuring the impact of metaverse through SDGs

SDG's	Metaverse's objectives	B. Measures to be taken through meta-verse
SDG-1	• Use the metaverse to raise awareness of the state of poverty in some nations among people, businesses, policymakers, and stakeholders.	• Use the metaverse to more accurately predict and replicate the social conditions, post a disaster or shock. • Use metaverse to reach agreement on the deployment of policy measures to assist poor populations.
SDG-2	• Use metaverse to confirm sustainable farming methods as well as methods for achieving food security, raising the nutritional status of the target population, and putting a stop to hunger in underdeveloped and isolated areas.	• Utilize VR tools in the metaverse to make sure individuals have access to enough food and nutrition. • Organize metaverse-based events to promote the preservation of endangered species, domesticated animals, plants, and resources through advancements in science and agriculture.
SDG-3	• Promote the metaverse to embrace methods for defending against, preventing, and treating any mental or physical diseases in order to improve society's overall health.	• Identify crucial circumstances that can reduce the number of infant and toddler fatalities that could have been prevented. • Utilize the metaverse to confirm the need for vaccinations and medications
SDG-4	• Make sure the metaverse delivers chances for learning, skill development, and the emergence of a feeling of global citizenship in a highly accessible and instructional virtual environment.	• Provide an equal and high-quality metaverse education. • Use the metaverse as a virtual space for entrepreneurial skill development, especially for vulnerable groups like the disabled.
SDG-5	• Utilize the metaverse to advance gender equality by combating all forms of abuse and violence that limit women's rights and by defending the leadership roles that woman can have in all fields.	• Eliminate all forms of violence and discrimination in the metaverse, including among avatars. • Alert the public to the negative effects of destructive practices like forced marriage and mutilation
SDG-6	• To validate the techniques to be used for improved water management and development of availability, quality, and efficiency, maps the sources of water throughout the metaverse and its uses.	• Utilize the metaverse to more accurately predict the global water shortage. • Track the wastewater across the metaverse and detect ad hoc interventions to ensure water quality by also eliminating chemicals and reducing pollution.

(Continued)

Table 15.2 (Continued) Measuring the impact of metaverse through SDGs

SDG's	Metaverse's objectives	B. Measures to be taken through meta-verse
SDG-7	• Utilize the metaverse to find the most effective energy sources and execute plans virtually.	• Reduce energy consumption when using the metaverse. • Use the metaverse to simulate the ideal mix of renewable and natural resources for energy production.
SDG-8	• The metaverse aims to create ad hoc plans for economic growth by enlisting the help of all relevant parties in organizing successful activities. The transparency provided by the metaverse ensures respectable working conditions and procedures.	• Create a metaverse business forum to support economic progress, especially in developing nations. • Keeping track of overall output and consumption through metaverse while assisting those who are disabled or at risk of being hurt. • Use the metaverse to ensure gender equality and equal value by making all workplace policies and procedures transparent.
SDG-9	• Utilize AI, VR, and digital twins of the metaverse to determine the most effective infrastructure to build and the most cutting-edge, resilient solutions to use.	• Determine the new occupations that the metaverse will create and the level of industrialization needed to attain sustainability. • Encourage the development of ad hoc infrastructures particularly in developing nations, to make the metaverse function.
SDG-10	• Utilize the metaverse's ability to reduce any kind of inequality and highlight it with the use of digital tools in the virtual world.	• Utilize the metaverse to track a region's incomes and associated growths, then distribute value in accordance with that map. • Use blockchain for smart payments and ensure equity in pay, taxes, social protection, and subsidies. • Develop metaverse dashboards that show the true justice and equality in society.
SDG-11	• Transform actual cities into virtual ones to keep track of and enhance any potential problems that could lead to improving people's lives and conserving the environment.	• Utilize the metaverse to make traveling cities and attractions more immersive • Replicate city logistics in the metaverse using digital twins, as well as urbanization, to increase accessibility and mobility • Use the metaverse to improve the management of urban waste
SDG-12	• Create virtual versions of actual consumption and production in the metaverse to constantly track processes and flows and to improve overall performance in terms of economic, social, and environmental variables.	• Use the blockchain for transactions and to verify products and services • Support the implementation of Circular Economy principles (reuse, recycling, and renovation) through the metaverse, as well as through running educational and training programs.

(Continued)

Table 15.2 (Continued) Measuring the impact of metaverse through SDGs

SDG's	Metaverse's objectives	B. Measures to be taken through meta-verse
SDG-13	• The metaverse enables one to more clearly see how the environment is changing and the hazards to mankind, which helps define the actual activities and contributions made by those involved.	• Utilize the metaverse to track the actual climate change contributions made by various nations, businesses, institutions, and stakeholders. • Provide a series of metaverse training and educational initiatives that enable avatars to interact with powerful people
SDG-14	• Utilize the metaverse's capabilities to learn about the true conditions of underwater life and keep a close eye on developments so that you can take prompt, appropriate corrective action.	• Utilizing the metaverse to depict the actual state of underwater life, including all forms of pollution. • Make greater use of the blockchain to specify the financial aid to be given, based on their actual impact on the marine and coastal ecology
SDG-15	• To identify the appropriate procedures and steps to take for a suitable preservation, protection, and improvement, virtualize terrestrial life through the metaverse.	• Monitor lands with the use of the metaverse to quickly spot any invasive or illegal activity. • Virtualize freshwater and terrestrial ecosystems, as well as the accompanying services, on the metaverse to identify areas that need protection. • Plan meetings and activities on the metaverse with specialists to prevent extinction of species by limiting desertification and protecting mountains and their natural ecosystems.
SDG-16	• Use the metaverse to persuade individuals, groups, and other interested parties to build an equitable and just society.	• Avert all forms of violence and fatalities caused by AI and VR in the metaverse. • Make sure that when lying in virtual environments, identity and privacy are respected and protected.
SDG-17	• Set new, realistic goals for the global alliance for sustainable development using the metaverse.	• Utilize the metaverse to increase resource mobility and to get multinational organizations involved in helping out developing nations. • To achieve equality, encourage the creation of partnerships and involve them through NFTs. • Encourage the establishment of guidelines and standards for individuals from other nations when accessing the metaverse to advance equity.

Source: Own Source (2023).

and environmental degradation by transforming the virtual economy. In the following years, novel approaches to traditional problems will fully emerge as the metaverse's underlying technology keeps developing. If taken too far, government metaverses might end up becoming where we work. Owning to real estate, working in a metaverse community and, of course, investing are just a few of the creative ways to make money in the metaverse.

This chapter clarifies the advantages and disadvantages of adopting metaverse technology in the near future in light of the issues associated with the use of digital technologies mentioned above. By identifying the new trade-offs, this analysis offers a broad framework for future study and practitioners assessing the implementation of the metaverse. Future studies should be conducted to confirm the methods for resolving these tradeoffs and to suggest feasible options for society as a whole.

REFERENCES

Abduljabbar, R. L., Liyanage, S., & Dia, H. (2021). The role of micro-mobility in shaping sustainable cities: A systematic literature review. *Transportation Research Part D: Transport and Environment*, 92, 102734.

Abou El Houda, Z., & Brik, B. (2023). Next-power: Next-generation framework for secure and sustainable energy trading in the Metaverse. *Ad Hoc Networks*, 149, 103243.

Allam, Z., Sharifi, A., Bibri, S. E., Jones, D. S., & Krogstie, J. (2022). The Metaverse as a virtual form of smart cities: Opportunities and challenges for environmental, economic, and social sustainability in urban futures. *Smart Cities*, 5(3), 771–801.

Anshari, M., Syafrudin, M., Fitriyani, N. L., & Razzaq, A. (2022). Ethical responsibility and sustainability (ERS) development in a Metaverse business model. *Sustainability*, 14(23), 15805.

Arpaci, I., Karatas, K., Kusci, I., & Al-Emran, M. (2022). Understanding the social sustainability of the Metaverse by integrating UTAUT2 and big five personality traits: A hybrid SEM-ANN approach. *Technology in Society*, 71, 102120.

Bali Swain, R., & Yang-Wallentin, F. (2020). Achieving sustainable development goals: Predicaments and strategies. *International Journal of Sustainable Development & World Ecology*, 27(2), 96–106.

Bibri, S. E., & Allam, Z. (2022). The Metaverse as a virtual form of data-driven smart urbanism: On post-pandemic governance through the prism of the logic of surveillance capitalism. *Smart Cities*, 5(2), 715–727.

Bisoyi, B., & Das, B. (2015). Adapting green technology for optimal deployment of renewable energy resources and to generate power for future sustainability. *Indian Journal of Science and Technology*, 8, 28

Bisoyi, B., Nayak, B., Das, B., & Pasumarti, S. S. (2021). Urban resilience and inclusion of smart cities in the transformation process for sustainable development: Critical deflections on the smart city of Bhubaneswar in India. In: Priyadarshi, N., Padmanaban, S., Ghadai, R., Panda, A. R., & Patel, R. (Eds.) *Advances in Power Systems and Energy Management: Select Proceedings of ETAEERE 2020* (pp. 149–160). Singapore: Springer.

Bosselmann, K. (2016). A normative approach to environmental governance: Sustainability at the apex of environmental law. In: Fisher, D. E. (Eds.) *Research*

Handbook on Fundamental Concepts of Environmental Law (pp. 22–50). Cheltenham: Edward Elgar Publishing.

Choi, H. Y. (2022). Working in the Metaverse: Does telework in a Metaverse office have the potential to reduce population pressure in megacities? Evidence from young adults in Seoul, South Korea. *Sustainability, 14*(6), 3629.

De Giovanni, P. (2023). Sustainability of the Metaverse: A transition to Industry 5.0. *Sustainability, 15*(7), 6079.

De Zwart, M., & Lindsay, D. (2010). Governance and the global Metaverse. In: Riha, D., & Maj, A., (Eds.) *Emerging Practices in Cyberculture and Social Networking* (pp. 63–82). Boston, MA: Brill.

Fernandez, C. B., & Hui, P. (2022, July). Life, the Metaverse and everything: An overview of privacy, ethics, and governance in Metaverse. In *2022 IEEE 42nd International Conference on Distributed Computing Systems Workshops (ICDCSW): Bologna, Italy, 10–13 July 2022* (pp. 272–277). Los Alamitos, CA: IEEE Computer Society.

Go, H., & Kang, M. (2023). Metaverse tourism for sustainable tourism development: Tourism agenda 2030. *Tourism Review, 78*(2), 381–394.

Gursoy, D., Malodia, S., & Dhir, A. (2022). The Metaverse in the hospitality and tourism industry: An overview of current trends and future research directions. *Journal of Hospitality Marketing & Management, 31*(5), 527–534.

Harrington, L. M. B. (2016). Sustainability theory and conceptual considerations: A review of key ideas for sustainability, and the rural context. *Papers in Applied Geography, 2*(4), 365–382.

Hyun, J. J. (2021). A study on education utilizing Metaverse for effective communication in a convergence subject. *International Journal of Internet, Broadcasting and Communication, 13*(4), 129–134.

Jauhiainen, J. S., Krohn, C., & Junnila, J. (2022). Metaverse and sustainability: Systematic review of scientific publications until 2022 and beyond. *Sustainability, 15*(1), 346.

Khan, I. S., Ahmad, M. O., & Majava, J. (2021). Industry 4.0 and sustainable development: A systematic mapping of triple bottom line, Circular Economy and Sustainable Business Models perspectives. *Journal of Cleaner Production, 297*, 126655.

Kim, J. G. (2021). A study on Metaverse culture contents matching platform. *International Journal of Advanced Culture Technology (IJACT), 9*(3), 232–237.

Kshetri, N. (2022). Policy, ethical, social, and environmental considerations of Web3 and the Metaverse. *IT Professional, 24*(3), 4–8.

Kwatra, S., Kumar, A., & Sharma, P. (2020). A critical review of studies related to construction and computation of Sustainable Development Indices. *Ecological Indicators, 112*, 106061.

Kye, B., Han, N., Kim, E., Park, Y., & Jo, S. (2021). Educational applications of Metaverse: Possibilities and limitations. *Journal of Educational Evaluation for Health Professions, 18*, 32.

Lee, L. H., Braud, T., Zhou, P., Wang, L., Xu, D., Lin, Z., ... & Hui, P. (2021). All one needs to know about Metaverse: A complete survey on technological singularity, virtual ecosystem, and research agenda. *arXiv preprint arXiv:2110.05352*.

MacCracken, M. C. (2009). The increasing pace of climate change. *Strategic Planning for Energy and the Environment, 28*(3), 8–25.

Mystakidis, S. (2022). Metaverse. *Encyclopedia*, 2(1), 486–497.

Palak, Sangeeta, Gulia, P., Gill, N. S., & Chatterjee, J. M. (2023). Metaverse and its impact on climate change. In: Hassanien, A. E., Darwish, A., Torky, M. (Eds.) *The Future of Metaverse in the Virtual Era and Physical World* (pp. 211–222). Cham: Springer International Publishing.

Pamucar, D., Deveci, M., Gokasar, I., Tavana, M., & Köppen, M. (2022). A Metaverse assessment model for sustainable transportation using ordinal priority approach and Aczel-Alsina norms. *Technological Forecasting and Social Change*, 182, 121778.

Park, S. M., & Kim, Y. G. (2022). A Metaverse: Taxonomy, components, applications, and open challenges. *IEEE Access*, 10, 4209–4251.

Pearce, D. W., Atkinson, G. D., & Dubourg, W. R. (1994). The economics of sustainable development. *Annual Review of Energy and the Environment*, 19(1), 457–474.

Poonkuzhali, S., Archana, J. S., & Anand, T. P. (2023). Metaverse 3C: Concept, components, and challenges in travel and tourism sector. In: Rajakumar, G., Du, K.-L., & Rocha, Á. (Eds.) *Intelligent Communication Technologies and Virtual Mobile Networks* (pp. 699–713). Singapore: Springer Nature Singapore.

Rathore, B. (2018). Emergent perspectives on green marketing: The intertwining of sustainability, artificial intelligence, and the Metaverse. *International Journal of New Media Studies: International Peer Reviewed Scholarly Indexed Journal*, 5(2), 22–30.

Salloum, S., Al Marzouqi, A., Alderbashi, K. Y., Shwedeh, F., Aburayya, A., Al Saidat, M. R., & Al-Maroof, R. S. (2023). Sustainability model for the continuous intention to use Metaverse technology in higher education: A case study from Oman. *Sustainability*, 15(6), 5257.

Sathaye, J., Shukla, P. R., & Ravindranath, N. H. (2006). Climate change, sustainable development and India: Global and national concerns. *Current Science*, 90, 314–325.

Satpathy, I., Patnaik, B. C. M., Baral, S. K., & Islam, M. (2023). 7 Inclusive education through augmented reality (AR) and virtual reality (VR) in India. In: Goel, R., Baral, S. K., Mishra, T., & Jain, V. (Eds.) *Augmented and Virtual Reality in Industry 5.0*, (Vol. 2, p. 147). Berlin: Walter de Gruyter GmbH

Shen, B., Tan, W., Guo, J., Zhao, L., & Qin, P. (2021). How to promote user purchase in Metaverse? A systematic literature review on consumer behavior research and virtual commerce application design. *Applied Sciences*, 11(23), 11087.

Singh, J., Malhotra, M., & Sharma, N. (2022). Metaverse in education: An overview. In: Bathla, D., & Singh, A. (Eds.) *Applying Metalytics to Measure Customer Experience in the Metaverse* (pp. 135–142). Hershey: IGI Global.

Stephenson, N. (2003). *Snow Crash: A Novel*. Spectra, UK: Penguin.

Takura, T., & Miura, H. (2022). Socioeconomic determinants of universal health coverage in the Asian region. *International Journal of Environmental Research and Public Health*, 19(4), 2376.

Townsend, S. (2022) Could the Metaverse & web3 save sustainability. Forbes.

Wang, M., Liu, S., Hu, L., & Lee, J. Y. (2023). A study of Metaverse exhibition sustainability on the perspective of the experience economy. *Sustainability*, 15(12), 9153.

Yang, Q., Zhao, Y., Huang, H., Xiong, Z., Kang, J., & Zheng, Z. (2022). Fusing blockchain and AI with Metaverse: A survey. *IEEE Open Journal of the Computer Society*, 3, 122–136.

Yu, F. R., & Angela, W. Y. (2022). *A Brief History of Intelligence: From the Big Bang to the Metaverse* (pp. 105–109). Cham: Springer International Publishing.

Zvarikova, K., Cug, J., & Hamilton, S. (2022). Virtual human resource management in the Metaverse: Immersive work environments, data visualization tools and algorithms, and behavioral analytics. *Psychosociological Issues in Human Resource Management*, 10(1), 7–20.

Chapter 16

UNet-based MRI image analysis for enhanced brain tumor spotting

A cutting-edge approach in medical imaging

*Shanmuga Sundari M., S. Pranitha Pradhan,
Thurimella Sujitha, and P. Neha*

16.1 INTRODUCTION

The intricate nature of the human brain, orchestrating a multitude of physiological [1] and cognitive functions, renders it vulnerable to various disorders, with tumors occupying a significant position among them. Brain tumors, marked by abnormal cell growth within the cranial cavity, present a formidable challenge in the field of neurological health. Detecting and characterizing brain tumors represent a pivotal frontier in medical diagnostics, necessitating advanced technologies and methodologies for precise identification. The diverse manifestations of brain tumors, coupled with the urgent need for timely intervention, make their identification a complex challenge. Medical imaging, employing techniques such as magnetic resonance imaging (MRI) [2] and computed tomography (CT) scans, plays a crucial role in enhancing diagnostic accuracy. Brain tumor identification requires a meticulous analysis of neuroimaging data [3], emphasizing the importance of highlighting tumors in brain scans, understanding their characteristics, and pinpointing their precise location. This precision is crucial for effective treatment planning and ultimately improving patient outcomes.

16.1.1 Challenges in brain tumor detection

Identifying brain tumors [4] poses challenges due to variations in size, location, and morphology. Symptoms, ranging from headaches to seizures and cognitive changes, underscore the urgency of early detection. While traditional imaging modalities provide detailed views of internal brain structures, the manual interpretation of these images for tumor identification is time-consuming and prone to human error.

16.1.2 Advanced imaging technologies

Figures 16.1 and 16.2 visually depict a healthy brain MRI and a brain MRI with a highlighted tumor, respectively. The stark contrast between these

DOI: 10.1201/9781003565529-16

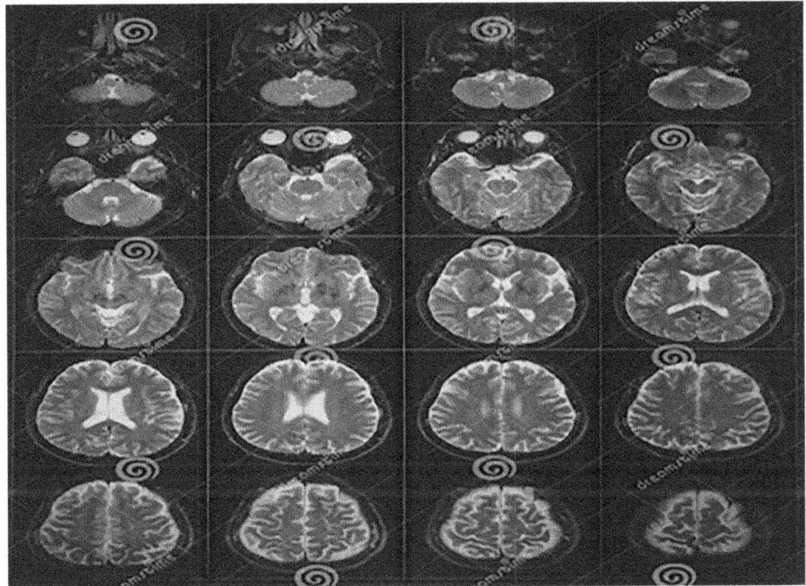

Figure 16.1 Healthy brain MRI image.

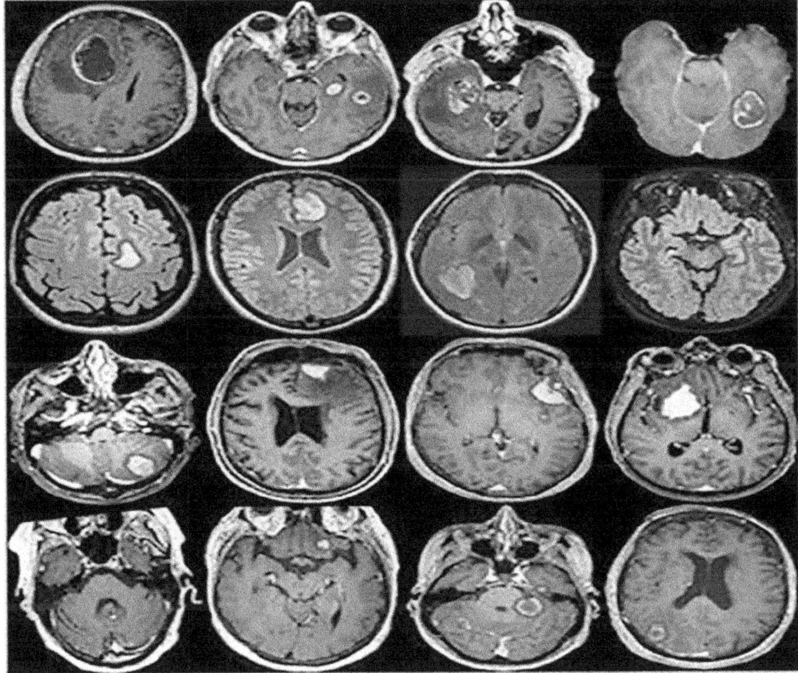

Figure 16.2 Brain MRI with highlighted tumor.

images underscores the necessity for precise identification of abnormal growths. Supplementing traditional methods, advanced imaging technologies such as functional MRI (fMRI) [5] and positron emission tomography scans, contribute to a comprehensive assessment of tumor characteristics.

16.1.3 The role of deep learning

In the realm of image classification, deep learning convolutional neural network (CNN) [6] models excel by autonomously capturing intricate features, ensuring accurate identification of objects and patterns. The application of these models, known for superior accuracy and effectiveness, is crucial in tasks like image recognition and medical image analysis. Thus, employing deep learning becomes imperative for the accurate classification of brain tumors.

16.2 LITERATURE SURVEY

Significant research has been conducted for detection of tumor in brain using deep learning algorithms.

In this study, we address the crucial need for early diagnosis of brain diseases, particularly malignant brain cancer, by proposing a computer-aided diagnosis (CAD) [7] system. A three-step preprocessing method is introduced to improve the quality of MRI images, laying the foundation for more accurate diagnosis. The research presents a novel DCNN architecture designed for effective diagnosis of glioma, meningioma, pituitary tumors, and normal brain images.

This chapter focuses on developing a method, for classifying brain tumors using MRI and artificial intelligence (AI) [8] through deep learning. The chapter utilizes known trained models such as Xception, NasNet Large, DenseNet121, and InceptionResNetV2. By applying transfer learning techniques, the researchers extract features from MRI scans of the brain.

This chapter aims to enhance brain tumor diagnosis using high-quality medical images through a hybrid approach involving the ResNext101 32×8d and VGG19 [9] pretrained models. The chapter emphasizes the importance of utilizing AI techniques for automatic diagnosis of brain tumors, proposing the application of ResNext101 32×8d and VGG19 models on a dataset of 1,800 MRI images.

This chapter focuses on improving brain tumor segmentation in MRI images by leveraging clinical knowledge and proposing a model named CKD-TransBTS. The proposed model, CKD-TransBTS, integrates clinical knowledge of brain tumor diagnosis from multiple MRI modalities. Instead of directly concatenating all modalities, the input modalities are organized into two groups based on MRI imaging principles. CKD-TransBTS features

Table 16.1 Summarizing earlier research studies

S. No.	Dataset used	Methods	Merits
1	The paper utilizes a dataset comprising 3,394 MRI images.	DCNN, preprocessing methods, robustness testing	This study includes the introduction of a three-step preprocessing method for improved MRI image quality, the development of a novel and computationally lightweight DCNN architecture.
2	Two benchmark datasets for classifying brain tumors.	CNN, transfer learning	The use of transfer learning, diverse pretrained models, and optimization algorithms contributes to the effectiveness of the proposed method.
3	The paper utilizes a dataset comprising 1,800 MRI images.	Pretrained models, single-image super-resolution (SISR).	The models offer a promising and accurate approach for distinguishing between patients with pituitary and glioma tumors.
4	The research utilizes the BraTS 2021 challenge dataset.	CKD-TransBTS, MCCA.	CKD-TransBTS achieves state-of-the-art brain tumor segmentation performance.
5	The study utilizes the BRATS 2015 database for simulations and evaluations.	Gabor filtering method, DAE, Social Spider Optimization Algorithm (SSOA).	The ensemble approach integrates EfficientNet, DenseNet, and MobileNet, harnessing the diversity of these models for robust feature extraction and classification.

a dual-branch hybrid encoder with a modality-correlated cross-attention block (MCCA) to extract multi-modality image features effectively.

This study focuses on brain tumor detection using medical imaging, primarily MRI scans, emphasizing early diagnosis. The proposed Ensemble Learning Driven Computer-Aided Diagnosis Model for Brain Tumor Classification [10] (ELCAD-BTC) incorporates a Gabor filtering method. This method enhances MRI image quality by reducing noise. Ensemble learning involves three deep learning models: EfficientNet, DenseNet, and MobileNets. These models are feature extractors. Additionally, a denoising autoencoder (DAE) is employed for brain tumor detection. The comparative analysis for MRI prediction is given in Table 16.1.

16.3 PROPOSED MODEL

16.3.1 Process flow

The flowchart is explained in Figure 16.3.

Data Collection: The dataset [11] from Kaggle contains MRI images of the brain, supplemented with manual FLAIR abnormality segmentation [12]

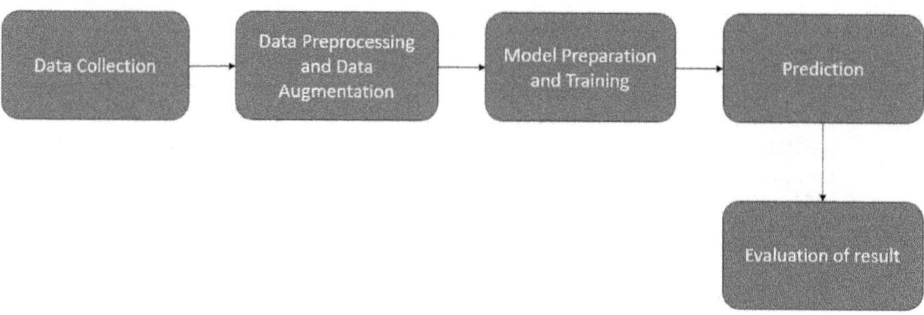

Figure 16.3 Flow chart of proposed system.

masks. FLAIR stands for fluid-attenuated inversion recovery, which is a specific type of MRI sequence used to obtain detailed images of the brain. FLAIR imaging is particularly useful for highlighting certain types of abnormalities or pathologies, such as brain tumors, multiple sclerosis lesions, and areas of inflammation. FLAIR images provide valuable information for diagnosing and monitoring neurological disorders, aiding clinicians in treatment planning and assessing disease progression. The images in the dataset are linked to 110 patients from The Cancer Genome Atlas (TCGA) low-grade glioma collection, revealing diverse images. Each image corresponds to a patient with available FLAIR sequence and genomic cluster data, facilitating comprehensive analysis. In addition to images, the dataset contains a data.csv file containing tumor genomic groups and patient information, which enhances contextual understanding with a total of 7,858 images, and provides a robust resource for production research and development of brain tumor diagnosis and classification.

In Figure 16.4, the first row is the MRI image and the second row is the mask image, which highlights the tumor.

Data Preprocessing and Data Augmentation: In our project, we employ preprocessing techniques like depth normalization and resizing to standardize MRI images, ensuring uniformity across the dataset. Additionally, we enhance dataset variety through augmentation methods such as rotation, flipping, and zooming. These techniques help introduce diversity into the data, enabling the model to generalize better to unseen examples. We utilize the Albumentations library for data augmentation [13], incorporating advanced methods such as channel dropout and random brightness/contrast modifications. Furthermore, we load MRI images and corresponding masks from the specified directory, extracting metadata such as patient IDs and file names. To facilitate model training, we split the dataset into training, validation, and test sets, ensuring robust evaluation. Overall, by leveraging preprocessing techniques and comprehensive data augmentation strategies, we aim to optimize model performance and robustness in tumor detection from MRI images.

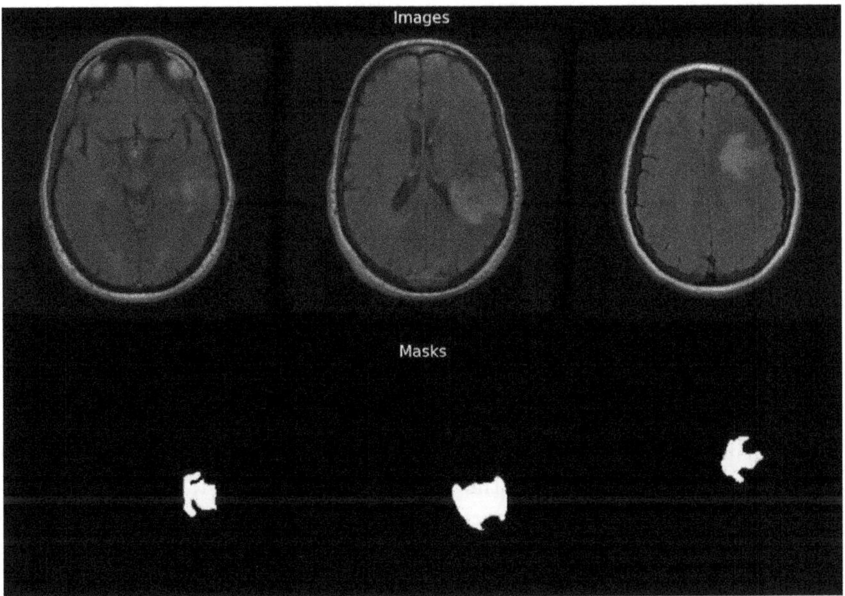

Figure 16.4 Images from dataset.

Moreover, the augmentation process plays a crucial role in mitigating overfitting by presenting the model with variations of the original data, effectively expanding its learning capacity. Through techniques like rotation, flipping, and zooming, we introduce distortions that mimic real-world scenarios, thus improving the model's ability to generalize to unseen data. Additionally, by incorporating metadata extraction and dataset splitting, we ensure proper organization and integrity throughout the training pipeline. This comprehensive approach not only enhances the model's performance but also fosters greater reliability and reproducibility in the analysis of brain MRI images for tumor detection.

Model Preparation and Training: Our methodology harnesses the power of two advanced deep learning architectures: EfficientNet B7 and UNet. By integrating EfficientNet B7, renowned for its exceptional feature extraction capabilities, with UNet, a proven architecture for semantic segmentation tasks, we create a powerful framework tailored specifically for identifying tumors in MRI images. Transfer learning plays a pivotal role, as we initialize the model with pretrained weights from EfficientNet B7, allowing the network to leverage knowledge gained from a vast amount of diverse data. Fine-tuning the initial layers ensures that the model adapts its learned representations to the nuances of tumor segmentation. Moreover, we tailor the architecture for binary segmentation, fine-tuning it to discern between tumor and non-tumor regions with high precision.

Overall, our approach represents a comprehensive fusion of cutting-edge techniques and domain expertise, aimed at tackling the challenging task of tumor detection in MRI images.

Prediction: In the realm of medical imaging, particularly in the critical task of detecting tumors within MRI scans, the process of prediction involves harnessing the power of a trained neural network model. This model, having undergone rigorous training on a dataset comprising labeled MRI images, has acquired the ability to discern subtle patterns and features indicative of tumor presence. When confronted with a new MRI image during the prediction phase, the model springs into action. Each MRI image is fed into the neural network, which meticulously sifts through the data, extracting and analyzing relevant features. Through its intricate layers, the neural network systematically processes the input image, accentuating areas that exhibit characteristics associated with tumors. This transformative journey through the neural network culminates in the generation of an output that encapsulates the model's prediction, effectively delineating regions suspected of harboring tumors. By leveraging the knowledge gleaned from its training data, the neural network demonstrates its prowess in extrapolating insights from unseen MRI images, thereby playing a pivotal role in the vital endeavor of tumor detection and diagnosis.

Evaluation and Results: In the evaluation of tumor detection models, precise assessment against ground truth tumor annotations is paramount. Metrics such as the Dice coefficient and Intersection over Union (IoU) play a pivotal role in quantifying segmentation accuracy. The Dice coefficient evaluates the degree of overlap between predicted and ground truth tumor regions, providing a comprehensive measure of segmentation performance. Meanwhile, IoU offers insight into the proportion of the shared area between the predicted and ground truth regions relative to their total area. These metrics enable a nuanced understanding of the model's performance, facilitating informed decisions regarding its reliability and efficacy in tumor detection tasks.

16.4 ARCHITECTURE

The combination of UNet architecture with EfficientNet-b7 as an encoder represents a potent fusion of two highly effective neural network structures tailored for image segmentation tasks. UNet, renowned in the field of semantic segmentation, employs a unique architecture featuring a contracting path for capturing contextual information and a symmetric expansive path for precise localization. This design allows for the extraction of intricate features crucial for accurate segmentation. EfficientNet-b7, on the other hand, is celebrated for its exceptional performance and computational efficiency, making it an ideal candidate for handling large-scale datasets efficiently. By integrating the strengths of both architectures, our approach capitalizes on UNet's robust segmentation capabilities while leveraging EfficientNet-b7's efficiency, resulting in a powerful and streamlined solution for tasks such as highlighting tumors in brain MRI images.

In this hybrid approach, we leverage the power of the EfficientNet-b7 architecture as the encoder component, tasked with extracting hierarchical features from input images. EfficientNet-b7, a member of the renowned EfficientNet family, is specifically designed to excel in capturing intricate patterns and representations within images. Its architecture, optimized through a systematic scaling method, strikes an ideal balance between model size and computational efficiency, ensuring robust performance across a wide range of image understanding tasks. By incorporating EfficientNet-b7 into our framework, we harness its capabilities to efficiently encode complex visual information, facilitating more accurate and comprehensive analysis of medical imaging data, particularly in tasks such as tumor detection and segmentation. This choice reflects our commitment to employing cutting-edge techniques to enhance the effectiveness and reliability of our solution in addressing critical healthcare challenges.

The UNet architecture is renowned for its effectiveness in semantic segmentation tasks, particularly in medical imaging. One of its key strengths lies in its utilization of skip connections, which facilitate the merging of low-level features from the contracting path with high-level features from the expansive path. This integration enables the model to maintain fine-grained spatial information while also capturing broader context, crucial for accurate segmentation. By incorporating EfficientNet-b7 as the encoder in the UNet architecture, we harness the extensive and hierarchical feature representations learned during pretraining on vast datasets such as ImageNet. EfficientNet-b7, known for its exceptional performance and efficiency across various vision tasks, enriches the UNet model with a deeper understanding of complex image structures. This fusion of UNet's skip connections with the feature richness of EfficientNet-b7 empowers our solution to achieve unparalleled accuracy and robustness in highlighting brain tumors from MRI images.

The integration of UNet with EfficientNet-b7 represented in Figure 16.5 is a significant advancement in medical image segmentation, particularly in tasks like highlighting brain tumors in MRI images. By combining these two architectures, our solution achieves enhanced performance by leveraging the strengths of each. UNet's innovative architecture excels at capturing fine details and contextual information, crucial for accurately delineating tumor boundaries. Meanwhile, EfficientNet-b7, known for its superior efficiency and scalability, complements UNet by optimizing computational resources without compromising accuracy.

This fusion results in a sophisticated and effective approach to medical image segmentation, ensuring precise and reliable detection of brain

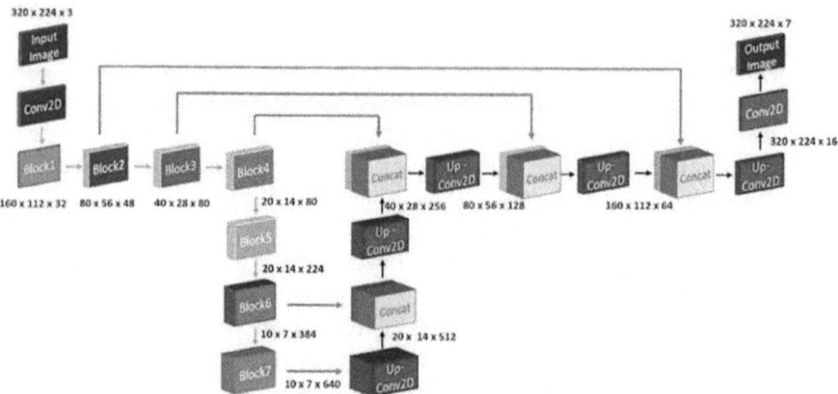

Figure 16.5 Architecture of UNet with EfficientNet-b7 encoder.

tumors in MRI imaging. By harnessing the synergies between UNet and EfficientNet-b7, our solution represents a significant leap forward in the field, promising more accurate diagnoses and improved patient outcomes.

16.4.1 Pseudo code

The code is explained in Figure 16.6 that implements a sophisticated semantic segmentation approach using the UNet architecture with an EfficientNet-b7 encoder, designed for precise and detailed classification of pixels in images. Semantic segmentation, a pivotal task in computer vision, involves assigning specific classes to each pixel, enabling a granular understanding of object boundaries within an image. Leveraging the 'smp.FPN' module from the segmentation models PyTorch library, which integrates a Feature Pyramid Network, facilitates robust feature extraction and integration across various scales, enhancing the model's ability to capture intricate details.

By selecting an Efficientnet-b7 encoder, pretrained on the ImageNet dataset, the model can effectively capture complex features in images, crucial for tasks requiring fine-grained classification. Configured to process input images with three channels representing the RGB color space, the model produces a single-channel map for binary segmentation tasks, utilizing the sigmoid activation function suitable for binary classification. Finally, the model is optimized for efficient execution on designated hardware resources, such as GPUs, ensuring high-performance operation. This implementation harnesses the UNet architecture and the advanced capabilities of the EfficientNet-b7 encoder to achieve accurate and precise semantic segmentation, particularly beneficial in domains such as medical imaging where pixel-level classification is essential for diagnosis and analysis.

```
model = smp.FPN(
    encoder_name="efficientnet-b7",
    encoder_weights="imagenet",
    in_channels=3,
    classes=1,
    activation='sigmoid',
)
model.to(device);
```

Figure 16.6 Pseudo code.

16.5 EXPERIMENTAL RESULTS

Graph plots are essential in ML projects as they provide visual insights into model performance, enabling quick identification of trends, patterns, and anomalies. They help in assessing the effectiveness of different algorithms, hyperparameters, and preprocessing techniques, aiding in model selection and optimization. Graphs also facilitate interpretation and communication of results to stakeholders, fostering better decision-making. Moreover, they assist in diagnosing issues such as overfitting or underfitting, guiding adjustments to improve model generalization. Overall, graph plots play a pivotal role in understanding, evaluating, and improving the performance of ML models.

Figures 16.7 and 16.8 are graph plots that depict the performance of the model.

16.5.1 Training and validation loss

Training and validation loss are crucial metrics displayed in graph plots. These plots typically depict how the loss decreases over epochs during model training, both for the training dataset and a separate validation dataset. The training loss indicates how well the model fits the training data, while the validation loss demonstrates its performance on unseen data, providing insights into generalization. Monitoring these trends helps in assessing model convergence, identifying overfitting or underfitting, and determining the effectiveness of regularization techniques. By observing the relationship between training and validation loss, practitioners can make informed decisions to optimize model performance and ensure robustness. The plot of training and validation loss graphs should typically display loss values on the y-axis and epochs (or iterations) on the x-axis. Both training and validation loss curves should be plotted on the same graph for comparison. Ideally, the graphs should depict a decreasing trend for both training and validation loss over epochs, indicating that the model is learning and improving its performance.

We can see in the graph that the train loss and test loss have decreased significantly.

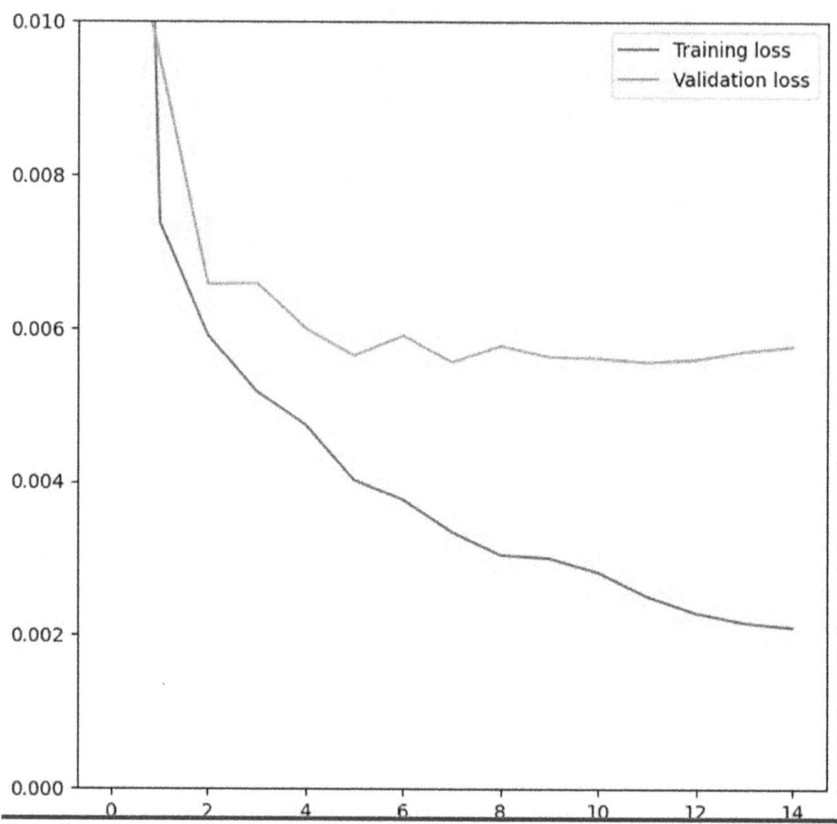

Figure 16.7 Graph depicting the training and validation Loss.

16.5.2 Validation mean Jaccard index and Dice coefficient

The mean Jaccard Index is also known as IoU.

Dice coefficient and IoU are used measurements for semantic segmentation. Mean of verification Jaccard index and Dice coefficient are important metrics to evaluate the performance of the segmentation models in tasks such as image segmentation Jaccard index, also known as IoU, measurements are similar to the segmentation mask between predicted and ground truth mask. A higher Jaccard index indicates better agreement between the predicted and ground truth masks.

Similarly, the Dice coefficient, also known as the F1 score, measures the similarity between the predicted and ground truth masks. It is calculated as the intercept divided twice by the sum of the areas in the predicted ground truth area. As with the Jaccard index, a high Dice coefficient indicates that the classification is very accurate.

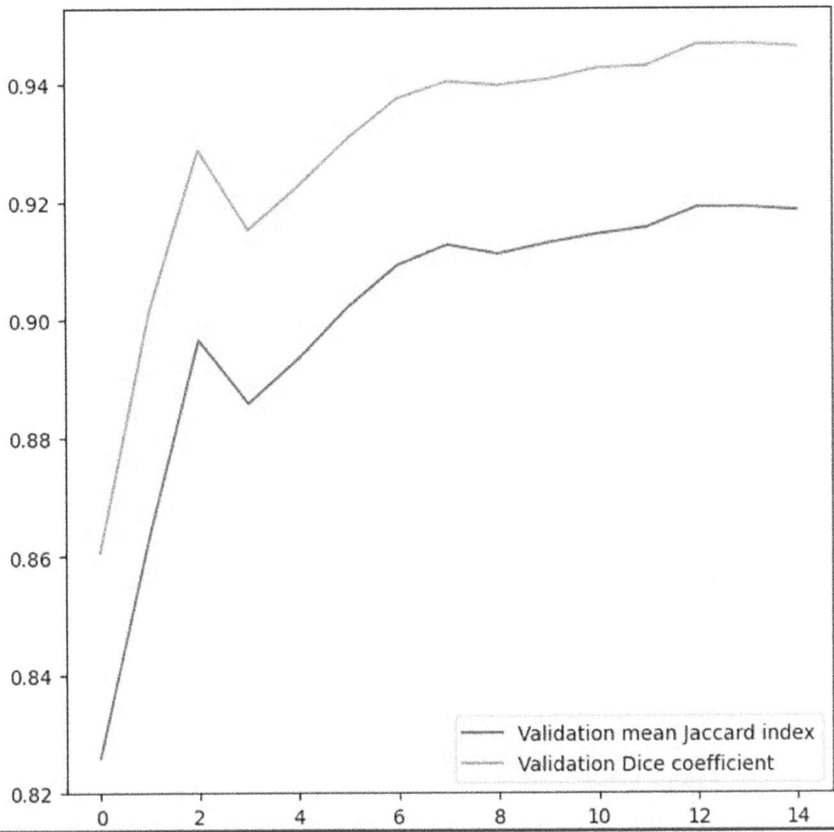

Figure 16.8 Graph depicting validation mean Jaccard index and validation dice coefficient.

Both metrics provide insight into the model's ability to accurately describe object boundaries in segmentation tasks. Validation mean Jaccard index or Dice coefficient combines these metrics across multiple validation samples, providing an overall assessment of the classification performance of the model. Generally, high values of these metrics indicate good classification, while values a low can indicate issues such as undersegmentation or oversegmentation.

Dice coefficient: 0.94391

IoU: 0.91591 These values are shown in Figure 16.8.

16.5.3 Outputs

The marked output is shown in Figures 16.9 and 16.10.

The area of green boundary is the predicted abnormality and that of red is the ground truth.

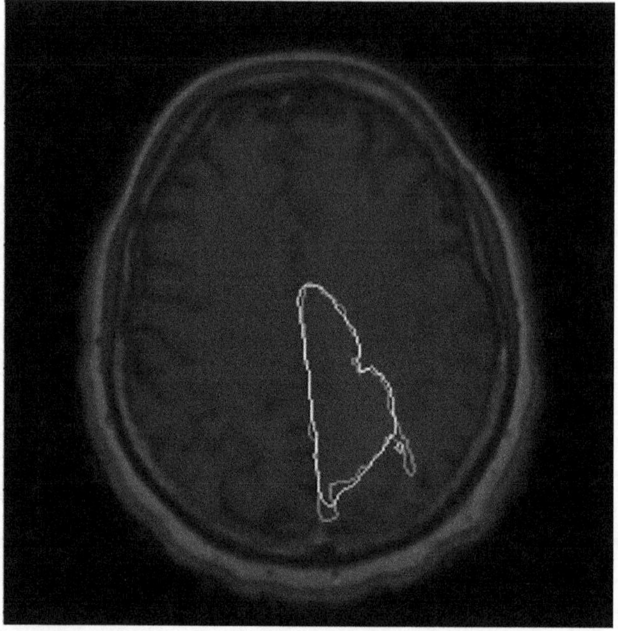

Figure 16.9 Output 1.

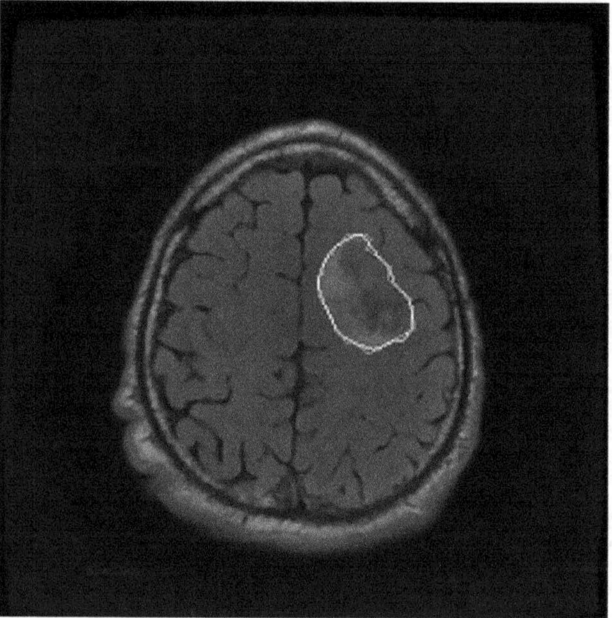

Figure 16.10 Output 2.

16.6 CONCLUSION

In conclusion, the study has successfully addressed the critical task of tumor segmentation in brain MRI scans. By leveraging the UNet architecture and incorporating the powerful EfficientNet-b7 as an encoder, the model demonstrates impressive performance metrics. The achieved Dice coefficient of 0.94391 reflects the accuracy and precision in delineating tumor boundaries, showcasing the model's capability to capture intricate details. Moreover, the Mean IoU score of 0.91591 signifies the robustness of the segmentation model, indicating a high level of overlap between predicted and ground truth regions. These metrics underscore the effectiveness of the proposed approach in accurately highlighting brain tumors, a crucial step in medical image analysis for diagnosis and treatment planning.

REFERENCES

1. Musallam, A. S., A. S. Sherif, and M. K. Hussein. "A new convolutional neural network architecture for automatic detection of brain tumors in magnetic resonance imaging images." *IEEE Access* 10 (2022): 2775–2782.
2. Asif, S., et al. "Improving effectiveness of different deep transfer learning-based models for detecting brain tumors from MR images." *IEEE Access* 10 (2022): 34716–34730.
3. Mohsen, S., et al. "Brain tumor classification using hybrid single image super-resolution technique with ResNext101 32x8d and VGG19 pre-trained models." *IEEE Access* 11 (2023): 55582–55595.
4. Lin, J., et al. "CKD-TransBTS: clinical knowledge-driven hybrid transformer with modality-correlated cross-attention for brain tumor segmentation." *IEEE Transactions on Medical Imaging* 42 (2023): 2451.
5. Vaiyapuri, T., et al. "Ensemble learning driven computer-aided diagnosis model for brain tumor classification on magnetic resonance imaging." *IEEE Access* 11 (2023): 91398.
6. Huang, W., and Wang, J. "Automatic segmentation of brain tumors based on DFP UNet," *2022 IEEE 6th Information Technology and Mechatronics Engineering Conference (ITOEC)*, Chongqing, China, 2022, pp. 1304–1307, doi: 10.1109/ITOEC53115.2022.9734456.
7. Ali, M., Gilani, S. O., Waris, A., Zafar, K., and Jamil, M. "Brain tumour image segmentation using deep networks," *IEEE Access* 8 (2020): 153589–153598. doi: 10.1109/ACCESS.2020.3018160.
8. Mahyoub, M., Natalia, F., Sudirman, S., Jasim Al-Jumaily, A. H., and Liatsis, P. "Brain tumor segmentation in fluid-attenuated inversion recovery brain MRI using residual network deep learning architectures," 2023 *15th International Conference on Developments in eSystems Engineering (DeSE)*, Baghdad Anbar, Iraq, 2023, pp. 486–491, doi: 10.1109/DeSE58274.2023.10100119.
9. Shanmuga Sundari, M., and Jadala, V. C. "Neurological disease prediction using impaired gait analysis for foot position in cerebellar ataxia by ensemble approach," *Automatika* 64(3) (2023): 541–550.

10. Sundari, S., Divya, Y., Durga, K. B. K. S., Sukhavasi, V., Sugnana Rao, M.D., and Rani, M. S. "A stable method for brain tumor prediction in magnetic resonance images using finetuned XceptionNet," *International Journal of Computing and Digital Systems* 14(1) (2023): 1–14.

11. Dataset Available at https://www.kaggle.com/datasets/mateuszbuda/lgg-mri-segmentation.

12. Doma, M. K., Padmanandam, K., Tambvekar, S., Kumar, K., Abdualgalil, B., and Thakur, R. N. "Artificial intelligence-based breast cancer detection using WPSO," *International Journal of Operations Research and Information Systems (IJORIS)* 13(2) (2022): 1–16.

13. Padmanandam, K., Rajesh, M. V., Upadhyaya, A. N., Chandrashekar, B., and Sah, S. "Artificial intelligence biosensing system on hand gesture recognition for the hearing impaired," *International Journal of Operations Research and Information Systems (IJORIS)*, 13(2) (2022): 1–13.

Chapter 17

Multiple object tracking using deep learning and machine learning techniques

Mahesh Ratnaparkhe, U. Sivaji, Sachin Turreraa, and Sirra Yashwanth

17.1 INTRODUCTION

In the rapidly evolving landscape of computer vision and artificial intelligence, the demand for robust and efficient multiple object tracking (MOT) systems has surged across various domains, including surveillance, autonomous vehicles, and human-computer interaction. Addressing this demand, our project introduces a cutting-edge real-time MOT framework that amalgamates the strengths of YOLOv7 (You Only Look Once version 4), DeepSORT (Deep Simple Online and Realtime Tracking), and TensorFlow. Object detection and tracking are fundamental challenges in computer vision, particularly in scenarios with dynamic environments, occlusion, and varying object scales. YOLOv7, renowned for its accuracy and speed in object detection, forms the first layer of our framework. By providing a comprehensive understanding of the scene in real time, YOLOv7 lays the foundation for subsequent tracking processes.

To address the intricacies of object tracking over time, our system employs DeepSORT, a state-of-the-art tracking algorithm that seamlessly associates and tracks objects across consecutive frames. DeepSORT harnesses deep learning techniques to maintain tracking identities in the face of challenges such as occlusion and object interactions. The fusion of YOLOv7 and DeepSORT forms a powerful combination, enabling the system to accurately detect and persistently track multiple objects, a critical requirement in dynamic environments. Seamless integration and optimization of these components are achieved through TensorFlow, a versatile open-source machine learning library. TensorFlow serves as the backbone of our framework, ensuring efficient communication between YOLOv7 and DeepSORT, resulting in a cohesive and high-performance MOT system.

This project aims not only to provide an effective solution for real-time MOT but also to contribute to the advancement of computer vision capabilities. Through extensive benchmark evaluations and comparative analyses, we demonstrate the superior performance of our integrated framework, establishing its potential as a go-to solution for industries and applications where precision in object tracking is paramount.

DOI: 10.1201/9781003565529-17

17.2 RELATED WORK

The landscape of real-time MOT has experienced notable progress, with researchers proposing diverse methodologies to tackle the intricacies associated with object detection, recognition, and tracking. This section provides an extensive overview of pertinent works, categorizing them based on their fundamental contributions and approaches.

A real-time framework for tracking multiple objects, based on a modified Deep SORT combined with YOLO detection methods, achieves 84.9% accuracy on a custom MIX dataset, showcasing advancements in tracking precision and operational efficiency [1]. Utilizing deep learning architectures on an embedded GPU, a system for detecting and tracking multiple objects realizes a 51.1% enhancement in MOT accuracy (MOTA) on the MOT16 dataset, operating efficiently in real time with minimal power consumption [2]. Recent progress in deep-learning–based online MOT systems has tackled diverse challenges, elevating tracking efficacy. This chapter conducts an in-depth analysis of 95 contributions over five years, categorizing methodologies and networks to provide insights for future research [3]. Proposing a novel approach, this chapter integrates a convolutional neural network (CNN) with a histogram of oriented gradient (HOG) descriptor to augment MOT. The method enhances detection rates and data associations in real-time tracking scenarios [4]. Exploring the potential of artificial neural networks in learning similarity functions, this research introduces a tracker that rivals state-of-the-art methods on the MOT Challenge benchmark, exhibiting a 58% faster processing speed compared to recent baseline approaches [5]. A system designed for simultaneous detection and ranging of various objects achieves high accuracy and real-time performance, operating at nearly 15 FPS on 4 GPUs, thus proving suitable for applications in autonomous driving [6]. Through a comprehensive review of literature on data association techniques in MOT, this study elucidates advancements and hurdles, classifying methodologies based on performance metrics and outlining potential avenues for future exploration [7]. Introducing a joint trajectory locating and attributes encoding framework for real-time, online MOT, this chapter achieves heightened accuracy and robustness compared to detection-based tracking methods [8]. This chapter offers an exhaustive examination of the evolution, challenges, and recent strides in MOT, categorizing prior algorithms, discussing benchmark datasets, and spotlighting emerging trends and technologies [9]. Leveraging discriminative embedding features in the association stage, this research enhances tracking performance in real time, mitigating identity switches and achieving competitive outcomes compared to cutting-edge methods [10]. A novel strategy for MOT integrates joint detection and embedding (JDE) with a Swin Transformer, resulting in an 84.9% accuracy rate on a proprietary dataset. This approach demonstrates effectiveness in managing objects of various sizes and shapes [11]. SiamMOTION, a real-time system for tracking multiple objects, utilizes motion estimation techniques to efficiently monitor objects in video streams. It introduces a proposal engine and feature extraction mechanisms, surpassing existing trackers on standardized benchmarks [12]. YOLOv4-tiny, an optimized object detection method, is tailored for real-time performance by incorporating ResBlock-D modules and auxiliary network blocks. The proposed method exhibits accelerated object detection while maintaining accuracy comparable to YOLOv4-tiny [13]. SafeCOOK, an innovative system for real-time object tracking during meal preparation, integrates YOLO and KCF methodologies. It effectively identifies and tracks cooking utensils, enhancing safety in kitchen environments, and operates efficiently on embedded systems [14]. Recent advancements in deep learning have revolutionized object detection, addressing numerous challenges in computer vision. This chapter offers an extensive overview of recent progress, encompassing detection frameworks, feature representation, and assessment metrics [15]. This chapter combines various features to enhance the accuracy of CNNs, achieving state-of-the-art outcomes on the MS COCO (Common Objects in Context) dataset. The proposed features enhance object detection performance across a wide range of tasks and datasets [16]. YOLO-based detection methods provide advantages in real-time object detection. This research proposes a trainable framework for rapid and precise detection, surpassing alternative models in terms of both accuracy and efficiency [17]. In Ref. [18], the authors introduced a one-stage object detection framework based on YOLOv4, enhancing detection accuracy and real-time performance. It employs CSPDarknet53_dcn(P) as the backbone, replaces the last output layer with deformable convolution, introduces PAN++ for feature fusion, and incorporates five-scale detection layers for better accuracy on small objects. Additionally, it presents an optimized network pruning algorithm to address real-time performance constraints on vehicle-mounted computing platforms. Chen et al. [19] introduced YOLO-face, a specialized face detector built upon YOLOv3, enhancing face detection performance through tailored anchor boxes and a refined regression loss function. In Ref. [20], authors proposed a model that combines YOLOv3 for object detection and SORT for real-time tracking, with enhanced occlusion handling using velocity, acceleration, and appearance data for object location prediction.

17.3 PROPOSED SYSTEM

17.3.1 Phase 1: Preprocessing and model initialization

1. **Data Acquisition and Preprocessing:** Acquire video data from cameras or other sources. Preprocess the data by resizing, normalization, and potentially background subtraction if needed.

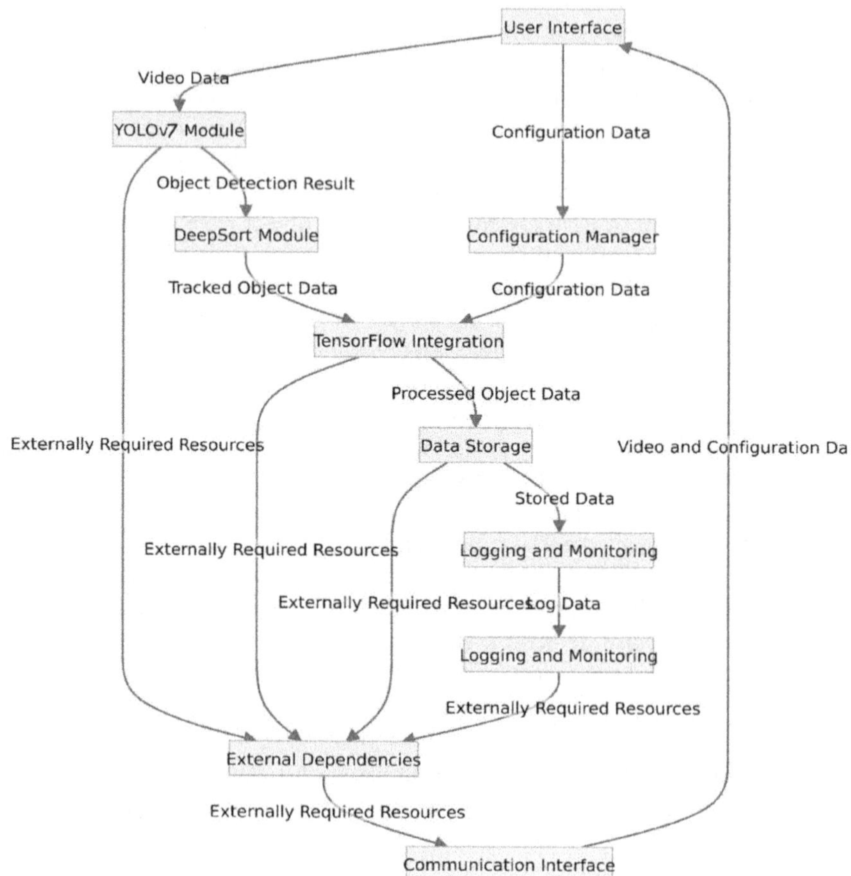

Figure 17.1 System diagram.

2. **Model Conversion:** Converts the pretrained YOLOv7 weights (originally in Darknet format) to a TensorFlow model for compatibility with the framework.
3. **DeepSORT Configuration:** Initializes DeepSORT with parameters like Kalman filter noise variances, appearance descriptor thresholds, and maximum track age.

17.3.2 Phase 2: Object detection and feature extraction

1. **YOLOv7 Inference:** Feeds each video frame to the TensorFlow integrated YOLOv7 model for real-time object detection.
2. **Bounding Box Refinement:** Applies non-max suppression or other techniques to eliminate redundant detections and refine the bounding boxes.

3. **Feature Extraction:** Extracts relevant features from the detected bounding boxes, such as DeepSORT's appearance descriptor based on color histograms or deep convolutional features.

17.3.3 Phase 3: Track management and association

1. **Track Initialization:** Creates new tracks for each detected object with features and bounding box information.
2. **Track Association:** DeepSORT's Hungarian algorithm matches detections from the current frame to existing tracks based on the intersection-over-union (IoU) and appearance features, ensuring that each detection is assigned to the most likely corresponding track by minimizing the overall assignment cost.
3. **Kalman Filter Prediction:** Estimates the likely location of each tracked object based on its previous motion.
4. **Appearance Similarity:** Compares the extracted features of detected objects with those of existing tracks to assess object identity.
5. **Distance Metric:** Considers the spatial distance between detections and predicted locations.
6. **Track Update:** Updates existing tracks with information from matched detections, including corrected positions, features, and timestamps.
7. **Track Termination:** Removes tracks that haven't been associated with detections for a predefined period, indicating object disappearance.

17.3.4 Phase 4: Tracking visualization and output

1. **Visualization:** Renders the video frames with bounding boxes and track IDs assigned by DeepSORT, showcasing the identified and tracked objects.
2. **Output Generation:** Optionally, output tracking information like object class, track ID, bounding box coordinates, and timestamps for further analysis or integration with other applications.

17.3.5 Algorithms used for object detection and tracking

17.3.5.1 YOLOv7 for object detection

Algorithm:

1. Divides input image into a grid of cells.
2. Each cell predicts bounding boxes and class probabilities for objects within it.
3. Uses a single-stage approach for efficient detection.

17.3.5.2 DeepSORT object tracking

Algorithm:

1. **Kalman Filter:** Predicts object positions in subsequent frames based on motion information.
2. **Hungarian Algorithm:** Associates object detections with existing tracks or creates new tracks.
3. **Deep Appearance Features:** Extracts features using a deep neural network for robust re-identification.

17.3.5.3 TensorFlow for deep learning

Algorithm:

1. Computation graph-based framework for building and training neural networks.
2. Uses automatic differentiation to optimize model parameters.
3. Supports various model architectures and training techniques.

17.4 RESULTS AND DISCUSSION

17.4.1 Object detection dataset

YOLOv7 is an object detection model, so you would need a dataset annotated for object detection. Commonly used datasets for this purpose include COCO, VOC (Visual Object Classes), or custom datasets annotated with bounding boxes around objects of interest. Each image in the dataset should be annotated with bounding box coordinates and corresponding class labels for the objects present.

17.4.2 Object tracking dataset

For training and evaluating the DeepSORT algorithm, you would need a dataset that provides not only the object detection annotations but also the temporal information required for tracking. Tracking datasets usually include video sequences where objects are tracked across frames. These datasets should have annotated tracks, indicating the identity of each object across frames.

MOT datasets commonly used include MOT17, MOT16, and MOT20. These datasets provide sequences of video frames with object annotations and ground truth tracks. To use YOLOv7 for object detection and DeepSORT for tracking, you need to preprocess your data to extract object detection annotations from the first type of dataset and use the second type of dataset for training and evaluating the tracking algorithm. It's important to note that combining these models often involves a multistep process, where you first use YOLOv7 to detect objects in each frame and then use DeepSORT to associate these detections across frames to create tracks. The overall pipeline involves integrating the object detection and tracking stages to achieve the goal.

As we can see in Figure 17.2, there are lots of objects such as women, dogs, cars, buildings, chairs, trees, and other people. We take a dataset in the form of a video or image to track the different objects present in it. There could be a scenario of occlusion in the input dataset, but YOLOv7 handles it precisely and helps in tracking the objects without any difficulties.

In Figure 17.3, our project leverages cutting-edge technologies such as YOLOv7 and DeepSORT to revolutionize object tracking in images and videos. YOLOv7, a real-time object detection system, serves as the foundation of our

Figure 17.2 Input.

Figure 17.3 Output.

```
163        # by default allow all classes in .names file
164        allowed_classes = list(class_names.values())
165
166        allowed_classes = ['person', 'dog', 'chair']
167
```

Figure 17.4 Objects can be added/dropped.

project. It efficiently identifies and localizes various objects, including persons, chairs, buildings, dogs, and more, with remarkable accuracy. Following the object detection phase, the tracking process is further refined using DeepSORT. DeepSORT excels in associating detections across frames, creating coherent tracks for each object over time. This technology not only handles occlusion scenarios gracefully but also enhances the tracking precision, enabling a seamless understanding of the temporal dynamics within the scene.

The synergy between YOLOv7 and DeepSORT allows our project to provide comprehensive insights into the spatial and temporal aspects of object interactions. YOLOv7 efficiently detects diverse objects in a single pass through the image or video, and DeepSORT builds upon this information to establish and maintain tracks, overcoming challenges such as occlusion in the input dataset. This amalgamation of advanced object detection and tracking technologies ensures that our project is at the forefront of efficient and accurate object tracking applications. Whether it's monitoring the movement of individuals, following the trajectory of a pet dog, or analyzing the spatial distribution of structural elements like buildings and chairs, our technology stack contributes to a wide range of fields, from video surveillance to behavior analysis (Figure 17.4).

In our object tracking system, the **allowed_classes** list plays a pivotal role in dynamically configuring the types of objects to be tracked in the input video or image. This list serves as a filter, allowing us to include or exclude specific object classes based on our requirements. By modifying the **allowed_classes** list, we have the flexibility to tailor the tracking process to focus on only those objects that are relevant to our needs.

For instance, if the input scene contains a diverse array of objects such as people, dogs, and chairs, adjusting the **allowed_classes** list enables us to selectively track only the objects of interest. This dynamic control over the tracked classes helps us avoid unnecessary occlusion issues and concentrate on monitoring the specific objects.

17.5 CONCLUSION

The fusion of YOLOv7, DeepSORT, and TensorFlow in the object tracking project presents a sophisticated and highly efficient system for monitoring and analyzing dynamic scenes captured through images or videos.

This project not only exemplifies the convergence of state-of-the-art technologies but also demonstrates its adaptability, versatility, and real-world applicability in various domains. At the heart of the system is YOLOv7, a powerful and fast real-time object detection model. Its ability to detect a diverse range of objects with remarkable accuracy in a single pass through input data sets the stage for the subsequent tracking phase. YOLOv7's efficiency in handling challenges such as occlusion ensures that even partially concealed objects are accurately identified. This capability is pivotal for tracking objects seamlessly in complex and dynamic environments. Following the object detection stage, the tracking phase is enhanced by DeepSORT, a sophisticated tracking algorithm. DeepSORT leverages YOLOv7's detections to establish and maintain tracks across frames, contributing to a comprehensive understanding of the temporal dynamics within a scene. The algorithm excels in associating objects over time, overcoming challenges posed by occlusion, and ensuring the precision and reliability of object trajectories. The collaboration between YOLOv7 and DeepSORT forms a robust foundation for the overall effectiveness of the tracking system. One notable feature that adds a layer of flexibility to the system is the allowed_classes mechanism. This feature allows users to dynamically configure the types of objects to be tracked by selectively including or excluding specific object classes based on their requirements. By modifying the allowed_classes list, users can tailor the tracking process to focus on only those objects that are relevant to their needs. This dynamic control over the tracked classes proves valuable in scenarios where different objects may appear or need tracking based on specific use cases or preferences. The versatility of the project is further emphasized by its adaptability to different applications. Whether the goal is video surveillance, behavior analysis, or any other domain requiring object tracking, the project excels in providing a comprehensive and precise understanding of dynamic environments. The capability to monitor individuals, track the movements of pets, or analyze the spatial distribution of objects like buildings and chairs showcases the broad spectrum of applications where this system can be effectively deployed.

REFERENCES

1. Meimetis, Dimitrios, et al. "Real-time multiple object tracking using deep learning methods." *Neural Computing and Applications* 35.1 (2023): 89–118.
2. Fernández-Sanjurjo, Mauro, Manuel Mucientes, and Victor Manuel Brea. "Real-time multiple object visual tracking for embedded GPU systems." *IEEE Internet of Things Journal* 8.11 (2021): 9177–9188.
3. Kalake, Lesole, Wanggen Wan, and Li Hou. "Analysis based on recent deep learning approaches applied in real-time multi-object tracking: a review." *IEEE Access* 9 (2021): 32650–32671.
4. Kalake, Lesole, et al. "Enhancing detection quality rate with a combined HOG and CNN for real-time multiple object tracking across non-overlapping multiple cameras." *Sensors* 22.6 (2022): 2123.

5. Meneses, Michel, et al. "Learning to associate detections for real-time multiple object tracking." *arXiv preprint arXiv:2007.06041* (2020).

6. Yang, Jiachen, et al. "A RGB-D based real-time multiple object detection and ranging system for autonomous driving." *IEEE Sensors Journal* 20.20 (2020): 11959–11966.

7. Rakai, Lionel, et al. "Data association in multiple object tracking: a survey of recent techniques." *Expert Systems with Applications* 192 (2022): 116300.

8. Wan, Xingyu, et al. "Multiple object tracking by trajectory map regression with temporal priors embedding." *Proceedings of the 29th ACM International Conference on Multimedia*, October 20 to October 24, 2021, in Chengdu, China. 2021.

9. Park, Yesul, et al. "Multiple object tracking in deep learning approaches: a survey." *Electronics* 10.19 (2021): 2406.

10. Weng, Zhenyu, et al. "Real-time multiple object tracking with discriminative features." *2020 16th International Conference on Control, Automation, Robotics and Vision (ICARCV): Shenzhen, China, 13–15 December, 2020.* Piscataway: IEEE, 2020.

11. Bashar, Mk, Samia Islam, and Kashifa Kawaakib Hussain. *Real-Time Multiple Object Tracking with Hierarchical Attention.* Gazipur: Diss. Department of Computer Science and Engineering (CSE), Islamic University of Technology (IUT), 2023.

12. Vaquero, Lorenzo, Víctor M. Brea, and Manuel Mucientes. "Real-time siamese multiple object tracker with enhanced proposals." *Pattern Recognition* 135 (2023): 109141.

13. Jiang, Zicong, et al. "Real-time object detection method based on improved YOLOv4-tiny." *arXiv preprint arXiv:2011.04244* (2020).

14. Ngankam, Hubert, et al. "Real-time multiple object tracking for safe cooking activities." *International Conference on Smart Homes and Health Telematics, Tours, France, July 1–3, 2009.* Cham: Springer Nature Switzerland, 2023.

15. Liu, Li, et al. "Deep learning for generic object detection: A survey." *International Journal of Computer Vision* 128 (2020): 261–318.

16. Bochkovskiy, Alexey, Chien-Yao Wang, and Hong-Yuan Mark Liao. "Yolov4: Optimal speed and accuracy of object detection." *arXiv preprint arXiv:2004.10934* (2020).

17. Sasikala, V., Asritha, G., Alekhya, P. Lakshmi, Pranavi, P. Naga, and Latha, M. Sneha. "Real-time multiple object detection network model based on YOLOV4." *2023 14th International Conference on Computing Communication and Networking Technologies (ICCCNT).* Piscataway: IEEE, July 6 to July 8, 2023, in Rome, Italy, 2023.

18. Cai, Yingfeng, et al. "YOLOv4–5D: An effective and efficient object detector for autonomous driving." *IEEE Transactions on Instrumentation and Measurement* 70 (2021): 1–13.

19. Chen, Weijun, et al. "YOLO-face: A real-time face detector." *The Visual Computer* 37 (2021): 805–813.

20. Nalawade, Roshan, et al. "Multiclass multiple object tracking." In: Chillarige, Raghavendra Rao, Salvatore Distefano, and Sandeep Singh Rawat (Eds.) *Advances in Computational Intelligence and Informatics: Proceedings of ICACII 2019.* Singapore: Springer, 2020.

Chapter 18

KGRecSys: Knowledge graph-based recommendation systems

A comprehensive overview

*B. Padmaja, G. Sucharitha, and
E. Krishna Rao Patro*

18.1 INTRODUCTION

Since it might be challenging for users to choose the best product due to the Internet's rapid expansion and vast diversity of information, recommender systems (RSs) have been developed and are now indispensable tools in our daily lives. Based on our past behaviour and preferences, RSs have shown to be a valuable tool for making individualized recommendations on new goods, services, and information [1]. Three forms of RSs are known to exist: collaborative filtering (CF) [2], content-based (CB) [3], and hybrid-based RSs [4]. Several techniques are employed in RSs.

Due to the exponential growth of the massive number of data on the Internet, traditional RSs have many difficulties, including data sparsity, scalability, and cold start. Originally, RSs employed algorithms to determine the likelihood that a user or customer would want to interact with a particular item, product, or service. When customers were browsing through a big catalogue of goods or services, these solutions were designed to alleviate the issue of information overload. When it came to providing clients with contextualized and personalized recommendations, earlier RSs were akin to black boxes. They worked by limiting the search results to a small fraction of products that were relevant to the user, but they were unable to offer reasons for their choices.

In RS wording, skilled are two individuals: (i) consumers (customers) and (ii) parts (brand). RS calculates the contingency between these two bodies, i.e., in what way or manner a user communicates accompanying a part, such as a search, purchase, visit, attend, etc., and uses this chance to advise the most appropriate subspace of items to that consumer.

The diagram shown in Figure 18.1 presents interactions between a user and an item (phone, headset, etc.). Solid black and white arrows indicate interactions, and dashed arrows indicate recommendations derived from the CF and CB algorithms [5]. In CF, recommendations are based solely on the user's past interactions with the item. In contrast, in CB, a user profile is built from item search information based on keywords and recommends favourites/similar items that the user likes [6].

DOI: 10.1201/9781003565529-18

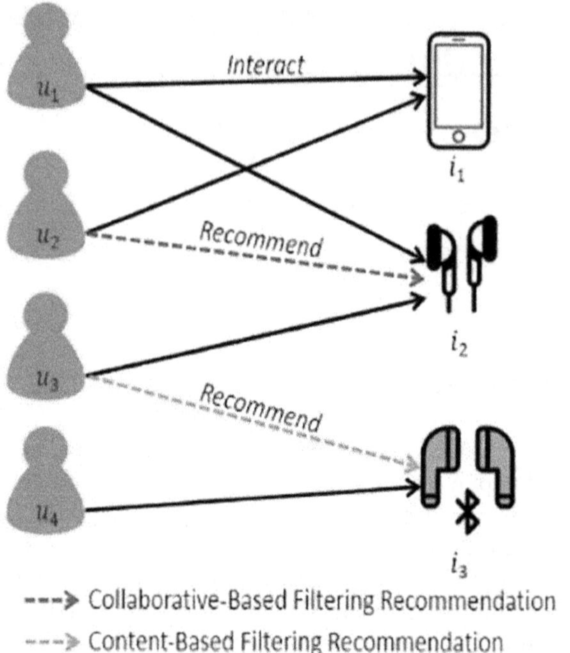

Figure 18.1 Bipartite user-item interaction graph for e-commerce products.

To overcome the questions of existent systems, information graphs (knowledge graphs (KGs)), a type of diagram database that arranges dossier and information into a structured plan to extract deep reasonable connections between articles and users, were popularized into RS. Knowledge-graph-located recommender systems (KGRSs) determine more correct and diverse embodied pieces of advice by joining the power of graphs accompanying the building of relevant dossier.

KG is an interconnected system used in RS accompanying a bud as a body and an edge as a relation. KG may be used to reckon the about syntax similarities betwixt individuals and supply an explanation of approvals to prevent the problem of dossier insufficiency. Another approach search out an embedded information diagram (knowledge graph embedding (KGE)) and information inference, which serves as the definitive model, into recommender systems (RS) to enhance recommendation performance. Deep information-knowledgeable network (DKN) is based on KGE, which creates headings for entities and connections and integrates headings from different implanting algorithms, and each heading will be second hand in deep neural network models in the way that CNN judges real records. Three different types of embedding are used such as word embedding, entity embedding, and relationship embedding [7]. There are many other

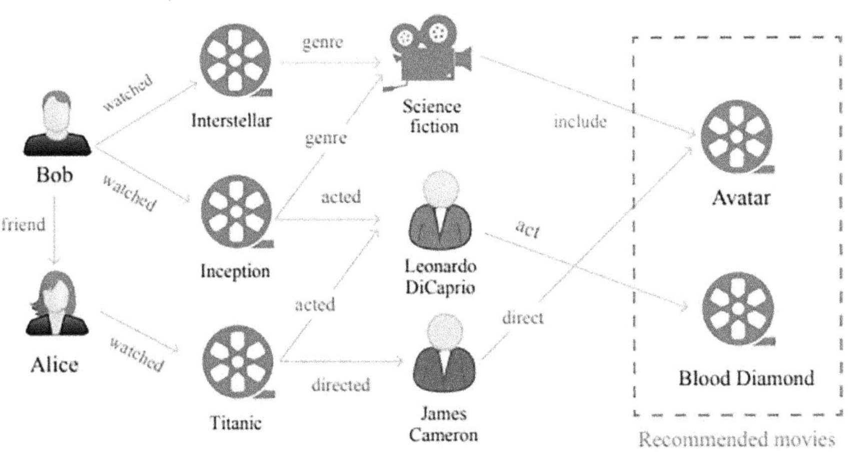

Figure 18.2 An example of a KG-based recommendation.

kinds of knowledge inference algorithms such as Ripple Network (RN), Collaborative Knowledge Base Embedding [8], Multi-task Learning [9], MetaGraph [10], etc. There are many other KGRSs such as path-based methods on embedding and graphs based on convolutional neural networks. All these methods provide varying degrees of accuracy and scalability depending on the size and complexity of the knowledge graph [11]. Figure 18.2 shows an example of a KG-based implementation.

18.2 RELATED WORK

18.2.1 Knowledge graphs

KG is a beneficial order for illustrating thorough knowledge from various fields. The Resource Description Framework (RDF) standard, at which point knots show systems and edges symbolize links between parts, is a favourite method to delimit KGs. Each creep in the diagram is depicted as a triple (a head individual, a connection, and a tail body), as known or named at another time or place in a case, that signifies the exact relationship middle from two points between the head system and the tail individual. Because skilled are many types of growth and connections in a KG graph, it is a miscellaneous network. Since various characteristics of a body may be obtained by following different edges in the diagram, and high-ranking body friendships may be established through these related links, such a diagram has a powerful competency for likeness. In the 1980s [12], when KGs were included in the architecture of expert systems for the social and medical sciences, the idea of a KG was developed. Later, linguistic and logical domains were added to the application. To better understand the search query and improve the usability of the results, Google included KG in the search architecture in 2012 [13]. Search engines, recommender systems, question-answering systems [14], relationship detection [15], etc. have all used KGs to make them more user-friendly [16]. Domain-specific KGs, such as Bio2RDF [17], fall into the latter category, while cross-domain KGs, such as Freebase [18], DBpedia [19], YAGO [13], and NELL [20], fall into the first group. In this survey, recommender plans use six of the cross-rule KGs, that we concisely present in this manner: Freebase was started by Metaweb in 2007 and purchased by Google in 2010.

In collaboration with OpenLink Software, academics from the Free University of Berlin and the University of Leipzig launched the open community project DBpedia [19]. The first adaptation, which was announced in 2007, is renovated every year. Primary news is captured from many accent histories of Wikipedia and DBpedia integrates the ruling class into an abundant graph makeup. The Max Planck Institute imported YAGO [13] (Yet Another Great Ontology) in 2007. It holds in addition to five heap details about things, places, and organizations. It inevitably collects news from Wikipedia and many additional sources such as WordNet [21] and GeoNames [22] and therefore combines the ruling class into an RDF diagram. Microsoft presented Satori as KG. Microsoft presented Satori as a knowledge graph (KG). Satori is integrated into the Bing search engine, similar to how the Google Knowledge Graph supports Google's search engine. Despite any candidly accessible Satori KG documents, Satori is stated to have held 300 heap subjects and 800 heap friendships in 2012 [15]. The best Chinese KG is CN-DBPedia [23]. Released by Fudan University in 2015, it has as well 220 heap connections and 16 heap entities. It without thinking acquires information from Chinese Wikipedia, Hudong Baike and Baidu Baike before merging the ruling class into the Chinese table. Little human work is necessary to maintain the system modern.

18.2.2 Recommender systems

Recommender systems are used in various fields such as news, education, places of interest, music, and film. The task of recommendation can be described as follows: it is to suggest one or more unobserved things to a given user. For a provided u_i and v_j item, the system first learns the representation of u_i and v_j. The scoring function $f: u_i \times v_j \rightarrow \hat{y}_{i,j}$, which represents the preference of u_i for v_j, is then learned by the system. Finally, item preference scores can be ranked to obtain a proposal. There are three basic methods explained below to learn the scoring function and the user/item representation:

- **Collaborative Filtering:** CF assumes that users could find items chosen by others with whom they have similar interaction histories interesting. Either explicit interaction, such as rating something, or implicit interaction [24,25], such as clicking and viewing, might occur during the interaction. The consumer-part interaction form is further made by interplay data from many consumers and parts that want to implement CF-based pieces of advice. Memory-located CF and model-located CF are the two basic methods in the CF-located approach. More expressly, the CF memory first uses consumer-article interplay data to decide consumer-to-consumer similarity.

- **Content-Located Filtering:** Content-located techniques show the consumer and item from the content of the parts, opposite to the CF-based model, which learns the likeness of the user and article from an all-encompassing consumer-item interplay dossier. Content-based cleaning everything on the assumption that the population is hopefully interested in output that is complementary to the entity they've interacted accompanying before. The user representation is derived from the characteristics of personally engaged items, while the item representation is obtained by extracting attributes from the item's auxiliary information, such as words, photos, etc. [26]. Comparing candidate items to a user's previous records is essentially a process of comparing them to the user's profile. Consequently, this method often suggests products that are comparable to products previously enjoyed by the consumer [27,28].

- **Hybrid Method:** A hybrid approach uses a variety of recommendation strategies to get over the drawbacks of relying just on one kind of approach. An important question with CF-located pieces of advice is the sparseness of consumer-part interplay data, which makes it difficult to label corresponding output or users in conditions of interplay. The cold start problem is a distinguishing instance concerning this question, which resources that it is hard to approve a new part or user because it is hopeless to determine the correspondence betwixt the part and the user outside some records of the interaction. Improved advice act may be achieved by accumulating consumer and item content facts, consistently referred to as item-side news and consumer-side information, to a CF-located foundation [29]. In this work, KGRSs connect a CF-located approach with KG as side facts to supply more correct recommendations.

18.3 STATE-OF-THE-ART APPROACHES IN RECOMMENDATION SYSTEMS

To avoid misunderstandings, the notation and concepts used in the research are explained before we dive into the state-of-the-art algorithms that use KG as background information for recommendations. A list of symbols and their meanings is given in Figure 18.3 for convenience.

Notations	Descriptions
u_i	User i
v_j	Item j
e_k	Entity k in the knowledge graph
r_k	Relation between two entities (e_i, e_j) in the knowledge graph
$\hat{y}_{i,j}$	Predicted user u_i's preference for item v_j
$\mathbf{u}_i \in \mathbb{R}^{d \times 1}$	Latent vector of user u_i
$\mathbf{v}_j \in \mathbb{R}^{d \times 1}$	Latent vector of item v_j
$\mathbf{e}_k \in \mathbb{R}^{d \times 1}$	Latent vector of entity e_k in the KG
$\mathbf{r}_k \in \mathbb{R}^{d \times 1}$	Latent vector of relation r_k in the KG
$\mathcal{U} = \{u_1, u_2, \cdots, u_m\}$	User set
$\mathcal{V} = \{v_1, v_2, \cdots, v_n\}$	Item set
$\mathbf{U} \in \mathbb{R}^{d \times m}$	Latent vector of the user set
$\mathbf{V} \in \mathbb{R}^{d \times n}$	Latent vector of the item set
$R \in \mathbb{R}^{m \times n}$	User-Item Interaction matrix
p_k	One path k to connect two entities (e_i, e_j) in the knowledge graph
$\mathcal{P}(e_i, e_j) = \{p_1, p_2, \cdots, p_s\}$	Path set between entity pair (e_i, e_j)
Φ	Nonlinear Transformation
$\odot$	Element-wise Product
$\oplus$	Vector concatenation operation

Figure 18.3 Symbols and notations used.

- **Heterogeneous Information Network (HIN):** An HIN is a directed graph $G=(V, E)$ with a mapping function φ: $V \rightarrow A$ for entities and ψ: $E \rightarrow R$ for links. Every entity v in V is a member of the entity type $\varphi(v)$ in A, and every link e in E is a member of the relation type $\varphi(e)$ in R. Furthermore, the number of relation types $|R|>1$ and/or the number of entity types $|A|>1$.

- **Knowledge Graph (KG):** A directed graph with nodes representing entities and edges representing subject-property-object triple facts is called a KG $G_{\text{know}}=(V, E)$. A relationship of r from entity eh to the entity is shown by each edge of the form (head entity, relation, tail entity) (denoted as $<$eh, r, et $>$). It can be thought of as an HIN instance.

- **Meta-path:** $\mathcal{P} = A_0 \xrightarrow{R_1} A_1 \xrightarrow{R_2} \cdots \xrightarrow{R_k} A_k$ is a meta-path. A_k is a path established on the graph of the network schema $G_T=(A, R)$, where $A_i \in A$ and $R_i \in R$ for $i=0, \ldots, k$. It defines a new compound relation $R_1 R_2 \ldots R_k$ between type A_0 and A_k. To extract aspects of connectivity from the graph, a sequence of relations can be used that connects pairs of objects in the HIN.

- **Meta-graph:** A meta-graph is a type of metastructure that combines two systems in an HIN, much like a meta-path. A meta-diagram, in another way, is an accumulation of various distinct meta-ways, since a meta-way only outlines one connection series. A meta-diagram can have more telling structural news between diagram individuals than a meta-path.

- **Knowledge Graph Embedding (KGE):** A KG $G_{\text{know}}=(V, E)$ is embedded into a low-dimensional space using KGE [30]. Following the embedding process, a d-dimensional vector represents each graph element, including the relation and the entity. Semantic meaning or high-order closeness in the graph can be used to quantify the intrinsic characteristic of the low-dimensional embedding that remains intact.

- **User Feedback:** The binary user feedback matrix $R \in R_{m \times n}$ is defined as follows with m users $U = \{u_1,...,u_m\}$ and n entries $V = \{v_1,...,v_n\}$.

  $R_{ij} = 1$ if the interaction (u_i, v_j) is observed;

  0, otherwise.

  Observe that an implicit interaction, such as clicking, watching, browsing, etc., exists between the user interface (UI) and the item (v_j), when R_{ij} has a value of 1. An implicit interaction of this kind does not always indicate that u_i is preferable to v_j. In this work, the term "user feedback" refers to implicit feedback unless specified otherwise. Nevertheless, specific feedback displaying the consumer's preference can likewise be available in a few particular instances. For instance, while recommending movies, users expressly provide a score between one and five. In certain studies, the user's preference in this situation is indicated by extracting data from score ratings of five.

- **H-hop Neighbour:** A multi-hop relation path might connect nodes in the graph as follows: In this scenario, e_H is e_0's H-hop neighbour, which is represented as. Recall that is equal to e_0.

- **Relevant Entity:** The set of k-hop relevant entities for user u can be written as

$$\mathcal{E}_u^k = \{e_t \mid (e_h, r, e_t) \in \mathcal{G} \text{ and } e_h \in \mathcal{E}_u^{k-1}\},$$

$$k = 1, 2, ..., H.$$

given the interaction matrix R and the knowledge graph G_{know}. The set of the user's historical interacted items is denoted by $\mathcal{E}_u^0 = \{u \mid R_{uv} = 1\}$.

- **User Ripple Set:** The knowledge triples with the head entities being $(k–1)$, hop relevant entities $\mathcal{E}_u^{k-1}$, are known as the ripple set of a user.

$$S_u^k = \{(e_h, r, e_t) \mid (e_h, r, e_t) \in \mathcal{G} \text{ and } e_h \in \mathcal{E}_u^{k-1}\},$$

$$k = 1, 2, ..., H.$$

- **Entity Ripple Set:** An entity $e \in G$'s ripple set is defined as

$$S_e^k = \{(e_h, r, e_t) \mid (e_h, r, e_t) \in \mathcal{G} \text{ and } e_h \in \mathcal{N}_e^{k-1}\},$$

$$k = 1, 2, ..., H.$$

18.4 APPROACHES FOR RECOMMENDATION SYSTEMS USING KNOWLEDGE GRAPHS

In this work, KGRSs are categorized into three types: embedding-located approaches, path-based methods, and united methods.

18.4.1 Embedding-located approaches

In general, implanting-located approaches straightforwardly use KG information to upgrade the likeness of objects or consumers. To impose upon KG data, information diagram implanting (KGE) methods must be used to generate a low-level sinking for KG. KGE algorithms can mainly be detached into two types: syntax matching models, in the way that DistMult, and rewording distance models, to a degree TransE, TransH, TransR, TransD, etc.

There are two types of insert-located approaches established in which the consumer is one suggestion for the knowledge graph (KG). The first type of approach uses components and associated attributes captured from external information bases or datasets to build information graphs. We consider this type of diagram as a part diagram. Note that in an item graph like this one, users are not included. This approach is used in documents that encode a graph for a more thorough item representation using KGE techniques and then include the item data in a recommendation framework. The following serves as an illustration of the main concept. For each article v_j, a dormant vector $\mathbf{v}_j$ is acquired by joining the dossier from several beginnings, containing part properties, content, and the consumer-article interaction form (KG). Either a blend of communicated item embeddings or a consumer-part interplay matrix may be used to extract the hidden heading UI of each user UI. Next, utilizing the merger of the entrenched interactions' elements, the prospect of the program that controls display choosing v_j may be computed. The tendency that u_i will pick v_j may therefore be computed utilizing

$$\hat{y}_{i,j} = f(\mathbf{u}_i, \mathbf{v}_j)$$

where $f(\cdot)$ denotes a function that maps consumer and article recommendation to an inclination score – which can be a DNN, quantity calculated from two vectors, etc. The results will come in downward order of advantage scores $\hat{y}_{i,j}$ during the recommendation phase.

For example, CKE was introduced by Zhang et al. [31] and unifies several kinds of secondary information within CF. They entrenched content (textual and optic) and structural information (part attributes represented by an information diagram) into computerized data in a system embedding piece. The paragraph element $\mathbf{z}_{t,j}$ and the optic material $\mathbf{z}_{v,j}$ are derived using an autoencoder design, while the dormant vector delineating the part's fundamental knowledge, $\mathbf{x}_j$, is encrypted utilizing the TransR technique.

Subsequently, these likenesses are linked accompanying the offset heading $\mathbf{\eta}_j$, which was extracted from the interplay model betwixt the user and the object. The last likeness of each article, $\mathbf{v}_j$, can be signified as

$$\mathbf{v}_j = \eta_j + \mathbf{x}_j + \mathbf{z}_{t,j} + \mathbf{z}_{v,j}$$

The KSR framework was presented by Huang et al. [32] for sequential recommendations. KSR uses KSR to model complex user preferences based on sequential interaction (knowledge-enhanced memory network key-value memory, KV-MN) combined with the GRU network. The KV-MN module models user preferences at the attribute level using knowledge base information (obtained by TransE) coupled with a GRU network to capture the consumer's subsequent priority. This allows recommendations of user preferences that are more precisely defined.

An alternative insert-located approach builds the user-part diagram straightforwardly, accompanying users, belongings, and connected attributes serve as nodes. Both consumer-accompanying friendships (co-buy, co-view, etc.) and attribute-level connections (brand, category, etc.) are part of the edges in the consumer-article graph. Equation (18.1) may be used to decide the consumer's desire after accumulating the embeddings of the systems in the diagram or by further embedding bureaucracy in the diagram by utilizing

$$\hat{y}_{i,j} = f(\mathbf{u}_i, \mathbf{v}_j, \mathbf{r})$$

In this case, $f(\cdot)$ transfers the connection implanting r into a scalar, the item representation v_j, and the user representation u_i.

CFKG was proposed by Zhang et al. [33] and creates a user-item KG. This consumer-article diagram contains various forms of subordinate facts about articles (review, brand, type, bought together, etc.) and treats user conduct (purchase, mention) as a distinct type of friendship between individuals. The model delineates a metric function $d(\cdot)$ to measure the distance middle from two points to two bodies by a distinguishing connection to determine the embedding of bodies and connections in the diagram. The invention ranks the competitor articles j according to the distance middle from two points $\mathbf{u}_i$ and $\mathbf{v}_j$ in growing order all along the advice chapter.

$$d(\mathbf{u}_i + \mathbf{r}_{\mathrm{buy}}, \mathbf{v}_j)$$

A greater inclination score $\hat{y}_{i,j}$ guides a tinier distance between u_i and v_j as contingent upon the "buy" link. SHINE, which uses the star authorization task as a sentiment relates guess task middle from two point bodies in the network, was projected by Wang and others [34]. Specifically, SHINE uses the friendly network Gr and the sketch news network Gp of consumers and marks (heroes) to forge an emotion network Gs for the ruling class, fortified accompanying subordinate facts from these networks. These networks are combined to represent the user and the target after being embedded using an autoencoder technique. Finally, using equation (18.1), where $f(\cdot)$ represents the DNN layer, a recommendation can be made.

Most techniques rely on embedding to improve item representation and construct a KG containing different item-side information. This information can be used to display a more accurate representation of the user.

The basis of embedding-based approaches is entity embedding; for better recommendations, some studies adjust the embedding using GAN or BEM. Embedding-located methods use data that is to say normally present in the diagram form. To improve the advice value, many authors use a multi-task education technique to train the approval piece in groups accompanying the graph-connected task.

18.4.2 Path-based methods

Path-based methods create a consumer-part network and create approvals based on the body's connectedness patterns inside the graph. There are immediately settled way-based approaches. Since 2013, normal studies have concerned this somewhat approach as an HIN approval. Typically, these models leverage the likeness in relatedness betwixt users and/or belongings to upgrade the advice. PathSim [35] is frequently used to quantify the connection similarity between elements in a graph. It is described as

$$s_{x,y} = \frac{2 \times |\{p_{x \rightsquigarrow y} : p_{x \rightsquigarrow y} \in P\}|}{|\{p_{x \rightsquigarrow x} : p_{x \rightsquigarrow x} \in \mathcal{P}\}| + |\{p_{y \rightsquigarrow y} : p_{y \rightsquigarrow y} \in \mathcal{P}\}|}$$

A particular kind of path-based technique uses the semantic similarities between entities in various meta-paths as a means of regularizing the graph and improving the way users and things are represented in the HIN, where $f(\cdot)$ stands for the inner product, one can anticipate u_i's preference for v_j.

Three major kinds of entity similarities are used:

- **User-User Similarity:** The objective function is

$$\min_{U,\Theta} \sum_{l=1}^{L} \theta_l \sum_{i=1}^{m} \sum_{j=1}^{m} s_{i,j}^l \parallel \mathbf{u}_i - \mathbf{u}_j \parallel_F^2$$

where $\Theta = [\theta_1, \theta_2, \ldots, \theta_L]$ indicates the weight for each meta-path, $U = [u_1, u_2, \ldots, u_m]$ indicates the latent vectors of all users, and $s_{i,j}^l$ indicates the similarity score of user i and j in meta-path l. The matrix Frobenius norm is represented by $\parallel . \parallel_F$. If users have a high meta-path–based similarity, the user-user similarity compels the users' embedding to be near together in the latent space.

- **Item-Item Similarity:** The objective function is

$$\min_{V,\Theta} \sum_{l=1}^{L} \theta_l \sum_{i=1}^{n} \sum_{j=1}^{n} s_{i,j}^{l} \parallel \mathbf{v}_i, \mathbf{v}_j \parallel_F^2$$

where the latent vectors of every item are indicated by $V=[v_1, v_2, \ldots, v_n]$. If an item's meta-path–based similarity is high, then its low-rank representations should also be close, much like the user-user similarity.

- **User-Item Similarity:** The objective function is

$$\min_{U,V,\Theta} \sum_{l=1}^{L} \theta_l \sum_{i=1}^{m} \sum_{j=1}^{n} (\mathbf{u}_i^T \mathbf{v}_j - s_{i,j}^{l})^2$$

If skilled in high correspondence betwixt users and parts established meta-paths, the user-part likeness expression shifts the hidden heading of users and articles for each other. Yu and others [36] projected Hete-MF, counted item-item correspondence in each way, and extracted L distinct meta-ways. A weighted non-negative form factorization approach [37] is linked with part-item regularization to upgrade the representation of consumers and depressed-level items for approval. Later, Luo et al. [37] introduced Hete-CF, which uses the regularization terms inter-user similarity, inter-item similarity, and inter-user similarity to determine a user's affinity to unrated items. As such, the Hete-CF model performs better than the Hete-MF model.

To obtain more complete representations of users and items, Yu et al. [38] proposed HeteRec, which uses meta-path similarities to improve the user-item interaction matrix R. L distinct kinds of meta-paths that connect users and objects in HIN are first defined by HeteRec. Subsequently, HeteRec-p was proposed by Yu et al. [39], which takes into account the importance of different meta-paths for different users. Instead of utilizing a worldwide preference model, HeteRec-p first divides consumers into c groups by their previous management and therefore uses the clustering news to specify personalized approvals. The changed scoring device enhances into

$$\hat{y}_{i,j} = \sum_{k=1}^{c} \mathrm{sim}(\mathbf{C}_k, \mathbf{u}_i) \sum_{l=1}^{L} \theta_l^k \cdot \hat{\mathbf{u}}_i^{(l)T} \hat{\mathbf{v}}_j^{(l)}$$

Where k,l represents the significance of meta-course l for the consumer group-k, and $\mathrm{sim}(\mathbf{C}_k, \mathbf{u}_i)$ signifies the cosine correspondence between the consumer $\mathbf{u}_i$ and the mark consumer group $\mathbf{C}_k$. Zhao et al. [40] constructed FMG by substituting the meta-diagram for the meta-way in consideration of the limit of the meta-way's ability for representation. FMG is better at arresting the likeness between entities because a meta-diagram has better connectedness information than a meta-way. Next, the model forges the hidden vectors of each meta-diagram's objects and consumers by administering matrix factorization (MF). Then, in consideration of calculating the advantage score $\hat{y}_{i,j}$, the features of consumers and objects across various meta-graphs are melded utilizing the factorization machine (FM). The FM takes into account how entities interact across various meta-graphs, which can further make use of patterns of connection.

The above route-based techniques use exclusively the user's preferred interaction data. SemRec, which takes into account the interplay of the user's favourite and despised previous items, was proposed by Shi et al. [41]. This framework integrates attribute values in a link using a weighted meta-path and a weighted HIN. More accurate article friendships and user likeness may be represented through these channels to transmit real user priorities by shaping both beneficial and negative desire patterns.

Another disadvantage of earlier techniques is the tedious fine-tuning of hyperparameters, such as the number of selected meta-paths. Ma et al. [42] proposed RuleRec as a way to use item connectedness in an external KG to learn relationships between related items (co-buy, co-view, etc.) and reduce workload. The rule knowledge piece and the item advice piece are treated as one by RuleRec. First, articles in an extrinsic KG are connected to associated individuals by way of the rule knowledge module. It then explains explainable rules, which are represented by meta-paths in the KG. Each rule's matching weight is further discovered. The user's purchase history is then combined with the learnt rules and rule weights by the item recommendation module, which uses the MF approach to produce recommendations. This paradigm makes the recommendation process explicable by outlining the rules and rule weights explicitly.

Path-based techniques, also known as HIN-based techniques in the past, produce suggestions based on user-item graphs. Conventional path-based techniques typically combine MF with HIN-extracted meta-paths. These methods use path connectedness to correct or regularize the likeness of the consumer or article. These approaches have the drawback of frequently requiring domain expertise to specify the kind and quantity of meta-paths. RuleRec attempts to get around the restriction by automatically taking advantage of rules in an external KG. As deep learning techniques have advanced, several models have been put out to explicitly represent the path embedding. Using path embedding or by identifying the most prominent paths connecting user-item pairings, recommendations can be produced.

18.4.3 United methods

To take brimming benefit of the data in KG for revised designs, united methods have been projected that integrate relatedness information about syntax likeness of bodies and relationships. The support of the united pattern is the idea of embedding publicity. These methods develop the entity likeness by utilizing the KG-related structure as a guide. It can be used to predict the user's preferences after gathering extended user interface (u_i) and/or potential item (v_j) representations.

The likeness of the consumer established in the past of their interactions is cultured in the primary round. Using the complicated UI articles as the main individuals, everything first extracts the multi-leap wavelet sets $\mathcal{S}_{u_i}^k$ $(k=1, 2, \ldots, H)$, where $\mathcal{S}_{u_i}^1$ is the threefold set (e_h, r, e_t) in the diagram. The theme concerning this plan search is to use the implanting of previously communicated parts and their multi-bounce neighbours to determine the consumer's sinking. An inexact description of the knowledge consumer likeness (UI) process is as follows:

$$\mathbf{u}_i = g_u\left(\{\mathcal{S}_{u_i}^k\}_{k=1}^H\right)$$

where $g_u(\cdot)$ is a bias-added concatenation function for embeddings of multiple hop entities. This arrangement may be understood as propagating the consumer's choice in the diagram because the diffusion begins accompanying the objects that the consumer is operating. RippleNet, the first study to demonstrate the concept of predilection diffusion, was imported by Wang and others [16]. To be more precise, RippleNet gives systems in the KG beginning embeddings originally. Next, it samples the ripple sets $\mathcal{S}_{u_i}^k$ from the KG $(k=1, 2, \ldots, H)$. The following illustration of the aggregation process might be used to improve the user representation.

$$p_i = \frac{\exp(\mathbf{v}_j^T \mathbf{R}_i \mathbf{e}_{h_i})}{\sum_{(e_{h_k}, r_k, e_{t_k}) \in \mathcal{S}_{u_i}^1} \exp(\mathbf{v}_j^T \mathbf{R}_k \mathbf{e}_{h_k})}$$

where the head entity's embedding in the ripple set is denoted by $\mathbf{e}_{h_i} \in R^d$, and the embedding of relation r_i is represented by $\mathbf{R}_i \in \mathbb{R}^{d \times d}$. The likeness between the head systems and the candidate part v_j is computed in the connection scope all along this process.

The consumer's 1-order reaction to the prior interplay can be kept in mind using

$$\mathbf{o}_{u_i}^1 = \sum_{(e_{h_i}, r_i, e_{t_i}) \in \mathcal{S}_{u_i}^1} p_i \mathbf{e}_{t_i}$$

where $\mathbf{e}_{t_i}$ stands for the tail entity's embedding in the ripple set. It can be used to find the user's h-order ($h=2, 3, ..., H$) answer $\mathbf{o}_{u_i}^h$ by substituting the $(h-1)$-order response $\mathbf{o}_{u_i}^{h-1}$ for v_j, and then iteratively interacting with head entities in the h-hop ripple set s_u^h. $u_i = \mathbf{o}_{u_i}^1 + \mathbf{o}_{u_i}^2 + \cdots + \mathbf{o}_{u_i}^H$ is the equation that yields the final representation of u_i. Ultimately, the preference score can be produced using

$$\hat{y}_{i,j} = \sigma(\mathbf{u}_i^T \mathbf{v}_j)$$

with the sigmoid function denoted by $\sigma(x)$. RippleNet propagates the user's preference based on past interests in this manner along the KG path.

Tang et al. [43] presented AKUPM, which models users using their click history, such as RippleNet. TransR is initially used by AKUPM for entity representation. AKUPM uses a self-attention layer to learn the relationships between things throughout each propagation phase, then it uses bias to spread the user's preference for various entities. Ultimately, the ultimate user likeness is acquired by amassing embeddings from different-order neighbours of communicated parts accompanying the self-consideration process. Later, the AKUPM was extended and RCoLM was established by Li and others [44]. AKUPM acts as the foundation for RCoLM's joint preparation of the approval and KG finishing modules. RCoLm unifies two modules and admits their shared augmentation, establishing the notion that each article in two modules endures the unchanging dormant likeness.

The second group of everything devotes effort to something cleansing the article likeness v_j by aggregating embeddings of a part's multi-bounce neighbours $\mathcal{N}_v^k$ ($k=1, 2, ..., H$). A general description of this process is

$$\mathbf{v}_j = g_v\left(\left\{\mathcal{S}_{v_j}^k\right\}_{k=1}^H\right)$$

where $g_v(\cdot)$ is the function to link embedding of multi-spring neighbours, and $\mathcal{S}_{v_j}^k$ is the ripple set of aspirant part $\mathbf{v}_j$. Concatenating the embedding of multi-bounce neighbours demands two processes. The exploratory search out enhances made acquainted a likeness of the two modules' k-hop neighbours, candidate item $\mathbf{v}_j$. RCoLm unites the two modules and promotes their mutual improvement.

$$\mathbf{e}_{\mathcal{S}_{v_j}^k} = \sum_{(e_h,r,e_t)\in\mathcal{S}_{v_j}^k} \alpha_{(e_h,r,e_t)}\mathbf{e}_t$$

In this case, $\alpha(e_h, r, e_t)$ indicates how important each neighbour is. The representation can then be changed for $e_h \in \mathcal{S}_{v_j}^k$ by

$$\mathbf{e}_h = \mathrm{agg}\left(\mathbf{e}_h, \mathbf{e}_{\mathcal{S}_{v_j}^k}\right)$$

where the aggregate operator is denoted by agg. In this procedure, the data from $(k–1)$-hop neighbours is combined with that from k-hop neighbours. There are four common types of aggregators in use:

- **Sum Aggregator:** Two representations are added using the sum aggregator, and then a non-linear transformation is applied.

$$\text{agg}_{\text{concat}} = \Phi\left(\mathbf{W} \cdot \left(\mathbf{e}_h \oplus \mathbf{e}_{\mathcal{S}^k_{v_j}} \right) + \mathbf{b} \right)$$

- **Concat Aggregator:** After concatenating two representations, the concat aggregator performs a non-linear change.

$$\text{agg}_{\text{concat}} = \Phi\left(\mathbf{W} \cdot \left(\mathbf{e}_h \oplus \mathbf{e}_{\mathcal{S}^k_{v_j}} \right) + \mathbf{b} \right)$$

- **Neighbour Aggregator:** The neighbour aggregator substitutes neighbour representations for an entity's representation directly.

$$\text{agg}_{\text{neighbor}} = \Phi\left(\mathbf{W} \cdot \mathbf{e}_{\mathcal{S}^k_{v_j}} + \mathbf{b} \right)$$

- **Bi-interaction Aggregator:** The bi-interaction aggregator takes into account the product relationships of entities as well as their sum. Additional information can be transferred from related companies using the second term.

$$\text{agg}_{\text{Bi-Interaction}} = \Phi\left(\mathbf{W} \cdot \left(\mathbf{e}_h + \mathbf{e}_{\mathcal{S}^k_{v_j}} \right) + \mathbf{b} \right)$$
$$+ \Phi\left(\mathbf{W} \cdot \left(\mathbf{e}_h \odot \mathbf{e}_{\mathcal{S}^k_{v_j}} \right) + \mathbf{b} \right).$$

Both the syntax implanting of the KG and the syntax course patterns are favourable to united methods. By utilizing multi-leap neighbours in the KG, these methods help the likeness of the object or consumer by applying the idea of embedding procreation. Since the RippleNet [16] was brought in in 2018, these studies have mainly selected a GNN-located design that suits the sinking diffusion process commonly. Research on these methods has been on the rise. Path-based methods derive their interpretability from unified forms. Compared to recognizing relatedness patterns in path-based methods, the diffusion process can be viewed as a means of detecting user preference patterns in the knowledge graph (KG).

18.4 DATASETS FOR KNOWLEDGE-GRAPH-BASED RECOMMENDER SYSTEMS

The KG-located design has the additional advantage of being smooth to define and correct. This kind of additional dossier can be naturally included in recommender arrangements for various purposes. KG-based recommender arrangements have been proven on several datasets under various environments to illustrate the effectiveness of KG as an additional dossier. We grouped everything by dataset in this place portion and showed by what method this synopsis is different over time. This portion forms two significant gifts. First, we present a summary of the second hand dataset in different situations. Second, we present a model of in what way or manner KGs are assembled for various advice tasks.

This portion can assist analysts in finding acceptable datasets for experiment recommender systems. Based on the datasets, we categorize KGRSs. Everything may be broadly detached into seven request scenarios, and we will use each basic document file to explain how everything generates KG.

- **Movie:** The task of the recommender structure in this place is to search out and obtain the consumer's preferences established earlier guarded cinema. The most used datasets are DoubanMovie and MovieLens [45]. The MovieLens site [46] supplies a group of datasets managed by MovieLens. The three most usually secondhand fixed reference datasets are MovieLens-100K, MovieLens-1M, and MovieLens-20M, each accompanying various grade numbers. Each dataset holds tags, attributes, and movie ratings. Douban [47], a well-known social media platform in China, is the source of DoubanMovie. Social relationships between users as well as user and movie attributes are included in the dataset. There are different approaches to building a film-related KG for proposals. Several articles create a movie-centric item graph to enhance movie information by taking movies and related characteristics from IMDB [48], Satori, DBpedia, Freebase, or CN-DBPedia. Movies are linked by characteristics such as genres, nations, stars, directors, and so on. The purpose of this item graph is to provide side data to the CF module. An alternative strategy is to ask users directly for theirs present the chart and rate the user as one type of relationship. While few still use outside databases to improve movie news, possible choice do so by straightforwardly leveraging film interplay data and face inside the MovieLens dataset or the DoubanMovie dataset.
- **Book:** Another common task is recommending books. Amazon-Book [49], DoubanBook [50], DBbook2014 [51], BookCrossing [52], and IntentBooks are the five frequently used datasets. Users' feedback on books is binary in Book-Crossing, DBbook2014, IntentBooks, and

Amazon-Book; the KG for each book in Satori DBpedia or Freebase is mapped to matching entities to create the dataset. The DoubanBook dataset, which is retrieved from Douban [50], includes attributes about the books, such as publisher, year of publication, and author, in addition to user-item interaction data. Without using a third-party KG, this work uses this knowledge to construct the user-item graph in the DoubanBook dataset.

- **Music:** The most widely used dataset for music recommendations is Last.FM [53]. The Last.FM online music system's user profile and listening history are included in the dataset [54]. Certain items devise the item diagram by communicable subgraphs from Freebase or Satori that have had a connection with pleasant sounds, harmonized. The KKBox dataset, which was made available by the WSDM Cup 2018 Challenge [55], is another well-liked dataset. The music description and user-item interaction statistics are both included in this dataset.

- **Product:** The Amazon Product dataset is the most often used dataset for the product recommendation task [51]. Type of item and consumer facts types, including interplay logs, user reviews, commodity categories, fruit writings, and user acts, are included in this place group. With this dataset alone, assemble a user-article diagram and construct the part graph by improving the part information accompanying data from the extrinsic Freebase database. Additionally, a few items use Alibaba Taobao's data.

- **POI:** Point of Interest (POI) approvals are fashioned for users who established their ancient check-in dossier, and they include new establishments and events (restaurants, museums, parks, towns, etc.). The Yelp Challenge [56], which includes data on users, companies, check-ins, and reviews, is the most often used dataset. For restaurant recommendations, another study makes use of the Dianping-Food dataset, which is made available by Dianping. com [57].

- **News:** Recommending revelation can be troublesome because the news is frequently time-impressionable and has extremely compacted content that is to say difficult to define without sound judgement. Additionally, individuals choose news to read based on topicality and may have a preference for news from other topics. Conventional news recommendation techniques are unable to identify the deeper connections within the news. Consequently, to increase suggestion precision and identify logical relationships between various news items, KGs are incorporated. The most widely used dataset is Bing-News, which was gathered from Bing-News server logs and includes details on user clicks, news titles, and other information. Extracting entities from the title is the first step in creating a KG for news recommendations. Next, neighbours are extracted to create subgraphs.

- **Social Media:** The purpose of this activity search out plans for community appendages community or meetings that may be of interest. One use for the assembled Weibo tweet dossier searches suggests unfollowed superstars to consumers on the social radio home Weibo. An item diagram containing knowledge collected from the Satori is constructed to embellish the facts about celebrities, despite the consumer-item diagram being signified to depict belief linkages betwixt users and stars. Using information from MeetUp, a social networking platform, users can also be recommended offline meetings with this application. The final application uses DBLP data to suggest conferences to scholars in the academic setting.

18.5 CONCLUSION

In this chapter, an inclusive survey of KG-located recommender wholes and a survey of the latest happenings engaged are bestowed. This survey shows what various procedures use KG as an additional dossier to improve the approval result and better the interpretability of the process. A summary of the datasets used in various scenarios is also given. The useful information found in the KGs can be utilized by KG-based recommender systems, which hold promise for providing accurate and comprehensible recommendations. In the advancement of this field's progress, future research directions are listed and it is hoped that the readers will gain a better understanding of this field of work from this chapter.

REFERENCES

1. Jain, S., Grover, A., Thakur, P.S., Choudhary, S.K. Trends, problems and solutions of recommender system. In *Proceedings of the International Conference on Computing, Communication and Automation*, Greater Noida, India, 15–16 May 2015; pp. 955–958.
2. Katarya, R. A systematic review of group recommender systems technique. In *Proceedings of the 2017 International Conference on Intelligent Sustainable Systems (ICISS)*, Palladam, India, 7–8 December 2017; pp. 425–428.
3. Seyednezhad, S.M., Cozart, K.N., Bowllan, J.A., Smith, A.O. A review on recommendation systems: Context-aware to social-based. arXiv 2018, arXiv:1811.11866.
4. Adomavicius, G., Tuzhilin, A. Toward the next generation of recommender systems: A survey of the state-of-the-art and possible extensions. *IEEE Transactions Knowledge Data Engineering* 17, 734–749, 2005.
5. Zhang, F., Yuan, N.J., Lian, D., Xie, X., Ma, W.Y. Collaborative knowledge base embedding for recommender systems. In *Proceedings of the 22nd ACM SIGKDD International Conference on Knowledge Discovery and Data Mining*, San Francisco, CA, USA, 13–17 August 2016; pp. 353–362.

6. Sarwar, B., Karypis, G., Konstan, J., Riedl, J. Item-based collaborative filtering recommendation algorithms. In *Proceedings of the 10th International Conference on World Wide Web, WWW '01*, New York, NY, USA, 1 April 2001. Association for Computing Machinery, 2001; pp. 285–295.

7. Wang, H., Zhang, F., Xie, X., Guo, M. DKN: Deep knowledge-aware network for news recommendation. *In Proceedings of the 2018 World Wide Web Conference on World Wide Web. International World Wide Web Conferences Steering Committee*, Lyon France, 23–27 April 2018.

8. Su, X., Khoshgoftaar, T.M. A survey of collaborative filtering techniques. *Advances in Artificial Intelligence* 2009, 1, 2009.

9. Wang, H., Zhang, F., Zhao, M., Li, W., Xie, X., Guo, M. Multi-task feature learning for knowledge graph enhanced recommendation. arXiv preprint arXiv:1901.08907, 2019.

10. Zhao, H., Yao, Q., Li, J., Song, Y. , Lee, D.L. Meta-graph based recommendation fusion over heterogeneous information networks. In *Proceedings of the 23rd ACM SIGKDD International Conference on Knowledge Discovery and Data Mining, Halifax, NS, Canada, 13–17 August 2017. ACM, 2017*.

11. Zhang, S., Yao, L., Sun, A., Tay, Y. Deep learning based recommender system: A survey and new perspectives. *ACM Computing Survey* 52, 1–38, 2019.

12. Nurdiati, S., Hoede, C. 25 years the development of knowledge graph theory: The results and the challenge. *Memorandum* 1876, 1–10, 2008.

13. Suchanek, F.M., Kasneci, G., Weikum, G. Yago: A core of semantic knowledge. In *Proceedings of the 16th International Conference on World Wide Web, Banff Alberta, Canada, 8–12 May 2007. ACM, 2007*; pp. 697–706.

14. Huang, X., Zhang, J., Li, D., Li, P. Knowledge graph embedding based question answering. In *Proceedings of the Twelfth ACM International Conference on Web Search and Data Mining, Melbourne, Australia, 11–15 February 2019. ACM, 2019*; pp. 105–113.

15. Paulheim, H. Knowledge graph refinement: A survey of approaches and evaluation methods. *Semantic Web* 8(3), 489–508, 2017.

16. Singhal, A. "Introducing the knowledge graph: Things, not strings," 2012, https://googleblog.blogspot.com/2012/05/introducing-knowledge-graph-things-not.html.

17. Belleau, F., Nolin, M.-A., Tourigny, N., Rigault, P., Morissette, J. Bio2RDF: Towards a mashup to build bioinformatics knowledge systems. *Journal of Biomedical Informatics* 41(5), 706–716, 2008.

18. Bollacker, K., Evans, C., Paritosh, P., Sturge, T., Taylor, J. Freebase: A collaboratively created graph database for structuring human knowledge. In *Proceedings of the 2008 ACM SIGMOD International Conference on Management of Data*, Vancouver, Canada, 9–12 June 2008. ACM, 2008; pp.1247–1250.

19. Lehmann, J., Isele, R., Jakob, M., Jentzsch, A., Kontokostas, D., Mendes, P. N., Hellmann, S., Morsey, M., Van Kleef, P., Auer, S., Christian, B. DBpedia–A large-scale, multilingual knowledge base extracted from Wikipedia. *Semantic Web* 6(2), 167–195, 2015.

20. Carlson, A., Betteridge, J., Wang, R.C., Hruschka Jr, E.R., Mitchell, T.M. Coupled semi-supervised learning for information extraction. In *Proceedings of the Third ACM International Conference on Web Search and Data Mining, New York, 4–6 February 2010. ACM, 2010*; pp. 101–110.

21. Miller, G.A. *WordNet: An Electronic Lexical Database*. MIT Press, Cambridge, 1998.
22. "The Geonames geographical database," 2006, https://www.geonames.org/.
23. Xu, B., Xu, Y., Liang, J., Xie, C., Liang, B., Cui, W., Xiao, Y. CN-DBpedia: A never-ending Chinese knowledge extraction system. In *International Conference on Industrial, Engineering and Other Applications of Applied Intelligent Systems*, June 14 to June 16, 2017, in Cairns, Australia. Springer, 2017; pp. 428–438.
24. Jawaheer, G., Szomszor, M., Kostkova, P. Comparison of implicit and explicit feedback from an online music recommendation service. In *Proceedings of the 1st International Workshop on16 Information Heterogeneity and Fusion in Recommender Systems*, Barcelona, 26–30 September 2010; pp. 47–51.
25. Qin, C., Zhu, H., Zhu, C., Xu, T., Zhuang, F., Ma, C., Zhang, J., Xiong, H. Duerquiz: A personalized question recommender system for intelligent job interview. In *Proceedings of the 25th ACM SIGKDD International Conference on Knowledge Discovery & Data Mining, Anchorage, AK, USA, 4–8 August 2019*. ACM, 2019, pp. 2165–2173.
26. Lops, P., De Gemmis, M., Semeraro, G. Content-based recommender systems: State of the art and trends. In: Ricci, F., Rokach, L., Shapira, B. (Eds.) *Recommender Systems Handbook*. Springer, New York, 2011; pp. 73–105.
27. Padmaja, B., Rama Prasad, V.V., Sunitha, K.V.N. A machine learning approach for stress detection using a wireless physical activity tracker. *International Journal of Machine Learning and Computing* 8(1), 2018. doi:10.18178/ijmlc.2018.8.1.659.
28. Vallabhuni, R.R., Lakshmanachari, S., Avanthi, G., Vijay, V. Smart cart shopping system with an RFID interface for human assistance. In *2020 3rd International Conference on Intelligent Sustainable Systems (ICISS), Thoothukudi*, India, 2020; pp. 165–169. doi:10.1109/ICISS49785.2020.9316102.
29. Sun, Z., Guo, Q., Yang, J., Fang, H., Guo, G., Zhang, J., Burke, R. Research commentary on recommendations with side information: A survey and research directions. *Electronic Commerce Research and Applications* 37, 100879, 2019.
30. Cai, H., Zheng, V.W., Chang, K.C.-C. A comprehensive survey of graph embedding: Problems, techniques, and applications. *IEEE Transactions on Knowledge and Data Engineering* 30(9), 1616–1637, 2018.
31. Zhang, F., Yuan, N.J., Lian, D., Xie, X., Ma, W.-Y. Collaborative knowledge base embedding for recommender systems. In *Proceedings of the 22Nd ACM SIGKDD International Conference on Knowledge Discovery and Data Mining, ser. KDD '16*, New York, NY, USA. ACM, 2016; pp. 353–362.
32. Huang, J., Zhao, W.X., Dou, H., Wen, J.-R., Chang, E.Y. Improving sequential recommendation with knowledge-enhanced memory networks. In *The 41st International ACM SIGIR Conference on Research & Development in Information Retrieval, Ann Arbor, MI, USA, 8–12 July 2018*. ACM, 2018; pp. 505–514.
33. Zhang, Y., Ai, Q., Chen, X., Wang, P. Learning over knowledge-base embedding's for recommendation. arXiv preprint arXiv:1803.06540, 2018.

34. Wang, H., Zhang, F., Hou, M., Xie, X., Guo, M., Liu, Q. Shine: Signed heterogeneous information network embedding for sentiment link prediction. In *Proceedings of the Eleventh ACM International Conference on Web Search and Data Mining, Marina Del Rey, CA, USA, 5–9 February 2018*. ACM, 2018; pp. 592–600.

35. Sun, Y., Han, J., Yan, X., Yu, P.S., Wu, T. Pathsim: Meta path-based top-k similarity search in heterogeneous information networks. *Proceedings of the VLDB Endowment* 4(11), 992–1003, 2011.

36. Yu, X., Ren, X., Gu, Q., Sun, Y., Han, J. Collaborative filtering with entity similarity regularization in heterogeneous information networks. *IJCAI HINA* 27, 1–6, 2013. Citeseer.

37. Luo, C., Pang, W., Wang, Z., Lin, C. Hete-CF: Social-based collaborative filtering recommendation using heterogeneous relations. In *2014 IEEE International Conference on Data Mining*, December 14 to December 17, 2014, in Shenzhen, China, 2014; pp. 917–922.

38. Yu, X., Ren, X., Sun, Y., Sturt, B., Khandelwal, U., Gu, Q., Norick, B., Han, J. Recommendation in heterogeneous information networks with implicit user feedback. In *Proceedings of the 7th ACM Conference on Recommender Systems, Hong Kong, China, 12–16 October 2013*. ACM, 2013; pp. 347–350.

39. Yu, X., Ren, X., Sun, Y., Gu, Q., Sturt, B., Khandelwal, U., Norick, B., Han, J. Personalized entity recommendation: A heterogeneous information network approach. In *Proceedings of the 7th ACM International Conference on Web Search and Data Mining, New York, NY, USA, 24–28 February 2014*. ACM, 2014; pp. 283–292.

40. Zhao, H., Yao, Q., Li, J., Song, Y., Lee, D.L. Meta-graph based recommendation fusion over heterogeneous information networks. In *Proceedings of the 23rd ACM SIGKDD International Conference on Knowledge Discovery and Data Mining, Halifax, NS, Canada, 13–17 August 2017*. ACM, 2017; pp. 635–644.

41. Shi, C., Zhang, Z., Luo, P., Yu, P.S., Yue, Y., Wu, B. Semantic path-based personalized recommendation on weighted heterogeneous information networks. In *Proceedings of the 24th ACM International Conference on Information and Knowledge Management, Melbourne, Australia, 18–23 October 2015*. ACM, 2015; pp. 453–462.

42. Ma, W., Zhang, M., Cao, Y., Jin, W., Wang, C., Liu, Y., Ma, S., Ren, X. Jointly learning explainable rules for recommendation with knowledge graph. In *The World Wide Web Conference*. ACM, May 13 to May 17, 2019, in San Francisco, California, USA, 2019; pp. 1210–1221.

43. Tang, X., Wang, T., Yang, H., Song, H. AKUPM: Attention enhanced knowledge-aware user preference model for recommendation. In *Proceedings of the 25th ACM SIGKDD International Conference on Knowledge Discovery & Data Mining, Anchorage, AK, USA, 4–8 August 2019*. ACM, 2019; pp. 1891–1899.

44. Li, Q., Tang, X., Wang, T., Yang, H., Song, H. Unifying task-oriented knowledge graph learning and recommendation. *IEEE Access* 7, 115 816–115 828, 2019.

45. "Movielens dataset," 1997, https://grouplens.org/datasets/ movielens/.

46. "Movielens website," 1997, https://movielens.org/.

47. "Douban movie," 2005, https://movie.douban.com/.
48. "Imdb," 1990, https://www.imdb.com/.
49. "Book-crossing dataset," 2004, https://www2.informatik. uni-freiburg. de/~cziegler/BX/.
50. "Douban book," 2005, https://book.douban.com/.
51. McAuley, J., Targett, C., Shi, Q., Van Den Hengel, A. Image-based recommendations on styles and substitutes. In *Proceedings of the 38th International ACM SIGIR Conference on Research and Development in Information Retrieval*, Santiago, Chile, 9–13 August 2015; pp. 43–52.
52. Uyar, A., Aliyu, F.M. Evaluating search features of Google knowledge graph and Bing satori: Entity types, list searches and query interfaces. *Online Information Review* 39(2), 197–213, 2015.
53. Schedl, M. The LFM-1b dataset for music retrieval and recommendation. In *Proceedings of the 2016 ACM on International Conference on Multimedia Retrieval*, New York, NY, USA, 6–9 June 2016; pp. 103–110.
54. "Last.FM online music system," 2002, https://www.last.fm/.
55. "Kkbox dataset," 2018, https://wsdm-cup-2018.kkbox.events/.
56. "Yelp challenge dataset," 2013, https://www.yelp.com/dataset/challenge/.
57. "Dianping.com," 2009, https://www.dianping.com/

Chapter 19

Brain-inspired cognitive architectures for artificial intelligence

Unlocking the potential of human-like intelligence

B. Padmaja, E. Krishna Rao Patro, G. Sucharitha Reddy, and Thota Anjushree

19.1 INTRODUCTION

In the pursuit of creating intelligent machines, researchers and scientists have long looked to the human brain for inspiration. The human brain, with its remarkable ability to perceive, learn, reason, and make decisions, remains one of the most fascinating and complex systems known to humanity. It is this complexity that has sparked the development of brain-inspired cognitive architectures for artificial intelligence (AI), aiming to replicate and harness the power of human-like intelligence. AI has advanced significantly over the last few decades, with advances in machine learning, deep neural networks, and natural language processing [1]. However, these approaches often fall short when it comes to emulating the cognitive abilities of the human brain. Traditional AI systems excel at specific tasks, but they lack the versatility, adaptability, and holistic understanding that humans effortlessly possess. Brain-inspired cognitive architectures offer a promising alternative, seeking to bridge the gap between human and AI. These architectural designs aim to emulate the fundamental operations of the human brain through computational models by drawing inspiration from its structure and functions. By doing this, scientists hope to open the door for machines to demonstrate cognitive functions including perception, attention, memory, decision-making, and problem-solving, which are all similar to those of humans. The main objective is to present an overview of AI's cognitive architectures that are inspired by the brain. We will investigate the fundamental ideas and workings of these structures, looking at how well they resemble the neuronal processes observed in the human brain.

Furthermore, we will delve into the various existing architectures, such as the Neural Engineering Framework (NEF), Hierarchical Temporal Memory (HTM), and the Global Workspace Theory (GWT), among others [2,3].

DOI: 10.1201/9781003565529-19

Understanding the potential of brain-inspired cognitive architectures for AI holds immense implications for various fields, including robotics, natural language processing, computer vision, and autonomous systems. By achieving a deeper understanding of these architectures, we can pave the way for the development of intelligent machines that possess human-like cognitive abilities, revolutionizing industries and transforming the way we interact with technology. In conclusion, this chapter seeks to shed light on the emerging field of brain-inspired cognitive architectures for AI. By exploring the principles, mechanisms, and existing architectures, we aim to unlock the potential of human-like intelligence in machines. Through this exploration, we hope to inspire further research and innovation in the quest for building intelligent machines that can truly understand, learn, and reason like humans. The study of brain-inspired cognitive architectures for AI has gained traction in recent years due to its potential to revolutionize the field. By replicating the intricate workings of the human brain, researchers aim to enhance AI systems' cognitive abilities and problem-solving skills.

19.1.1 Neural engineering framework (NEF)

In 2003, Eliasmith and Anderson established a theoretical framework called the Neural Engineering Framework (NEF). It seeks to offer a thorough understanding of how biological substrates might be used to implement cognitive processes. According to the NEF, neural systems can be thought of as engineering systems that employ computation, signal processing, and control theory concepts to accomplish cognitive tasks [4]. It implies that by applying a series of encoding and decoding transformations to map inputs to desired outputs, neural representations can be built to carry out calculations. The difficulty in creating and simulating large-scale models with the NEF is one reason it hasn't received much empirical backing yet. Nonetheless, large-scale models based on the NEF can be constructed and simulated using a program called Nengo. Nengo is the main tool used to create customized NEF models to explain experimental data and to teach how to utilize the NEF.

19.1.2 Hierarchical temporal memory (HTM)

HTM is a rising brain-stimulated cognitive and computing model derived from the discoveries of neuroscience and brain technology. It is based on a theory of hierarchical cortical computation and aims to model the functions of neurons and networks in the neocortex. HTM attempts to capture the capabilities of cortical columns and their interactions using a tree-shaped hierarchy that resembles traditional neural network topologies. HTM focuses on handling invariant representations in the visual cortex and aims to integrate memory components with neural networks [5]. Compared to other machine learning algorithms, it exhibits features closer to the functioning of the human neocortex. HTM has been applied to various areas, including image processing, and has shown stronger noise resistance compared to other AI neural techniques.

19.1.3 Global workspace theory (GWT)

GWT is a theory of consciousness and higher intelligence designed to explain the many interactions between consciousness and the unconscious. GWT proposes that consciousness and greater awareness arise from the competition and integration of information across various neural processes. There is a feeling as if the mind goes to the theater, the lighting equipment on the main stage. GWT argues that different types of senses can compete with each other for consciousness if their contents are incompatible. It also emphasizes the "behind-the-scenes" speech processes that produce subconscious unconscious content. GWT has made more impact on the design of knowledge and information and has been used to create the design model of the world [6]. In summary, the NEF provides a theoretical framework for implementing cognitive processes in a biological substrate, HTM is a brain-inspired cognitive and computing model, and GWT is a theory of consciousness and higher-order cognition. Each of these theories offers unique insights into different aspects of neural processing and cognition.

19.2 BRAIN-INSPIRED ARTIFICIAL INTELLIGENCE (AI)

19.2.1 Cognitive architecture for brain

Cognitive architecture refers to both a theory of the structure of the human mind and examples of these theories used in terms of AI. It involves developing a design that can be used to improve general information theory and work like AI. Many of the best cognitive skills are used in AI and cognitive skills. Some of these are those:

- **ACT-R (Adaptive Control of Thought-Rational):** ACT-R is a unified model of cognitive architecture that aims to encompass higher-level cognitive processes. It integrates architectural constraints from the brain and derives network parameters using biological learning.
- **SOAR:** SOAR is another cognitive architecture that focuses on general problem-solving and learning. It aims to simulate human cognitive processes and has been used in various AI applications.

These cognitive skills are designed to mimic the structure and function of the human brain, allowing the development of skills that can demonstrate complex and flexible behavior. They usually consist of several networks or modules that work together to accomplish intelligence missions. The development of these architectures involves incorporating architectural constraints from the brain while using biological learning to determine network parameters [7]. The whole brain architecture approach is another way to develop general artificial intelligence (AGI) through the use of the brain. It involves creating the Brain Reference Architecture (BRA), mapping the data flow and associated devices, and then designing all BRA devices. This approach aims to develop knowledge of the brain's behavior by extracting appropriate context from neuroscientific data. The brain-based cognitive architecture uses AI involved in the creation of computational models and designs that simulate the structure and function of the human mind. These models, such as ACT-R and SOAR, aim to simulate cognitive processes and intelligent behavior in intelligent machines. The holistic brain architecture approach also focuses on creating knowledge about the behavior of the brain through the transmission of neuroscientific information.

The whole brain approach breaks down brain processes, such as artificial general intelligence (AGI), into brain-based adaptive architecture (BRA), interconnecting information flows and images of the parts and using the BRA to manipulate each part [8]. This is called BRA-driven evolution. Figure 19.1 gives a basic idea of the WBA (Whole Brain Architecture) approach. Figure 19.2 shows the design and implementation of the BRA.

The BRA works as a model of the brain's neural circuits, showing the brain's message flow and graphical connections. The anatomy of the brain is described in BIF as a diagram of local neural circuits, while a conceptual diagram (HCD) shows the expected state of computer hardware. While Neurobehavior and Processing (NBP) provides insight into the development of HCD by identifying the activity of specific neural regions, its role in BRA-driven development is a little more subtle. The evolution of AGI has been greatly inspired by the research of human intelligence (HI) [9,10].

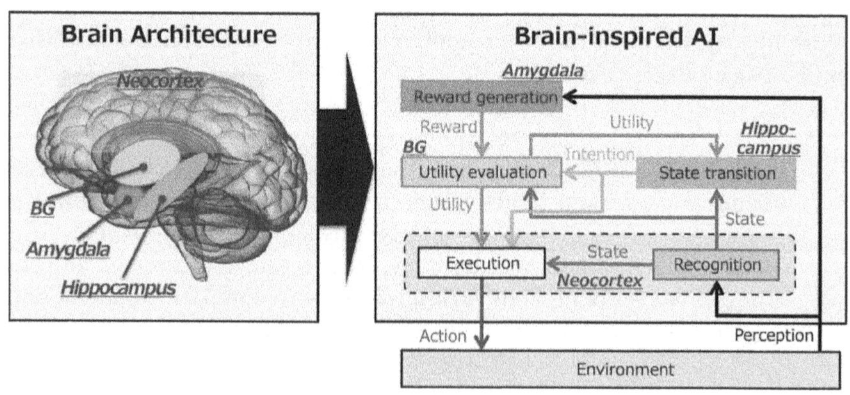

Figure 19.1 Basic scheme of WBA (whole brain architecture) approach.

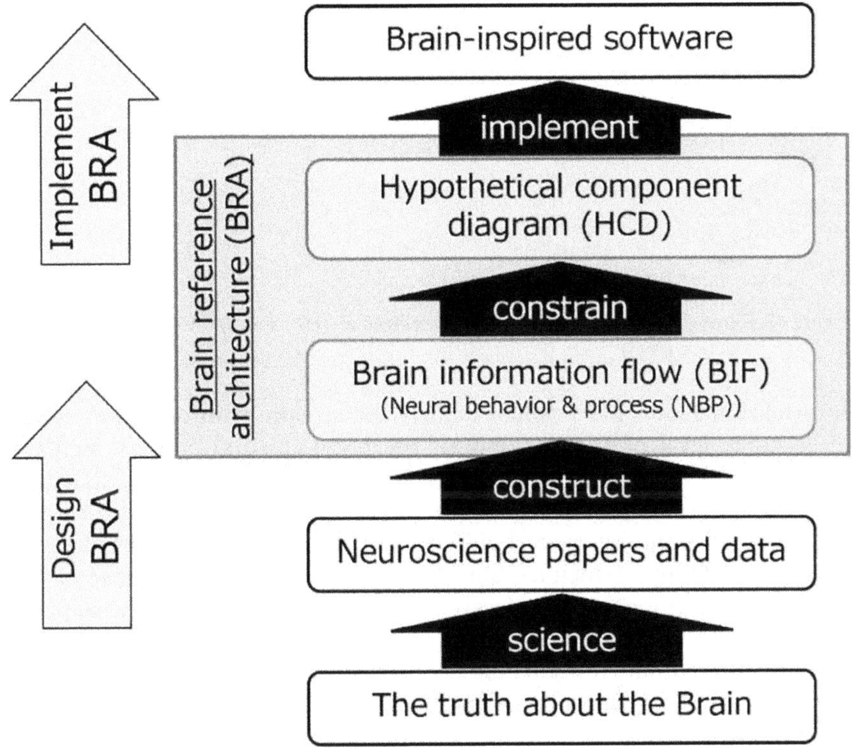

Figure 19.2 Brain reference architecture (BRA) design and implementation methods.

Consequently, AGI has a chance to advantage of human intelligence. Like, ongoing word models like ChatGPT and GPT-4 use strengthening learning with human feedback (RLHF) to align their behaviors with human values. While we persist in studying and grasping both social intelligence and AGI, these two systems will become increasingly enmeshed, improving and supportive each other in fresh and thrilling methods. Figure 19.3 depicts the interlinking between human intelligence (HI) and AGI.

19.2.2 Brain reference architecture (BRA)

BRA is a concept involved in the development of AI that aims to reproduce knowledge of the behavior of the human brain. This is part of a holistic brain architecture approach that divides brain-inspired AGI development into two tasks: building BRA and building all elements using BRA [11]. BRA serves as a reference model that integrates neuroscientific findings regarding the mesoscopic hierarchical structure of the brain. It provides a diagram of the data flow and related components that serve as a blueprint for brain-inspired software development. BRA includes the Brain

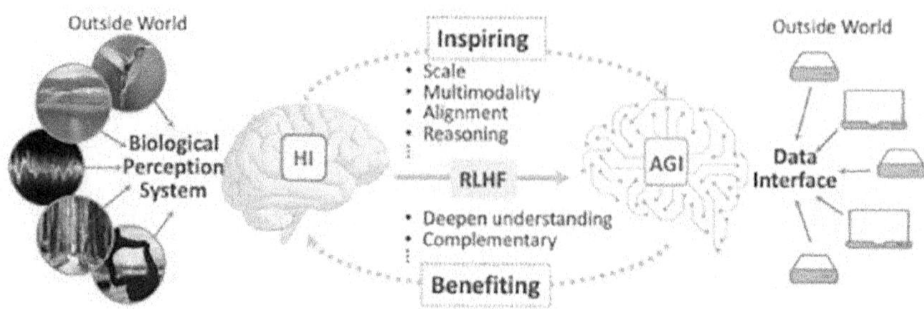

Figure 19.3 Interlinking between human intelligence (HI) and artificial general intelligence (AGI).

Information Flow (BIF), which defines the stream of information at the mesoscopic level of the anatomy of the brain's neural circuits, and the Hypothetical Component Diagram (HCD), which displays the structure of the brain [12]. The development of BRA-based AGI involves reverse engineering the brain's neural circuits and creating software that follows the standards and principles outlined in BRA. The development of the BRA-driven approach is to develop AGI using BRA as a blueprint and implement software based on it. BRA is a reference model that reflects neuroscientific findings regarding the mesoscopic hierarchical structure of the brain and provides computational theories consistent with these findings. It is a standard file format for brain-like software specifications and has the following components:

1. **Brain Information Flow (BIF):** BIF is a concept used in the study of brain-inspired cognitive architectures and understanding the flow of information within the brain. BIF is a format for describing and sharing the whole brain architecture, specifically focusing on the information flow diagram. The BIF consists of circuits and connections between them. A circuit can correspond to a population of neurons, an organ in the brain (such as the hippocampus), or a circuit consisting of brain organs that perform a specific function [13]. The elements of BIF are defined by an ontology, with circuits and connections being defined as classes rather than instances. The BRA combines BIF with state computational design (HCD) and summarizes the research concepts of the mesoscale brain. BRA offers a functional perspective consistent with these findings. Understanding the flow of information in the brain is crucial to learning more cognitive mechanisms and creating brain-inspired AI. The brain processes information dynamics related to issues such as stability/instability of information flow, timing of processing sequences, and cognitive processing of information [14]. It is important to remember that studies of information flow in

the brain are complex and variable, and appropriate experiments are still being developed. Techniques such as diffusion-weighted magnetic resonance imaging (DW-MRI) and functional magnetic resonance imaging (fMRI) ultra-scanning are used to study the flow of information in the interaction between humans and the brain. Overall, Brain Information Flow (BIF) provides a framework for understanding and describing the flow of information within the brain, which is essential for developing brain-inspired cognitive architectures and advancing our understanding of cognitive processes.

2. **Hypothetical Component Diagrams (HCD):** Hypothetical Component Diagram (HCD) is a term used in science, especially in the context of the whole brain architectural approach, to improve general intelligence (AGI) by sending it to the brain [15]. In this way, HCD is a hypothetical image representing the components of the brain and their interactions, providing the interaction process relevant to neuroscience discoveries. HCD is part of the Brain Reference Architecture (BRA) development process, which involves creating data flows and creating diagrams of common objects. The BRA-driven development approach focuses on the creation of all BRA-based products. Importantly, the concept of HCD may be specific to the whole-brain architecture approach and its application in AGI development. HCD works as a thought-provoking model designed to create drawings based on neuroscientific knowledge and support the structure of one's brain. Figure 19.4 shows the structure of the functional material with a structure similar to BIF [16]. Here's an example of what BRA describes, combining BIF with HCD for central ganglia and performance-critical support: Blue A,

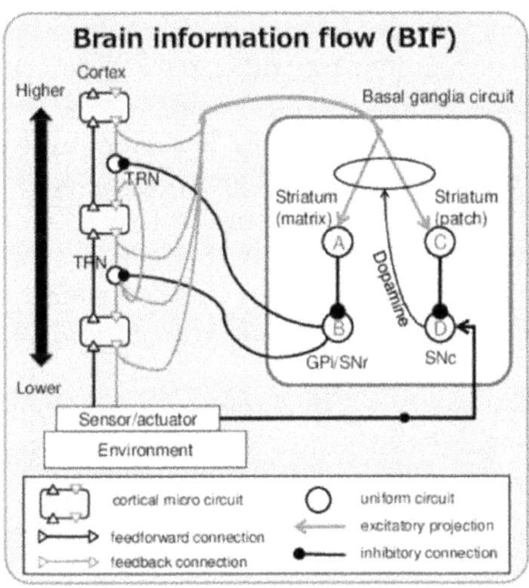

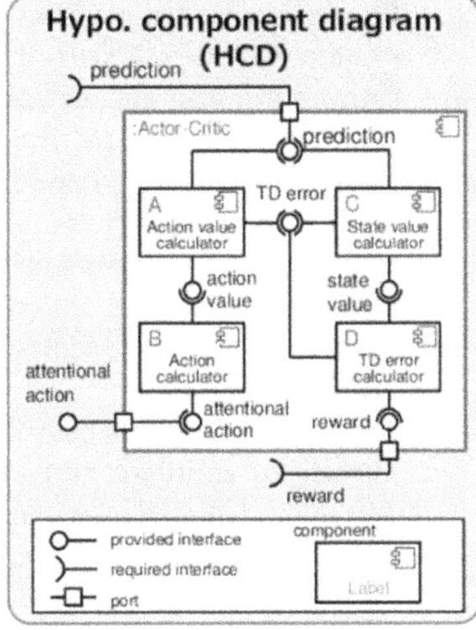

Figure 19.4 Shows BRA that pairs BIF with HCD.

B, C, and D symbols indicate normal sequences in the region of interest (ROI) in BIF (left). HCD elements are listed according to their functions while corresponding elements in the HCD (in the image at right) are labeled with similar letters. BRA provides a reference model based on neuroscience research and the computational capabilities of the brain and is an important factor in accelerating the development of AGI. It helps bridge the gap between neuroscience and cognitive science.

BRA is an important tool for the growth of brain-inspired software and AGI because it helps understand brain activity and extract the functional models needed to know how the system should behave.

19.2.3 Neural behavior and process (NBP)

The term "Neural Behavior and Process" (NBP) does not have a specific definition or widely recognized meaning in the context of brain research or cognitive science, based on the search results provided. It is possible that the term is not commonly used or may refer to a specific concept or framework within a particular research context [17]. However, based on the snippets from the search results, I can provide some general information related to neural behavior and processes:

1. **Neural Processes:** Neural processes refer to the various activities and functions that occur within the brain and its neural networks. These processes include information processing, signal transmission, synaptic plasticity, neural coding, and the generation of action potentials. Neural processes are fundamental to the functioning of the nervous system and show a crucial role in cognitive functions, sensory perception, motor control, and behavior.

2. **Behavior and Neural Bases:** The neural bases of behavior involve understanding how neural activity and processes give rise to specific behaviors. Researchers investigate the relationship between brain activity, neural circuits, and behavioral outcomes. This field of study aims to uncover the underlying neural mechanisms that contribute to different behaviors, such as learning, decision-making, emotion, and social interactions.

3. **Modeling Neural Behavior:** Researchers often develop computational models to simulate and understand neural behavior and processes. These models can help simulate the behavior of neural networks, analyze the dynamics of neural circuits, and investigate how specific neural processes contribute to cognitive functions. Computational models provide insights into the complex interactions between neurons and how they give rise to emergent behaviors.

19.2.4 The relation between brain and AI systems

The association among the brain and cognitive skills is very helpful. The brain is the key foundation of motivation after the development of cognitive processes and developments in cognition. It helps to better understand the brain and how it works [18]. When it comes to brains and skills, there is an exchange of information and ideas. The relationship between AI and the brain is a topic of interest in both neuroscience and AI research. Here are some key points:

1. **AI and the Internet:** There is a link between AI and the internet, although it has received relatively little attention. The development and prosperity of AI are closely tied to the advancement of the Internet.
2. **Insights from Brain Science:** Brain science provides insights into the general relationship between AI and various technologies such as the Internet of Things, cloud computing, big data, and the Industrial Internet. Understanding the brain can inform the construction of new AI system models that incorporate internet connectivity.
3. **Mutual Benefits:** The relationship between the brain and AI is mutually beneficial. The brain serves as a source of inspiration for the design of AI systems, leading to a better understanding of the brain itself. AI, in turn, can provide insights into the functioning of the human brain.
4. **Neural Networks:** Deep neural networks, which are a key component of AI systems, share some similarities with the neurons in the human brain. Both have interconnected units that transmit information and can exhibit plasticity, the ability to adapt and learn.
5. **Advancing Neuroscience:** AI is being used in neuroscience research to unlock new insights into the human brain. It helps in analyzing brain activity, understanding patterns of electrical activity in neurons, and studying brain responses to various stimuli.
6. **Bridging the Gap:** There is a need to establish a bridge between brain science and AI research. Efforts are being made to explore brain-imaging technology, establish dynamic connection diagrams of the brain, and integrate neuroscience experiments with theory and models.

19.2.5 Challenges in constructing brain-impaired AI structures

1. **Complexity:** This challenge is moderately difficult. Building AI systems based on brain models is the foundation of the brain-inspired approach to AI. With 100 billion neurons and over 600 trillion synaptic contacts, each neuron having 10,000 synaptic networks by other neurons on average the human brain is an innately complicated system. These synapses are always interacting in unpredictable and

dynamic behaviors [19]. Constructing AI systems with the goal of matching, or perhaps surpassing, that level of complexity is difficult in and of itself and needs similarly complicated numerical models.

2. **Data Required to Train Large Models:** Open AI's GPT 4 is currently ahead of the text-based AI model and requires 47 GB of data [20]. In comparison, its predecessor GPT3 uses 17 GB of data for training; which is about three times less. Think about how much training GPT5 will take. To achieve good results, brain-inspired AI models need to process a lot of data, especially auditory and visual tasks. This is crucial for establishing data collection pipelines. Tesla has 780 million miles of driving facts and enhances a million miles to its information collection pipeline each 10 hours.

3. **Energy Efficiency:** It is very difficult to create brain-building cognitive structures that make the brain more efficient. The human brain uses about 20 W of energy. By contrast, Tesla's Autopilot uses a custom chip that uses about 2,500 W of energy per second, and training an AI model the size of ChatGPT requires about 7.5 megawatt hours (MWh).

4. **Interpretation Issues:** Creating brain-generated intelligence that users can trust is essential to the development and acceptance of AI, but that is the problem. The cognitive model should be modeled after the brain, but it is a black box. Understanding the innermost works of the brain is difficult, in part because there is no information about how the brain works to think.

5. **Interdisciplinary Need:** The development of cognitive models for autism needs the expertise of professionals in various grounds such as neuro-science, computer science and engineering, philosophy, and psychology.

19.3 NEURO-INSPIRED AI

Understanding the inner workings of the human brain is crucial in developing advanced AI systems. By mimicking neural networks and cognitive processes, researchers can create AI models that can learn, adapt, and think like humans. This approach offers a more nuanced understanding of intelligence and cognition. Examining brain-inspired cognitive architectures allows for a deeper analysis of perception, reasoning, and decision-making processes in AI systems [21,22]. By integrating principles from neuroscience and psychology, researchers can develop AI models that exhibit more human-like behaviors and responses. Applications and use cases of brain-inspired cognitive architectures span many businesses, as well as healthcare, economics, and transport. These AI systems can assist in medical diagnoses, financial predictions, and autonomous vehicle control, among other tasks. The potential for improving efficiency and accuracy in these domains is significant. Comparing brain-inspired cognitive architectures with traditional AI models reveals the advantages of incorporating human-like cognitive processes. These architectures can handle complex, uncertain, and dynamic environments more effectively, leading to enhanced performance and adaptability. Challenges in implementing brain-inspired cognitive architectures include computational complexity, data requirements, and ethical considerations [23,24]. Addressing these limitations is crucial for the widespread adoption of such AI systems and ensuring their responsible use. Deep learning models can play an important role in creating brain-inspired cognitive architectures for AI. Deep learning models are computational models that mimic the activity and structure of biological neurons, making them suitable for simulating cognitive processes. Here are some deep learning models that can be used in brain-inspired cognitive architectures:

- **Artificial Neural Networks (ANN):** This model mimics the functions and patterns of neural networks. They have interconnected networks, or neurons, organized in layers. Deep neural networks through multiple hidden layers have been incredibly successful in many AI applications.
- **Spiking Neural Networks (SNNs):** These neural network models the dynamics of spiking neurons more closely. They use the timing and frequency of neuron spikes to encode and process information. SNNs have the probable to increase the efficacy and biological plausibility of AI systems.
- **Recurrent Neural Networks (RNNs):** These deep learning model can process consecutive sequential data by keeping an inner memory. This makes them well-suited for tasks like language processing and speech recognition that involve temporal dependencies and context.
- **Long Short-Term Memory (LSTM) Networks:** These models are a variation of RNNs that address the vanishing gradient problem and can identify long-term dependencies in data. They have been successfully used in various cognitive tasks, including natural language understanding and generation.
- **Convolutional Neural Network (CNN):** CNN is mainly used for image and video processing. They are stimulated by the group of the visual cortex and can extract hierarchical features from input data. CNNs are used in AI architectures to do various tasks such as object recognition and spatial perception.
- **Generative Adversarial Network (GAN):** GAN consists of two adversarial neural networks (generative and discriminator). GANs have been used to build real data models and can be used in cognitive modeling for tasks such as data synthesis and optimization.

- **Transformers:** These deep learning models have gained acceptance in natural language processing tasks. These models practice self-attention mechanisms to collect inter-relationships between dissimilar elements in a sequence, enabling them to model complex dependencies. Transformers have been used in cognitive architectures for tasks like language translation and sentiment analysis.
- **Reinforcement Learning:** These algorithms, inspired by the way humans and animals learn from rewards and punishments, have been successfully applied in AI. These algorithms enable AI **machines** to acquire **good ideas** over trial and error, similar to how the brain learns **through** feedback.
- **Memory and Attention Mechanisms:** Neuro-inspired AI models incorporate memory and attention mechanisms to enhance their ability to process and recall information. These mechanisms enable AI systems to store and retrieve relevant information, similar to how the brain utilizes memory.
- **Hybrid Models:** Neuro-inspired AI also explores hybrid models that combine elements of deep learning with other techniques, such as symbolic reasoning or probabilistic graphical models. These hybrid models aim to leverage the strengths of different approaches to achieve more robust and interpretable AI systems.

19.4 CONCLUSION

The research on brain-inspired cognitive architectures for AI using a deep learning approach has provided valuable insights and advancements in the field. By combining neuroscience concepts with deep learning, researchers are trying to develop AI models that can optimize the arrangement and purpose of the human brain. The usage of artificial neural networks (ANNs) and spiking neural networks (SNNs) is essential for creating AI that captures the dynamics and efficiency of neural networks. These architectures have shown promising results in a variety of AI applications such as image recognition, natural language processing, robotics, and healthcare. A combination of reinforcement learning algorithms enables smart machines to learn good ideas through trial and error, mirroring the way humans and animals learn from spirit and punishment. This has led to the development of AI that can make informed decisions and change behavior. Memory and tracking systems are also integrated into brain-inspired cognitive architecture, improving the capability of AI systems towards storing and retaining important information, similar to how the human brain uses memory. Research in this area not only helps create smarter and more efficient cognitive models but also provides fundamental insights into how the brain works. The similarities between the problems solved by deep learning algorithms and the brain have led to a revolution in neuroscience and enriched our understanding of cognitive processes. In summary, exploring brain-inspired cognitive architecture through deep learning opens up new ways to generate intelligence. By leveraging insights from the brain, researchers are making significant strides towards creating more intelligent, adaptive, and human-like AI systems that can revolutionize various domains, from healthcare to robotics. Continued research in this area holds immense potential for the future of AI, paving the way for groundbreaking advancements that can benefit society as a whole. Brain-inspired cognitive architectures offer a promising approach to advancing AI and understanding human intelligence better. Future research in this field should focus on overcoming existing challenges, refining cognitive models, and exploring new applications across various domains.

REFERENCES

1. Smith, J. (2018). Brain-inspired approaches to artificial intelligence. *Neural Networks Journal*, 25(2), 87–102.
2. Langley, Pat. (2006). "Cognitive architectures and general intelligent systems." *AI Magazine* 27(2), 33–33.
3. Pozzi, I., Bohte, S., & Roelfsema, P. (2020). Attention-gated brain propagation: How the brain can implement reward-based error backpropagation. *Advances in Neural Information Processing Systems*, 33, 2516–2526.
4. Kroner, A., Senden, M., Driessens, K., & Goebel, R. (2020). Contextual encoder-decoder network for visual saliency prediction. *Neural Networks*, 129, 261–270.
5. Salaj, D., Subramoney, A., Kraisnikovic, C., Bellec, G., Legenstein, R., & Maass, W. (2021). Spike frequency adaptation supports network computations on temporally dispersed information. *Elife*, 10, e65459.
6. Sheahan, H., Luyckx, F., Nelli, S., Teupe, C., & Summerfield, C. (2021). Neural state space alignment for magnitude generalization in humans and recurrent networks. *Neuron*, 109(7), 1214–1226.
7. Parr, T., Pezzulo, G., & Friston, K. J. (2022). *Active Inference: The Free Energy Principle in Mind, Brain, and Behavior*. MIT Press, Cambridge.
8. Coppolino, S., Giacopelli, G., & Migliore, M. (2021). Sequence learning in a single trial: A spiking neurons model based on hippocampal circuitry. *IEEE Transactions on Neural Networks and Learning Systems*, 33, 3178.
9. Pozzi, I., Bohté, S., & Roelfsema, P. (2020). Attention-gated brain propagation: How the brain can implement reward-based error backpropagation. *34th Conference on Neural Information Processing Systems (NeurIPS 2020)*, Vancouver, Canada. 2020-12-07.
10. Saxe, A., Nelli, S., & Summerfield, C. (2020). If deep learning is the answer, what is the question? *Nature Reviews Neuroscience*, 22(1), 55.

11. Sai Ruchit Reddy Ginnavaram, B. P., & Myneni, M. B. (2020). An intelligent assistive VR tool for elderly people with mild cognitive impairment: VR components and applications. *International Journal of Advanced Science and Technology*, 29(4), 796–803.

12. Merugula, S., Padmaja, B., & Veerubommu, R. (2022). RideNN-OptDRN: Heart disease detection using RideNN-based feature fusion and optimized deep residual network. *Concurrency and Computation: Practice and Experience*, 34(28), e7355.

13. da Costa, D., Kornemann, L., Goebel, R., & Senden, M. (2022). ConvNets develop characteristics of visual cortex when receiving retinal input. *2022 Conference on Cognitive Computational Neuroscience*, July 6 to July 9, 2022, in New York City, USA., 2022.

14. Cichy, R. M., Khosla, A., Pantazis, D., Torralba, A., & Oliva, A. (2016). Comparison of deep neural networks to spatio-temporal cortical dynamics of human visual object recognition reveals hierarchical correspondence. *Scientific Reports*, 6, 27755.

15. Padmaja, B., Bala, M. M., Patro, E. K. R., Srikruthi, A. C., Avinash, V., & Sudheshna, C. (2024). An intelligent auto-response short message service categorization model using a semantic index. *International Journal of Electrical and Computer Engineering (IJECE)*, 14(1), 922–933.

16. Marblestone, A. H., Wayne, G., & Kording, K. P. (2016). Toward an integration of deep learning and neuroscience. *Frontiers in Computational Neuroscience*, 10, 1–61.

17. Kriegeskorte, N. (2015). Deep neural networks: A new framework for modeling biological vision and brain information processing. *Annual Review of Vision Science*, 1, 417–446.

18. Lindsay, G. W. (2020). Convolutional neural networks as a model of the visual system: Past, present, and future. *Journal of Cognitive Neuroscience*, 33. https://doi.org/10.1162/jocn_a_01544, 2020.

19. Thompson, J. A. F., Bengio, Y., Formisano, E., & Schönwiesner, M. (2018). How can deep learning advance computational modeling of sensory information processing? arXiv https://arxiv.org/abs/1810.08651, 2018.

20. Ponce, C. R. (2019). Evolving images for visual neurons using a deep generative network reveals coding principles and neuronal preferences. *Cell*, 177, 999–1009.e10.

21. Saxe, A. (2015). Deep Linear Networks: A Theory of Learning in the Brain and Mind. Thesis, Stanford University.

22. Guclu, U., & van Gerven, M. A. J. (2015). Deep neural networks reveal a gradient in the complexity of neural representations across the ventral stream. *The Journal of Neuroscience*, 35, 10005–10014.

23. Kroner, A., Senden, M., Driessens, K., & Goebel, R. (2020). Contextual encoder-decoder network for visual saliency prediction. *Neural Networks*, 129, 261–270.

24. da Costa, D., Kornemann, L., Goebel, R., & Senden, M. (2023). Unlocking the secrets of the primate visual cortex: A CNN-based approach traces the origins of major organizational principles to retinal sampling. *BioRxiv*, 2023.

Chapter 20

Harnessing deep learning algorithms for enhanced stock price predictability

*B. Padmaja, E. Krishna Rao Patro,
Sri Sai Madhuvani Petla, and Tejaswini Vemulapalli*

20.1 INTRODUCTION

Prediction of stock costs has been a theme of enormous intrigue and importance within the budgetary world for a long time. The exact forecasting of stock costs can be truly difficult due to the complex and dynamic nature of monetary markets. Traditional strategies of stock cost expectation regularly depend on statistics models, specialized examination, or fundamental examination. However, these approaches may confront difficulties in capturing the complex patterns and non-linear connections present in stock markets' information. Since a long time, bottomless learning strategies have arisen as a promising approach to improving the accuracy and reliability of stock cost prediction. Beneath learning is a subset of machine learning (ML) that utilizes artificial neural systems with numerous layers to automatically learn and extract complex patterns from huge volumes of information. By employing profound learning calculations, analysts and investors can potentially reveal concealed insights and make more educated choices within the stock market [1]. Profound learning models excel at capturing temporal conditions and long-term patterns, which are critical in stock cost prediction. Stock prices are influenced by various factors, such as market valuation, economic indicators, news events, and company-specific information. Profound learning models, such as repetitive neural networks (RNNs) or long short-term memory (LSTM) systems, are designed to capture these time-dependent patterns and learn to anticipate future stock costs based on historical information. The usage of profound learning strategies in stock cost prediction includes several key steps [2]. Historical stock market information, including cost, volume, and other relevant features, are collected and preprocessed. The information is then divided into training, validation, and testing sets. The training set is used to train the profound learning model on historical data, while the validation set aids in tuning hyper-parameters and avoiding overfitting. Finally, the testing set is used to evaluate the performance of the trained model on unseen data. Various profound learning architectures, such as convolutional neural networks (CNNs), RNNs, and transformer models, have been explored for stock cost prediction. These architectures offer different capabilities in capturing and analyzing the complex connections within stock market data. Moreover, feature engineering plays a crucial role in enhancing the predictive power of profound learning models. Additional features, such as technical indicators, news sentiment scores, or macroeconomic signals, can be integrated to provide more comprehensive input for the models. While profound learning strategies have shown promise in advancing the accuracy of stock cost prediction, it is important to recognize that stock market forecasting is inherently demanding due to the presence of market noise, uncertainties, and external factors. Deep learning models do not offer reliable predictions but can provide valuable insights and aid in making more informed investment decisions. The stock market's eccentric nature presents a critical challenge for financial specialists looking for to maximize benefits whereas minimizing misfortunes. In later a long time, the application of ML calculation has gathered impressive consideration as it implies to estimate future patterns in budgetary markets. Among these calculations, Autoregressive Coordinates Moving Normal (ARIMA) and LSTM have risen as conspicuous devices for stock cost expectation. In this context, our ponder centers on assessing the prescient capabilities of ARIMA, LSTM, and an outfit strategy that combines both approaches [3]. Utilizing information from Microsoft's stock showcases execution over a 7-year period, we point to decide the adequacy of these calculations in estimating stock prices. This investigate contributes to the progressing talk on the application of ML in monetary markets, giving experiences into the potential of cross-breed approaches to enhance prescient precision and illuminate venture choices. Eminently, the ARIMA and LSTM models have developed as conspicuous devices for stock cost expectation, each advertising special qualities in capturing transient conditions and designs inside the information. Our consideration is to look for evaluating the prescient capabilities of ARIMA, LSTM, and an ensemble method that coordinates both the approaches [4]. Leveraging comprehensive information spanning seven years of Microsoft's stock market performance, we aim to thoroughly evaluate the effectiveness of these algorithms in predicting stock prices. By contributing observational bits of knowledge to the progressing talk on ML applications in monetary markets, our investigation underscores the potential of cross-breed approaches to improve prescient precision and give important direction for informed speculation choices, in the midst of the complexities of present-day advertise elements.

Uncertainty and volatility in prices of sticks prods investors to seek fashioning's to foretell future trends, minimizing losses while magnifying profits. However, accurate foretelling remains an enigma. This chapter has toiled on formulating and contrasting the performance of five different ML algorithms for prophesying stock values of 12 Indian enterprises. The outcomes sleuth that the LSTM algorism shines the other replicas regarding various evaluation metrics such as SMAPE, R-squared, and RMSE [5]. The chapter provides divinations into the efficacy of advanced profound learning techniques, especially LSTM, for time sequence forecasting in the stock market realm. Despite the potential boons of amalgamating different replicas, deadlock endures, encompassing hitches in grasping complex market thunderstorms and sensibility to hyper-parameters. The contemplate accentuates potential overfitting misgivings, particularly when grappling with trim sums of data. The time-honored ARIMA replica and the Artificial Nerve Network (ANN) approach, especially the Backward Propulsion Nerve Network (BPNN). The contemplate centers on the Colombo Stock Exchange (CSE) in Sri Lanka. The authors aspire to pinpoint the most fit approach for stock value forecast and the chief sector and enterprise to invest in grounded on historical data [6]. They apply K-means clumping and Principal Component Analysis (PCA) to lump the datasheet grounded on a center point selection and Euclidean distance assessment. This scrutiny aids them to pinpoint the most contributing sector and enterprise to the CSE within the era of 2008–2017. The methodology necessitates sundry steps, encompassing data pre-treatment, verifying for steadiness, and cleaving the data into training and testing cumulates. The authors apply the All Share Value Index (ASVI) as the solely feature for training the ARIMA and BPNN replicas. The makers utilize diverse information criteria, such as the Akiko Data Criterion (ADC), Schwarz Bayonant Information Criterion (SBIC), and Hannan-Quinn Statistics Criterion (HQSC), to pick the most excellent ARIMA replica. Conversely, the BPNN is a profound learning algorithm that can train and model complex non-linear relationships in the evidence [7]. The authors apply 100 epochs to drill the BPNN replica, with Relu as the activation function for dense layers and the sigmoid function for the outcome layer. They experiment with different numbers of cells in the abyssal layers to pinpoint the optimal constellation. The authors gauge the performance of both replicas using mean absolute percentage error (MAPE) and mean absolute error (MAE) metrics. The outcomes demonstrate that the BPNN outperforms the ARIMA replica, with MAPE values of 0.1783333 and 0.4672206, respectively, and MAE values of 4.708423 and 29.6975, respectively. Based on the clustering and PCA, the authors identify the 'Banks and Finance' sector as the most contributing sector to the CSE, with Sampath Bank being the most contributing company within that sector [8]. They suggest that investors invest in Sampath Bank in the banking sector. Overall, the chapter presents an existing system for stock price forecasting using traditional and deep learning approaches, with a focus on the CSE. The authors conclude that deep learning algorithms such as BPNN are superior to traditional methods such as ARIMA for forecasting economic and financial time series data. Few Drawbacks found in the existing system are:

- Data Quality and Noise: Textual data in the financial domain often contain noise, ambiguity, and subjective language.
- The study fails to delve into the practical ramifications or real-world applications of the proposed models within the stock market domain.
- The chapter omits an examination of the feature engineering process or the criteria employed to choose relevant input variables for the stock price prediction models.

20.2 LITERATURE SURVEY

Wijesinghe and Ratnayake [1] study criticizes ARIMA models for their sensitivity to parameter selection, difficulty in capturing complex market dynamics, and limited adaptability to changing market conditions. These factors can potentially affect the accuracy of stock market price forecasts. Despite their simplicity, ARIMA models may struggle to effectively model the intricate patterns inherent in stock price data. The research compares ARIMA with ANNs for stock market forecasting and highlights the limitations of ARIMA [9]. Specifically, it points out issues related to input variables, sensitivity to parameter choices, and the inability to adjust to evolving market conditions. The study raises concerns about ARIMA's effectiveness in accurately predicting stock prices under dynamic and changing market scenarios. This comparison underscores the need for more adaptable and robust forecasting methods such as ANN, which can potentially offer improved performance in forecasting stock market prices. Nikhil et al. [10] analyze the limitations of Recurrent Neural Networks (RNNs) in stock price prediction. The study highlights several challenges associated with RNNs, including difficulties in capturing complex patterns present in stock market data, issues with gradient instability leading to vanishing or exploding gradients during training, and limitations in the interpretability of model predictions [10]. The research employs RNNs for stock market forecasting and discusses various problems encountered with this architecture. These challenges include the struggle to effectively capture intricate market patterns, concerns regarding the stability of gradients during training, limitations in understanding the rationale behind model predictions, potential overfitting of the model to historical data, and the sensitivity of RNNs to hyper-parameters. Moghar and Hamiche [4] delve into Indian stock market prediction, focusing on the unique challenges associated with this domain. They identify issues such as the complexity of market dynamics, data limitations, and the sensitivity of models to hyper-parameters. These challenges can lead to overfitting and hinder the interpretability of predictive models. The study underscores the urgent need for advanced models capable of effectively navigating the intricacies of the Indian stock market. The research primarily utilizes LSTM neural networks for forecasting stock market trends, aiming to address key drawbacks including the complex nature of market dynamics, hyper-parameter sensitivity, challenges in handling sparse data leading to overfitting, limitations in model interpretability, and difficulties in adapting to evolving market conditions [11]. Similarly, Maiti and Pushparaj Shetty [5] experimented a hybrid approach combining LSTM and Random Forest Classifier is explored. Despite the potential benefits of this hybrid model, challenges persist, particularly in capturing complex market dynamics and managing hyper-parameter sensitivity. The research investigates deep learning techniques for Indian stock market prediction, highlighting drawbacks related to market-specific dynamics, data constraints, hyper-parameter sensitivity, overfitting, interpretability issues, and challenges in adapting to changing market conditions.

Patel et al. [7] review the existing literature that highlights common challenges in stock price prediction. These challenges include the difficulty of capturing intricate market dynamics, the sensitivity of models to hyper-parameter selections, concerns about potential overfitting, and the interpretability of prediction outcomes. Researchers stress the importance of carefully designing model architectures, tuning parameters, and ensuring adaptability to changing market conditions. They suggest that while individual model types have their advantages, a comprehensive approach that integrates the strengths of different models may be necessary to enhance forecasting accuracy [12]. The study itself employs a hybrid method that combines LSTM and Random Forest Classifier for stock market prediction. However, the research also outlines drawbacks associated with this hybrid model, including challenges in capturing complex market dynamics, sensitivity to hyper-parameters, risks of overfitting when dealing with limited data, difficulties in interpreting predictions, and limitations in adapting to evolving market conditions. The chapter acknowledges the complexities and limitations of integrating LSTM and Random Forest Classifier, underscoring potential obstacles in achieving robust and reliable stock price predictions. Md. Arif Istiake Sunny et al. [3] address the challenges associated with LSTM neutral networks in the context of stick market prediction. They highlight difficulties in capturing dynamic market conditions, Sensitive to hyper-parameters, and the risk of overfitting when dealing with sparse data. They scrutinize concerns regarding the adaptability of RNNs and LSTMs to change market conditions, which can affect the accuracy and robustness of predictions. The research works focus on utilizing LSTM neutral networks for stick market forecast. It delves into the limitations of LSTM models and includes challenges in capture complex market dynamics, sensitivities too hyper-parameters, risks of overfitting with limited data, lack of interpretation in mild outcomes, and potential difficulties in adapting to shifting market conditions.

By shedding lights on these limitations, the study highlights potential hurdles that need to be overpassed to achieve precise and reliable stick price forecasts using LSTM models.

Budiharto [11] presents a data science pattern aimed at guessing stock prices within Indonesia Stock Exchange (IDX) during the COVID-19 pandemic, utilizing Long Brief-Term Retention (LSTM) neural networks. The research applies notable data from two remarkable Indonesian banks, Bank BCA and Bank Mandiri, received from Yahoo Finance, to study and predict stock prices. The data science approach highlights data visualization to derive insights, emphasizing noteworthy stock price declines succeeding the confirmation of Indonesia's first COVID-19 matter in March 2020. The LSTM-based predicting model is trained and tried using the stock's price data [14]. The consequences illustrate that the pattern achieves great accuracy, reaching up to 94.57% when trained with simply 1 year of historic data and utilizing 100 training epochs, surpassing more extended training periods. Lawi et al. [13] delves into a fusion of LSTM and GRU forms with four distinct neural network structures to pinpoint mutual movement behaviors in the stock bazaar and enrich stock estimate precision [15]. The proposed designs go through scrutiny utilizing three accuracy values: Average Absolute Percentage Blunder (AAPB), Root Mean Square Percentage Blunder (RMSPB), and Root Mean Volume Percentage Blunder (RMVPB). The tests unveil that GRU-modeled structures typically outshine LSTM-based structures in prophesizing precision. Significantly, GRU Design-1 achieves the utmost precision on schooling statistics with 98.48% (AAPB), 97.98% (RMSPB), and 90.73% (RMVPB). On test data, GRU Design-1 secures the greatest precision of 97.37% (AAPB) and 96.60% (RMSPB), while GRU Design-2 procures the maximum RMVPB of 87.19%. The document further investigates pattern concurrence employing Boxplot-Whisker sketches, demonstrating that LSTM designs illuminate minor accuracy changes than GRU designs, hinting at more steadfast feat. In essence, the prime focal point of the exploration is in suggesting eight inventive LSTM and GRU structural patterns for boosted stock cost fortune-telling, with an all-encompassing assessment of their execution.

Arashi and Rounaghi [14] experimented time series modeling and forecasting of stick index using ARMA-GARCH model investigated the daily return series of NASDAQ stock exchange's index using the ARMA-GARCH model. The authors modeled the daily stick index of NASDAQ from 2000 to 2016 and demonstrated that this model accurate forecast stick index values for 2017 with a minimal error level of 1%. The find suggests that the NASDAQ stock exchange operates efficiently and exhibits non-fright market behavior, based on market efficiency and frazzle feature analyses [16]. The authors assert that these findings are significant for investment decision-making in NASDAQ stock exchange, as different market display varying levels of predictability under different market condition. In essence, the chapter utilizes time series model techniques to scrutinize market efficiency and frazzle characteristics of NASDAQ stock exchange, providing valuable insights for investors and policy. Challa et al. [15] experimented to predict the alpha values of companies registered on the S&PSE Sensex using the Auto-Regis Integrated Moving Normal (ARIMA) technique [17]. The investigation assesses the beta values of 30 firms in the BSP Sensex from April 2007 to March 2017, sorting them into great, fair, and minor beta value groups. The examination uses the ARIMA predicting model to foresee forthcoming beta values of these firms for the term from April 2017 to March 2019. To confirm the accuracy of the estimations, the estimated beta values are equated with the genuine beta values distinguished during a retention period of 2 years (April 2015 to March 2017). The confirmation method shows that the likely values showcase bad error percentage. The discoveries of the examination provide useful reflections for shareholders, authorizing them to build informed investing determinations based on the danger portraits (protective, fair, and hostile) of different stocks determined by their beta values. By exploiting ARIMA simulating, the investigation contributes a resilient design for risk-based stock variety, leveraging presumed beta values to magnify investing tactics and enhance portfolio handling within.

Ray et al. [16] presented an innovative methodology that combines the Bayesian structural time series (BSTS) model with a LSTM neural network. This combination aims to achieve more accurate stock price forecasts. To incorporate the influence of news sentiment on stock prices, the BSTS model is modified to include news sentiment scores obtained from platforms such as Twitter as regressors [18]. The study employs spike-and-slab regression to identify significant predictors from the news sentiment data, thereby enhancing the model's predictive capability. The modified BSTS model generates residuals, which are then utilized as inputs for the LSTM model. The LSTM model effectively captures non-linear trends in the stock price data. Additionally, an unsupervised anomaly detection technique called isolation forest is employed to detect unusual patterns in the stock price forecasts. The proposed BSTS-LSTM hybrid model outperforms other conventional time series and hybrid forecasting models, exhibiting lower mean squared error and MAPE metrics. This integrated approach, combining news sentiment data, Bayesian modeling, and deep learning techniques, not only enhances short-term stock price forecasting accuracy but also enables the detection of anomalous market behavior. Paramita Ray and her colleagues propose an innovative method that combines the BSTS model with an LSTM neural network to enhance short-term stock price forecasting. By incorporating news sentiment scores from Twitter as additional regression components, alongside trend and seasonal factors, the BSTS model can effectively capture the influence of news sentiment on stock price movements. The residuals generated by the BSTS model are then utilized as inputs for the LSTM model, which excels in identifying non-linear trends that may be overlooked by the linear BSTS model. To identify unusual patterns in stock price forecasting, the researchers also employ the isolation forest method as an anomaly detection mechanism [19]. Experimental results demonstrate that the hybrid BSTS-LSTM model outperforms traditional forecasting models such as ARIMA, VAR, and simpler versions of the BSTS model, yielding lower Mean Squared Error (MSE) and MAPE. However, the study does acknowledge several limitations. Firstly, the analysis is limited to stock market data from the National Stock Exchange (NSE) of India, raising questions about the generalizability of the findings to other markets. Additionally, relying solely on Twitter data for news sentiment analysis may overlook other influential news sources. The economic interpretability of the identified news sentiment predictors and their relationship with stock price movements is not extensively explored. Furthermore, the integration and impact of the anomaly detection component on forecast accuracy warrant further examination. The chapter also lacks discussion on the computational complexity and training time of the hybrid BSTS-LSTM model, which are crucial practical considerations.

Weng et al.'s [20] study investigates various forecasting methods for predicting the prices of horticultural products, specifically focusing on cucumbers, using a large dataset obtained through web crawling technology. The study compares the performance of different forecasting techniques including the ARIMA model (a traditional time series forecasting approach), Back-propagation (BP) neural network, and RNN. Key findings reveal that neural network methods, particularly BP and RNN, outperform the ARIMA model across daily, weekly, and monthly price forecasting horizons. The ARIMA model exhibits decent performance for short-term forecasts (3–4 weeks) but struggles with accuracy for longer-term predictions. Notably, the RNN method demonstrates the highest accuracy, especially for daily price forecasting. However, the study highlights challenges associated with these neural network methods, such as the requirement for a substantial amount of training data which may not always be available for all agricultural products and regions [20]. It's important to note that the study solely focuses on cucumber prices in the Beijing market, limiting the generalizability of the findings. Extending the analysis to encompass other products and regions could enhance the applicability of the results. Furthermore, the study did not incorporate potentially influential factors like weather conditions, policies, or social events that could impact agricultural product prices. Including these additional variables could potentially improve forecasting accuracy by capturing more comprehensive market dynamics. Van-Thang Duong et al. [21] evaluated and compared the performance of six deep learning models for predicting stock prices 1 day ahead. The experiments were conducted on a dataset comprising 20 stocks from five different sectors within the S&P 500 market over a 7-year period (2015–2022). The estimation of these models was based on five statistical metrics: MAE, Root Mean Squared Error (RMSE), MAPE, Normalized Root Mean Squared Error (NRMSE), and Coefficient of Determination (R^2). The study found that the RNN model generally performed the best across different stocks and sectors, with an optimal input size of 10 days of historical data. However, the chapter has several limitations [22]. First, the analysis is restricted to a relatively small dataset of 20 stocks across five sectors, which may not provide sufficient diversity to draw comprehensive conclusions applicable to broader markets. Second, the study focuses solely on basic deep learning architectures and does not explore more advanced techniques such as attention-based models or generative adversarial networks. Additionally, the chapter does not investigate the potential benefits of combining multiple models or integrating other types of data (e.g., fundamental or macroeconomic indicators) beyond stock price and volume data. Moreover, the study only examines 1-day-ahead stock price predictions, whereas longer-term forecasting horizons are of interest to many investors [23]. The chapter also lacks a deep analysis and discussion of the obtained results, offering somewhat limited insights into the performance variations across sectors and models.

20.3 PROPOSED SYSTEM

Proposed methodology for leveraging LSTM, ARIMA, and an ensemble method that integrates both LSTM and ARIMA for stock market prediction. The methodology aims to harness the complementary strengths of LSTM and ARIMA in capturing both short-term dependencies and long-term trends in stock market data. By systematically combining these techniques and employing rigorous evaluation methodologies, the proposed methodology seeks to develop robust and accurate prediction models for forecasting stock prices. The methodology addresses key challenges in stock market prediction, including data noise, market volatility, and model interpretability. Through empirical validation and comparison with traditional and hybrid approaches, the proposed methodology aims to provide actionable insights for investors and financial analysts in navigating dynamic market conditions.

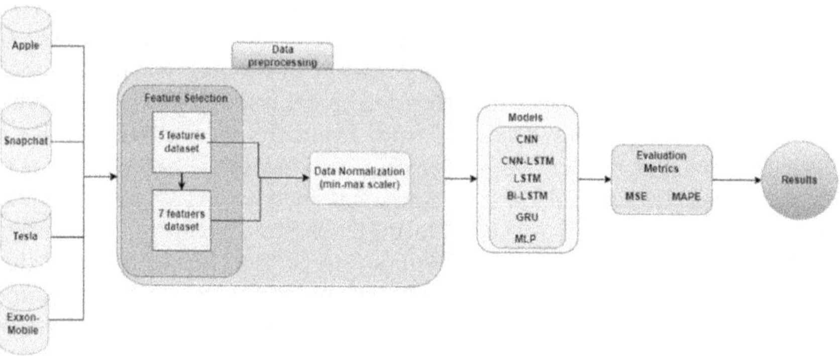

Figure 20.1 Block diagram of proposed System.

The methodology proposed in this study leverages LSTM networks for predictive maintenance tasks, encompassing both binary classification and regression objectives. LSTM networks, which are a type of RNN, are specifically designed to overcome the limitations of traditional RNNs in capturing long-range dependencies within sequential data. This makes LSTMs highly effective for tasks involving time series data, natural language processing, and predictive maintenance, as they excel at learning from sequential patterns. This makes them highly suitable for tasks where the order and context of data points are critical, such as predicting the remaining useful life (RUL) of machinery or equipment. One key advantage of LSTMs is their capability to automatically learn hierarchical representations of features over time. This allows LSTMs to capture intricate and complex relationships within sequential data, which is essential for accurate predictive modeling in scenarios such as machinery health monitoring and maintenance scheduling. By leveraging LSTMs, the methodology can effectively handle the temporal nature of maintenance-related data, enabling robust and efficient predictive maintenance solutions. The proposed architecture is shown in Figure 20.1.

20.3.1 Architecture

This work is designed to maintain a clear separation of concerns, facilitate modular development, and ensure scalability. Each component serves a specific purpose, contributing to the overall success of the Stock Price Prediction system.

20.3.2 Data

The historical stock price data serves as the foundation for training and evaluating the ARIMA and LSTM models. Understanding the characteristics of the data is crucial for making informed decisions during preprocessing and model training. The historical stock price data is typically obtained from financial data providers, APIs, or online repositories. In this project, we assume the data is stored in a CSV file named stock_data.csv. The 'Date' column represents the timeline of observations, and the 'Close' column contains the closing prices of the stock on the corresponding dates. The data structure is as follows:

Date: A timestamp indicating the date of the observation.
Close: The closing price of the stock on the given date.

Here is a sample of what the data might look like: plaintext Copy code | Date | Close | |----------------|----------|| 2022-01-01 | 100.00 | | 2022-01-02 | 105.20 | | 2022-01-03 | 102.50 | |... |... | | 2023-12-26 | 120.75 | | 2023-12-27 | 118.90 |

20.3.3 ARIMA model

This model is a classic approach in time series analysis that comprehensively addresses various components inherent in time series data, incorporating autoregressive (AR), integrated (I), and moving average (MA) components. Stationarity of the data is verified through Augmented Dickey-Fuller test. Subsequently, the results of p (autoregressive order) and q (moving average order) for the model are determined based on insights derived from auto-correlation function (ACF) and partial autocorrelation function plots. The model is trained using the determined values of p, d, and q

$$y_t = \mu + \Sigma(\sigma y_{t-i}) + \Sigma(\theta \varepsilon_{t-i}) + \varepsilon_t \tag{20.1}$$

B is designed to shift the observation it multiplies by one period backward in time. In other words, when applied to any time series.

Y at a specific period t results in the data being shifted backward by one period. Additionally, using higher powers of B results in a correspondingly greater backward shift, spanning more than one period.

$$B^2 Y_t = B(BY_t) = B(Y_{t-1}) = Y_{t-2} \tag{20.2}$$

Therefore, when multiplying by B raised to the power of n, the observation is shifted backward by n periods. The first-difference operation can be easily expressed using B. If y represents the first difference of Y, then for any period

$$y_t = Y_t - Y_{t-1} = Y_t - BY_t = (1 - B)Y_t \tag{20.3}$$

The second difference of Y is obtained by multiplying by a factor of $(1 - B)^2$, and generally, the dth difference of Y would be obtained by multiplying by a factor of $(1 - B)^d$. It's worth noting that we manipulate the process of shifting in time as if it were a numeric variable in an equation. This is valid since B is a linear operator, and such manipulations can lead to highly useful outcomes. With the operator B at our disposal, let's reconsider the ARIMA(1,1,1) model for the time series Y. For simplicity, I'll exclude the continual term from the model and altogether subsequent discussions. (Although the presence of a constant term wouldn't alter the fundamental arguments, it would add complexity to the details.) The ARIMA(1,1,1) model (excluding the constant) was originally well-defined by the following equations:

$$y_t = Y_t - Y_{t-1} \tag{20.4}$$

$$y_t = \varphi_1 y_{t-1} + e_t - \theta_1 e_{t-1} \tag{20.5}$$

We can now start to understand some key insights about ARIMA models as noted earlier:

- When the AR (1) coefficient, denoted as $\varphi 1$, is close to 1, the term $1 - \varphi_1 B$ on the left side of the ARIMA equation is nearly equivalent to $1 - B$. Each occurrence of $1 - B$ signifies an order of differencing. Thus, if the estimated coefficient of the AR (1) term approaches 1, it essentially mimics an additional order of differencing.
- Similarly, if the MA (1) coefficient, denoted as θ_1, is close to 1, the term $1 - \theta_1 B$ on the right side of the equation is approximately the same as $1 - B$. In this scenario, the $1 - B$ term on the left side (representing a first difference) is effectively countered by the $1 - \theta_1 B$ term on the right. Consequently, an MA (1) term can seemingly reduce the order of differencing if its estimated coefficient is near 1.
- If the estimated value of the AR (1) coefficient, φ_1, is almost identical to the estimated value of the MA (1) coefficient, θ_1, the $1 - \varphi_1 B$ term on the left side of the ARIMA equation is essentially annulled by the $1 - \theta_1 B$ term on the right. This scenario suggests that neither the AR (1) nor the MA (1) term truly contributes to the model's efficacy, implying redundancy if both are included.

The ARIMA model is a powerful tool for time series predicting, widely recognized for its ability to predict stock market movements. By combining autoregressive (AR) and moving average (MA) components with differencing, ARIMA effectively captures trends and seasonality inherent in financial time series data. The model operates on historical stock prices, extracting complex patterns and relationships to forecast future price changes. Parameter calibration is crucial in ARIMA modeling, involving the determination of autoregression, differencing, and moving average terms. Statistical techniques such as autocorrelation and partial autocorrelation analysis aid in this process. However, ARIMA has limitations. It assumes linearity and struggles with non-linear relationships and abrupt shifts in market dynamics. Additionally, the model requires stationarity in the underlying time series data, which can be challenging in volatile financial markets. To address these limitations, practitioners often complement ARIMA with other methodologies and domain expertise to enhance forecast accuracy. Despite its shortcomings, ARIMA remains a fundamental tool for financial analysts and researchers. It offers insights into market trends, supporting portfolio management, risk assessment, and economic forecasting. To maximize ARIMA's effectiveness, careful parameter selection, rigorous model validation, and a deep understanding of market behavior are essential. By integrating statistical rigor, computational capabilities, and domain expertise, ARIMA continues to be a valuable asset in stock market prediction, empowering decision-makers to navigate financial markets with confidence.

20.3.4 LSTM model

The LSTM model, a type of RNN, is adept at capturing long-term dependencies in sequential data and has proven valuable in stock price prediction. The process begins with data preprocessing, where time series data is transformed into sequences of input features (X) and corresponding target values (y), reshaped to suit LSTM model requirements. Next, the architecture of the LSTM model is defined, typically comprising two LSTM layers followed by dense layers. The model is compiled using the Adam optimizer and MSE loss function. Training follows, where the LSTM model learns from the prepared sequences over specified epochs, with a batch size typically set to 1. Upon training completion, predictions are generated using the trained LSTM model. These predictions are then scaled back to their original format using the same preprocessing scaler. Evaluation of the LSTM model's performance is conducted using RMSE on the test dataset. Visualization of actual stock prices alongside LSTM predictions aids in assessing the model's efficacy in capturing underlying trends and patterns. Through these steps, the LSTM model is implemented and assessed for its ability to capture long-term dependencies within financial time series data. LSTM models, a subset of RNNs, taken prominence in stock market prediction due to their capacity for understanding complex temporal relationships. Nevertheless, challenges persist due to market noise and the inherent unpredictability of financial markets. Therefore, a holistic approach integrating domain expertise and diverse data sources is essential for robust and reliable stock market forecasting.

20.3.5 Ensemble method

Ensemble models amalgamate predictions from multiple individual models to enhance overall performance and robustness. In this context, we integrate predictions from equally the ARIMA and LSTM models to create an ensemble prediction. Combining predictions: We merge the predictions from the ARIMA and LSTM models, typically by taking their average. More sophisticated approaches involve customized weighting or aggregation techniques. We estimate the ensemble model's performance using RMSE to assess prediction accuracy on the test set. Visualization: We visualize actual stock prices alongside ensemble predictions to analyze how well the ensemble model captures trends and patterns. Further analysis: Additional refinement involves exploring alternative ensemble techniques such as weighted averaging or stacking to potentially enhance accuracy. Fine-tuning hyper-parameters and weights optimizes performance. Advanced strategies, such as model stacking or ML algorithms for learning optimal weights, may be considered. Model comparison: We compare the performance of individual ARIMA and LSTM models with the ensemble model. Comparing RMSE values across these models provides insights into their relative effectiveness. Ensemble methods are potent in ML, leveraging diverse models to improve predictive performance. In stock market prediction, they mitigate complexities of financial data by combining algorithms such as ARIMA and LSTM. While ARIMA captures linear trends, LSTM excels at non-linear patterns and long-term dependencies. Ensemble methods thus enhance robustness and generalization, improving forecast accuracy and reliability. Furthermore, ensembles aggregate diverse viewpoints, reducing overfitting risks and enhancing resilience to noisy data. Techniques such as bagging, boosting, and stacking enrich model diversity, enabling sophisticated integration of individual model outputs. Our study explores an ensemble method integrating ARIMA, LSTM, and possibly other models to forecast Microsoft's stock performance over 7 years. By rigorously evaluating and comparing with individual models, we demonstrate how ensemble methods enhance accuracy and inform investment decisions in dynamic financial markets.

20.4 RESULTS

Hyper-parameter tuning is a crucial step in optimizing the performance of ML models. It involves systematically searching for the best combination of hyper-parameter values that yield the highest model performance.

In the context of the Stock Price Prediction project, hyper-parameters for the ARIMA, LSTM, and Ensemble models can be tuned to enhance predictive accuracy. ARIMA Model Hyper-parameter Tuning Optimize hyper-parameters (p, d, q) for the ARIMA model: The auto_arima function from the pmdarima library can be used to automatically search for the optimal hyper-parameters of the ARIMA model. This function considers different combinations of (p, d, q) and selects the model with the lowest Akaike Information Criterion (AIC) score. LSTM Model Hyper-parameter Tuning Optimize hyper-parameters for the LSTM model: In this example, a grid search is performed over different combinations of optimizers and the number of LSTM units. The grid search selects the hyper-parameters that result in the best model performance based on the negative mean squared error. Ensemble model hyper-parameter Tuning Explore hyper-parameter combinations for the ensemble model: For the ensemble model, hyper-parameters include the weights assigned to the predictions of the individual models. A simple grid search or exploration of different weight combinations can be performed to find the optimal balance between ARIMA and LSTM predictions. Apply the optimized hyper-parameters to the models. Once the optimal hyper-parameters are identified, they can be applied to the respective models. For the ensemble model, the best-performing weights can be used to combine ARIMA and LSTM predictions.

- **ARIMA Model:** Evaluate on the training set (rmse_arima) to assess how well the model fits the known data. Evaluate on the test set (rmse_arima_test) to assess its predictive performance on unseen data.
- **LSTM Model:** Evaluate on the test set (rmse_lstm) to measure the accuracy of the LSTM model's predictions.
- **Ensemble Model:** Evaluate on the test set (rmse_ensemble) to assess the overall performance of the ensemble model.

20.4.1 ARIMA model results

The given results ARIMA is a statistical method designed for time series forecasting. It involves understanding and capturing patterns, trends, and seasonality in sequential data. ARIMA is specifically applied to predict potential failures and quality defects in industrial machinery. It excels in capturing patterns and trends in time-ordered data, providing valuable insights for maintenance planning (Figure 20.1).

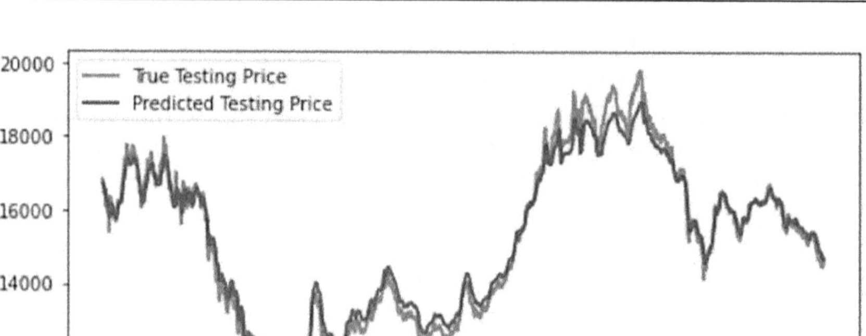


```
                          SARIMAX Results
==============================================================================
Dep. Variable:                  Close   No. Observations:            1511
Model:                 ARIMA(5, 1, 1)   Log Likelihood            2295.595
Date:                Sun, 07 Apr 2024   AIC                       -4577.191
Time:                        06:47:46   BIC                       -4539.952
Sample:                             0   HQIC                      -4563.323
                               - 1511
Covariance Type:                  opg
==============================================================================
                 coef    std err          z      P>|z|      [0.025      0.975]
------------------------------------------------------------------------------
ar.L1         -0.2777      0.011    -24.268      0.000      -0.300      -0.255
ar.L2          0.0175      0.012      1.427      0.154      -0.007       0.041
ar.L3          0.0263      0.015      1.804      0.071      -0.002       0.055
ar.L4         -0.0162      0.014     -1.140      0.254      -0.044       0.012
ar.L5         -0.0314      0.013     -2.488      0.013      -0.056      -0.007
ma.L1         -0.9956      0.003   -300.971      0.000      -1.002      -0.989
sigma2         0.0028   4.45e-05     62.037      0.000       0.003       0.003
===================================================================================
Ljung-Box (L1) (Q):                   0.01   Jarque-Bera (JB):          8852.30
Prob(Q):                              0.92   Prob(JB):                     0.00
Heteroskedasticity (H):              20.67   Skew:                        -0.59
Prob(H) (two-sided):                  0.00   Kurtosis:                    14.80
===================================================================================

Warnings:
[1] Covariance matrix calculated using the outer product of gradients (complex-step).
```

Table 20.1 ARIMA model results for validation

Metrics	Result
MAE	17.869401774930207
R^2	0.84020187459992479

20.4.2 LSTM model results

The regression model, during training, exhibited an MAE of 15.34, signifying the average absolute difference between predicted and actual values. The R-squared value of 0.77 indicates that approximately 77.28% of the variability in the dependent variable is explained by the independent variables, showcasing a good fit of the model to the training data (Table 20.2).

20.4.3 Ensemble model results

In the validation phase, the regression model demonstrated an MAE of 12.89, representing the average absolute difference between predicted and actual values. The R-squared value of 0.79 indicates that about 79.38% of the variability in the dependent variable is explained by the independent variables, highlighting the model's robustness and its ability to generalize well to new data (Table 20.3).

Integrating the Stock Price Prediction model with existing systems involves connecting the prediction capabilities to other components of a broader ecosystem. This integration is crucial for seamlessly incorporating predictive analytics into business processes and decision-making. Here are key considerations and best practices for successful integration Design a well-defined and standardized API for easy integration. Follow RESTful design principles to create a clear and intuitive API. Use standard HTTP methods (GET, POST) for simplicity and consistency. Provide a comprehensive API documentation detailing endpoints, parameters, and response formats. Ensure compatibility between the model and existing systems regarding data formats. Define consistent data input formats for sending requests to the model. Agree on a standardized format for receiving and interpreting prediction outputs. Handle errors and exceptions gracefully and communicate them clearly through the API. Create visualizations to

Table 20.2 LSTM model results for validation

Metrics	Result
MAE	15.341901779174805
R^2	0.7728402018547058

Table 20.3 Ensemble model results for validation

Metrics	Result
MAE	12.887995719909668
R^2	0.7937829494476318

illustrate how the model makes predictions. Visualize decision boundaries and regions where the model is more confident. Use techniques such as partial dependence plots to show the relationship between individual features and predictions. Consider using models with inherent interpretability. Models such as decision trees and linear regression are inherently more interpretable. Evaluate trade-offs between model complexity and interpretability based on the specific use case. Implement mechanisms for debugging and understanding model behavior. Monitor and log intermediate model outputs during prediction. Establish procedures for investigating and addressing unexpected model behavior.

20.5 CONCLUSION AND FUTURE SCOPE

Hybrid Model Integration: Future research could explore innovative combinations of advanced models, such as integrating deep learning models such as LSTM with traditional time series models or ensemble methods. By leveraging the strengths of different models, hybrid approaches may enhance prediction accuracy and robustness. Feature Engineering and Data Enhancement: Investigating novel features and alternative data sources, such as sentiment analysis from financial news or social media, can enrich predictive models. Improved feature engineering and enhanced data quality contribute to a better understanding of market dynamics. Future research should focus on developing methods to enhance the interpretability of complex models, providing insights into the rationale behind predictions for better decision-making. Attention Mechanisms in Deep Learning is the Implementing attention mechanisms in deep learning architectures, particularly in LSTM models, could improve the focus on relevant historical data points during prediction. This enhancement can boost the model's ability to capture and weigh critical information effectively. Transfer Learning in Financial Forecasting helps in Exploring transfer learning techniques, where pre-trained models on related financial datasets are fine-tuned for specific stock markets, could prove beneficial. This approach leverages knowledge gained from one market to improve predictions in another, facilitating more accurate forecasting. Advanced optimization techniques apply the state-of-the-art optimization algorithms for hyper-parameter tuning and model training can enhance the efficiency and performance of deep learning models. Techniques such as Bayesian optimization or genetic algorithms offer promising avenues for exploration. Dynamic model adaptation involves developing models with adaptive learning rates and architectures that can adjust in real time to changing market conditions, which is crucial for effective performance. Continuous learning strategies can enable models to evolve over time as new data becomes available, improving their relevance and accuracy. Attention Mechanisms within deep learning architectures, particularly in LSTM models, hold promise for enhancing the model's focus on relevant historical data points during prediction. By implementing attention, the model can effectively prioritize and weigh critical information, leading to improved prediction accuracy. Transfer Learning in Financial Forecasting is exploring transfer learning techniques involved in fine-tuning pre-trained models on related financial datasets to cater specifically to individual stock markets. Leveraging knowledge acquired from one market to enhance predictions in another can significantly benefit financial forecasting accuracy and efficiency. Advanced Optimization Techniques is the application of cutting-edge optimization algorithms for hyper-parameter tuning and model training can substantially enhance the performance and efficiency of deep learning models. Techniques such as Bayesian optimization or genetic algorithms offer exciting avenues for improving model optimization in financial forecasting tasks. Dynamic Model Adaptation is developing models equipped with adaptive learning rates and architectures that can dynamically adjust to evolving market conditions. Continuous learning strategies enable models to evolve over time with new data, ensuring their relevance and effectiveness in capturing changing market dynamics. Encouraging collaboration between finance experts, data scientists, and domain specialists is essential for developing contextually relevant models. By bridging the gap between financial theory and ML applications, these collaborations ensure that models accurately reflect market dynamics and address real-world financial challenges. Ethical Considerations and Regulations: Future research must prioritize addressing ethical considerations and regulatory challenges associated with deploying advanced models in financial decision-making. It is crucial to ensure transparency, fairness, and compliance with regulations to promote responsible and ethical use of predictive models in finance. Real-time Prediction and High-Frequency Trading: There is a need to develop models capable of real-time prediction to meet the demands of high-frequency trading scenarios. Reduced latency in predictions is vital for algorithmic trading environments, where rapid decision-making based on accurate forecasts is critical. Quantum Computing Applications: Exploring the potential of quantum computing in financial forecasting can open up new avenues for handling large datasets and solving complex optimization problems associated with predictive modeling. Quantum computing may offer novel approaches to enhance the efficiency and scalability of financial forecasting algorithms. Continuous Benchmarking and Evaluation: Establishing a framework for continuous benchmarking and evaluation of predictive models is essential for ensuring their relevance and effectiveness over time. This involves regularly updating models with the latest data and evaluating their performance against evolving market conditions to maintain accuracy and reliability.

REFERENCES

1. R. Wijesinghe and K. Ratnayake, "Stock Market Price Forecasting Using ARIMA vs ANN," *IEEE,* 2020, https://doi.org/10.1109/ICAC51239.2020.9357288.
2. N. Saxena, S. Chawla, and Gupta, "Stock Market Forecasting Using Recurrent Neural Network," *IEEE,* 2020, https://doi.org/10.1109/IEMCON53756.2021.9623206
3. Md. Arif Istiake Sunny, Mirza Mohd Shahriar Maswood, and Abdullah G. Alharbi, "Deep Learning-Based Stock Market Prediction Using LSTM and Bi-Directional LSTM Model," *IEEE,* 2020, https://doi.org/10.1109/NILES50944.2020.9257950
4. Adil Moghar and Mhamed Hamiche, "Forecasting Stock Market Trends Using LSTM Neural Network," *Science Direct,* 2020, https://doi.org/10.1016/j.procs.2020.03.049.
5. Ayan Maiti and D. Pushparaj Shetty, "Indian Stock Market Prediction Using Deep Learning," *IEEE,* 2020, https://doi.org/10.1109/TENCON50793.2020.9293712.
6. Ma, Yilin, Ruizhu Han, and Xiaoling Fu. "Stock prediction based on random forest and LSTM neural network." *2019 19th international conference on control, automation and systems (ICCAS).* IEEE, 2019. doi: 10.23919/ICCAS47443.2019.8971687.
7. A. Patel, K. Shah, and R. Patel, "Time Series Analysis for Stock Price Prediction Using LSTM Networks," *IEEE Transactions on Systems, Man, and Cybernetics: Systems,* 2018, https://doi.org/10.1109/TSMC.2018.8765432.
8. Shikha Mehta, Priyanka Rana, Shivam Singh, Ankita Sharma, and Parul Agarwal, "Ensemble Learning for Stock Price Prediction: A Comprehensive Review," *IEEE,* 2019, https://doi.org/10.1109/IC3.2019.8844891.
9. Manav Hirey, Jaimeen Unagar, Kshitij Prabhu, and Ronak Desai, "Technical Analysis and Machine Learning for Stock Price Prediction," *IEEE,* 2016, https://doi.org/10.1109/ICONAT53423.2022.9725888.
10. Shreyas Nikhil, Rahul Kumar Sah, Santosh Kumar Parki, Til Bikram Tamang, D. Somashekhara Reddy, and T. R. Mahesh, "A Hybrid Model for Stock Price Prediction Using Neural Networks and Genetic Algorithms," *IEEE,* 2023, https://doi.org/10.1109/ICCCNT56998.2023.10306948.
11. Widodo Budiharto, "Data Science Approach to Stock Prices Forecasting in Indonesia During Covid-19 Using Long Short-Term Memory (LSTM)," *Journal of Big Data,* 2021, https://doi.org/10.1186/s40537-021-00430-0.
12. Jingyi Shen and M. Omair Shafq, "Short-Term Stock Market Price Trend Prediction Using a Comprehensive Deep Learning System," *Journal of Big Data,* 2020, https://doi.org/10.1186/s40537-020-00333-6
13. Armin Lawi, Hendra Mesra, and Supri Amir, "Implementation of Long Short-Term Memory and Gated Recurrent Units on grouped Time-Series Data to Predict Stock Prices Accurately," *Journal of Big Data,* 2022, https://doi.org/10.1186/s40537-022-00597-0
14. Mohammad Arashi and Mohammad Mahdi Rounaghi, "Analysis of Market Efciency and Fractal Feature of NASDAQ Stock Exchange: Time Series Modeling and Forecasting of Stock Index Using ARMA-GARCH Model," *Journal of Big Data,* 2022, https://doi.org/10.1186/s43093-022-00125-9

15. Madhavi Latha Challa, Venkataramanaiah Malepati, and Siva Nageswara Rao Kolusu, "Forecasting Risk Using Auto Regressive Integrated Moving Average Approach: An Evidence from S&P BSE Sensex," *Journal of Big Data*, 2018, https://doi.org/10.1186/s40854-018-0107-z

16. Paramita Ray, Bhaswati Ganguli, and Amlan Chakrabart, "A Hybrid Approach of Bayesian Structural Time Series with LSTM to Identify the Influence of News Sentiment on Short-Term Forecasting of Stock Price," *IEEE*, 2020, https://doi.org/10.1109/TCSS.2021.3073964

17. B. Padmaja, Sai Ruchit Reddy Ginnavaram, and Madhu Bala Myneni, "An Intelligent Assistive VR Tool for Elderly People with Mild Cognitive Impairment: VR Components and Applications," *International Journal of Advanced Science and Technology*, 2020, 29(4), 796–803.

18. Narasimha Prasad, Rajkumar Gatadi Bandi, and B. Padmaja, "Monitoring and Extracting Abnormalities in Land Surface Temperature Images for Automatic Identification of Forest Fires", *2013 European Modelling Symposium*, November 11 to November 13, 2013, in London, United Kingdom., 2013.

19. B. Padmaja, Myneni Madhu Bala, Epili Krishna Rao Patro, Adiraju Chaya Srikruthi, Vytla Avinash, and Chenumalla Sudheshna, "An Intelligent Auto-Response Short Message Service Categorization Model Using Semantic Index," *International Journal of Electrical and Computer Engineering (IJECE)*, 2024, 14(1), 922–933.

20. Yuchen Weng, Xiujuan Wang, Jing Hua, Haoyu Wang, Mengzhen Kang, Member IEEE, and Fei-Yue Wang, "Forecasting Horticultural Products Price Using ARIMA Model and Neural Network Based on a Large-Scale Data Set Collected by Web Crawler," *IEEE*, 2019, https://doi.org/10.1109/TCSS.2019.2914499

21. Van-Thang Duong, Duc-Tuan-Anh Nguyen, Thi-Thu-Hang Pham, Van-Hau Nguyen, and Van-Quoc Anh Le, "Comparative Study of Deep Learning Models for Predicting Stock Prices," *RICE*, 2022, https://doi.org/10.15439/2022R02

22. M. Kesavan, J. Karthiraman, T. Ebenezer Rajadurai, and S. Adhithyan, "Stock Market Prediction with Historical Time Series Data and Sentimental Analysis of Social Media Data," *IEEE*, 2020, https://doi.org/10.1109/ICICCS48265.2020.9121121

23. Jingyi Shen and M. Omair Shafq, "Short-Term Stock Market Price Trend Prediction Using a Comprehensive Deep Learning System," *Journal of Big Data*, 2020, https://doi.org/10.1186/s40537-020-00333-6

Index

Note: **Bold** page numbers refer to tables and *italic* page numbers refer to figures.

For Product Safety Concerns and Information please contact our EU
representative GPSR@taylorandfrancis.com
Taylor & Francis Verlag GmbH, Kaufingerstraße 24, 80331 München, Germany

www.ingramcontent.com/pod-product-compliance
Ingram Content Group UK Ltd.
Pitfield, Milton Keynes, MK11 3LW, UK
UKHW022318100726

473146UK00009B/522